The World at Our Fingertips

The World at Our Fingertips

A Multidisciplinary Exploration of Peripersonal Space

Edited by

FRÉDÉRIQUE DE VIGNEMONT

Director of Research
Institut Jean Nicod
Department of cognitive studies
ENS, EHESS, CNRS, PSL University
Paris, France

ANDREA SERINO

Professor and Head of Myspace Lab
CHUV, University Hospital and University of Lausanne
Lausanne, Switzerland

HONG YU WONG

Professor and Chair of Philosophy of Mind and Cognitive Science
University of Tübingen
Tübingen, Germany

ALESSANDRO FARNÈ

INSERM Senior Researcher
ImpAct
Lyon Neuroscience Research Centre
Lyon, France

OXFORD
UNIVERSITY PRESS

OXFORD
UNIVERSITY PRESS

Great Clarendon Street, Oxford, OX2 6DP,
United Kingdom

Oxford University Press is a department of the University of Oxford.
It furthers the University's objective of excellence in research, scholarship,
and education by publishing worldwide. Oxford is a registered trade mark of
Oxford University Press in the UK and in certain other countries

Published in the United States of America by Oxford University Press
198 Madison Avenue, New York, NY 10016, United States of America

British Library Cataloguing in Publication Data

Data available

Library of Congress Control Number: 2020944520

ISBN 978-0-19-885173-8

DOI: 10.1093/oso/9780198851738.001.0001

Printed and bound by
CPI Group (UK) Ltd, Croydon, CR0 4YY

Contents

PART III. THE SPACE OF SELF AND OTHERS

Contributors

Adrian Alsmith, PhD
Lecturer in Philosophy (Mind and Psychology)
Department of Philosophy
King's College London
London, UK

Tommaso Bertoni, PhD Student
Lemanic Neuroscience Doctoral Program
UNIL Department of Clinical Neurosciences
and Department of Radiology
University Hospital and University of Lausanne
Lausanne, Switzerland

Elvio Blini, PhD
Postdoc Fellow
Department of General Psychology
University of Padova
Padova, Italy

Claudio Brozzoli, PhD
Researcher
Integrative Multisensory Perception Action &
Cognition Team (ImpAct)
INSERM, CNRS, and Lyon Neuroscience Center
Lyon, France

R.J. Bufacchi, PhD
Postdoc
Department of Neuroscience, Physiology and
Pharmacology
Centre for Mathematics and Physics in the Life
Sciences and Experimental Biology (CoMPLEX)
University College London
London, UK

Michela Candini, PhD
Postdoctoral Researcher
Department of Psychology
University of Bologna
Bologna, Italy

Justine Cléry, PhD
Post-Doctoral Associate
Robarts Research Institute and BrainsCAN
University of Western Ontario
London, ON, Canada

Yann Coello, PhD
Professor of Cognitive Psychology and
Neuropsychology
University of Lille
Lille, France

Frédérique de Vignemont, PhD
Director of Research
Institut Jean Nicod
Department of cognitive studies
ENS, EHESS, CNRS, PSL University
Paris, France

H.C. Dijkerman, DPhil
Professor of Neuropsychology
Helmholtz Institute, Experimental
Psychology
Utrecht University
Utrecht, The Netherlands

Alessandro Farnè, PhD
Director of Research
Integrative Multisensory Perception
Action & Cognition Team (ImpAct)
INSERM, CNRS, and Lyon Neuroscience
Center
Lyon, France

Francesca Frassinetti, MD, PhD
Full Professor
Department of Psychology
University of Bologna
Bologna, Italy

Matthew Fulkerson, PhD
Associate Professor
Department of Philosophy
UC San Diego
La Jolla, CA, USA

Michael S.A. Graziano, PhD
Professor
Psychology and Neuroscience
Princeton University
Princeton, NJ, USA

Fadila Hadj-Bouziane, PhD
Researcher
Integrative Multisensory Perception Action &
Cognition Team (ImpAct)
INSERM, CNRS, and Lyon Neuroscience Center
Lyon, France

Suliann Ben Hamed, PhD
Director of the Neural Basis of Spatial Cognition
and Action Group
Institut des Sciences Cognitives Marc Jeannerod
(ISCMJ)
Lyon, France

Tina Iachini, PhD
Professor
Department of Psychology
University of Campania Luigi Vanvitelli
Caserta, Italy

G.D. Iannetti, MD, PhD
Professor
Ianettilab, UCL
London, UK

Colin Klein, PhD
Associate Professor
School of Philosophy
The Australian National University
Canberra, ACT, Australia

Alisa Mandrigin, PhD
Anniversary Fellow
Division of Law and Philosophy
University of Stirling
Stirling, UK

Mohan Matthen, PhD
Professor
Philosophy
University of Toronto
Toronto, ON, Canada

W.P. Medendorp, PhD
Principal Investigator
Donders Institute for Brain, Cognition and
Behaviour
Radboud University
Nijmegen, Netherlands

Anders Pape Møller, PhD
CNRS Senior Researcher
Ecologie Systematique Evolution
PARIS, France

Jean-Paul Noel, PhD
Post-Doctoral Associate
Center for Neural Science
New York University
New York City, NY, USA

Matthew Nudds, PhD
Professor of Philosophy
Department of Philosophy
University of Warwick
Coventry, UK

George D. Park, PhD
Senior Human Factors Engineer
Systems Technology, Inc.
Hawthorne, CA, USA

Giuseppe di Pellegrino, MD, PhD
Full Professor
Department of Psychology
University of Bologna
Bologna, Italy

Catherine L. Reed, PhD
Professor of Psychological Science and
Neuroscience
Claremont McKenna College
Claremont, CA, USA

Andrea Serino, PhD
Professor and Head of Myspace Lab
CHUV, University Hospital and University of
Lausanne
Lausanne, Switzerland

Hong Yu Wong, PhD
Professor and Chair of Philosophy of Mind and
Cognitive Science
University of Tübingen
Tübingen, Germany

Wayne Wu, PhD
Associate Professor
Department of Philosophy and Neuroscience
Institute
Carnegie Mellon University
Pittsburgh, PA, USA

PART I
PERCEPTION PREDICTION ACTION

1

Peripersonal space

A special way of representing space

Frédérique de Vignemont, Andrea Serino, Hong Yu Wong, and Alessandro Farnè

1.1 Introduction

It is easy to believe that our representation of the world is structured under a binary mode: there is the self and then there is the rest of the world. However, one can question this conception in light of recent evidence on the existence of what is known as 'peripersonal space', which we may think of as something like a buffer zone between the self and the world. As a provisional definition, we can say that peripersonal space corresponds to the immediate surroundings of one's body. Nonetheless, it should be noted, even at this early stage, that the notion of peripersonal space does not refer to a well-delineated region of the external world with sharp and stable boundaries. Instead it refers to *a special way of representing* objects and events located in relative proximity to what one takes to be one's body. Indeed, as we shall see at length in this volume, peripersonal processing displays highly specific multisensory and motor features, distinct from those that characterize the processing of bodily space and the processing of far space. Depending on the context, the same area of physical space can be processed as peripersonal or not. This does not entail that any location in space can be perceived as peripersonal, but only that there is some degree of flexibility. Now, when some area of space is perceived as peripersonal, it is endowed with an immediate significance for the subject. Objects in one's surroundings are directly relevant to the body, because of their potential for contact. With spatial proximity comes temporal proximity, which gives rise to specific constraints on the relationship between perception and action. There is no time for deliberation when a snake appears next to you. Missing it can directly endanger you. You just need to act.

It may then seem tempting to reduce the notion of peripersonal space to the notion of behavioural space, a space of actions, but this definition is both too wide and too narrow. It is too wide because the space of action goes beyond peripersonal space. Actions can unfold at a relatively long distance: for instance, I can reach for the book at the top of the shelf despite it being relatively distant from my current location at the writing desk. On the other hand, the definition is too narrow because the space that surrounds us is represented in a specific way no matter whether we plan to act on it or not. There is thus something quite unique about peripersonal space, which requires us to explore it in depth.

Despite the intuitive importance of peripersonal space, it is only recently that the significance of the immediate surroundings of one's body has been recognized by cognitive science. The initial discovery that parietal neurons respond to stimuli 'close to the body' was made by Leinonen and Nyman (1979), but it was Rizzolatti and his colleagues who described the properties of premotor neurons specifically tuned to this region of space in 1981 and named

Frédérique de Vignemont, Andrea Serino, Hong Yu Wong, and Alessandro Farnè, *Peripersonal space* In: *The World at Our Fingertips*. Edited by Frédérique de Vignemont, Andrea Serino, Hong Yu Wong, and Alessandro Farnè, Oxford University Press (2021).
© Oxford University Press. DOI: 10.1093/oso/9780198851738.003.0001.

it 'peripersonal space' (Rizzolatti et al., 1981). Detailed neurophysiological exploration of this still unknown territory had to wait until the late 1990s, in monkeys (e.g. Graziano and Gross, 1993) and in human patients (e.g. di Pellegrino et al., 1997). Since then, there has been a blooming of research on peripersonal space in healthy human participants with the help of new experimental paradigms (such as the cross-modal congruency effect, hereafter 'CCE', Spence et al., 2004) and new tools (such as virtual reality, Maselli and Slater, 2014). For the first time, leading experts on peripersonal space in cognitive psychology, neuropsychology, neuroscience, and ethology are gathered in this volume to describe the vast number of fascinating discoveries about this special way of representing closeness to one's body along with their ongoing research. For the first time too, these empirical results and approaches are brought into dialogue with philosophy.

Our aim in this introduction is not to summarize the 18 chapters that this volume includes. Instead, we offer an overview of the key notions in the field and the way they have been operationalized. We then consider some of the implications of peripersonal space for fundamental issues in the philosophy of perception and for self-awareness.

1.2 Theoretical and methological challenges

What is peripersonal space? Surprisingly, perhaps, this is one of the most difficult questions that the field has had to face these past 30 years. To better understand the notion of peripersonal space, it is helpful to contrast it with notions that may be more familiar in the literature and that are closely related. In particular, we shall target the notions of personal space, reaching space and egocentric space. We shall then turn to the different experimental paradigms and the questions they leave open.

1.2.1 Personal space

Historically, the idea that space around the subject is not represented uniformly was first described in ethology and in social psychology. On this approach, the boundaries are defined exclusively in social terms and vary depending on the type of social interactions, whether negative or positive. The Swiss biologist Heini Hediger (1950), director of the Zurich zoo, first described how animals react in specific ways depending on the proximity of approaching predators. Each of the distances between the prey and the predator (flight, defence, and critical) is defined in terms of the specific range of actions that it induces. Animals also react differently to conspecifics depending on their proximity. Hediger, for instance, distinguishes between personal distance (the distance at which the presence of other animals is tolerable) and social distance (the distance at which one needs to be to belong to the group). Later, the social psychologist Hall (1966) introduces more distinctions: intimate space, in which we can feel the warmth of another person's body (up to 45 cm); personal space, in which we can directly interact with the other (up to 1.2 m); social space, in which we can work or meet together (up to 3.6 m); and public space, in which we have no involvement with other people. He concludes: 'Each animal is surrounded by a series of bubbles or irregularly shaped balloons that serve to maintain proper spacing between individuals' (Hall, 1966, p. 10). Hall's work linked these distinctions to key features of human societies and described how they shape social interactions, how they vary between cultures (e.g. European vs Asian), and how they influence the design of architectural spaces (e.g. houses, workplaces).

Although of great interest, the analysis of most of these bubbles remains purely descriptive. There is both the risk of unwarranted proliferation and that of undue conflation across concepts and terminology. Moreover, space is only considered in its social dimension. The definition of peripersonal space, by contrast, includes not only the proximity of individuals but of objects too. It would thus be a category mistake to confuse personal space and peripersonal space. Indeed, some studies point to dissociations between the two (Patané et al., 2017). Clearly marking the distinction between the two notions, however, should not prevent us from exploring their relationship.

1.2.2 Reaching space

Another notion that is often discussed together with peripersonal space is the notion of reaching space. It is functionally defined as the distance at which an object can be reached by the subject's hand without moving her trunk. The two notions are sometimes reduced one to the other, but we believe that they should be carefully distinguished. To start with, reaching space is typically larger than peripersonal space although they can spatially overlap. A second difference between the two notions is that reaching space refers to a unique representation that is shoulder-centred. By contrast, there are several distinct representations of peripersonal space, which are centred respectively on the hand, the head, the torso, and the feet. A third difference is at the neural level. Both the representations of reaching space and of peripersonal space involve fronto-parietal circuits but they include different brain areas. Reaching representation is associated with the dorsal premotor cortex, primary motor cortex, supplementary motor area (SMA), and parietal areas 5 and 7. On the other hand, peripersonal representation is associated with the ventral and dorsal premotor cortex, ventral intraparietal area, intraparietal sulcus (IPS), and putamen.

The final major difference concerns the functional role of these two representations. Most research in cognitive neuroscience has restricted their investigation to bodily movements such as reaching, grasping, or pointing. These movements allow us to act on the world, to explore it, and to manipulate objects. From an evolutionary point of view, these are the movements whose ultimate function is to find food and to eat. Reaching space is exclusively concerned by this type of movement. But there is another class of movements, possibly even more important, whose function concerns a different dimension of survival—namely, self-defence. These movements are sometimes summarized by the famous '3 Fs': freeze, fight, flight. One should not think, however, that we engage in protective behaviour only when there are predators. In everyday life, we avoid obstacles in our path, we retract our hand when coming too close to the fire, we tilt our shoulder when walking through a door, and so forth. Unlike the representation of reaching space, the representation of peripersonal space plays this dual role: to engage with the world but also to protect oneself from the world (de Vignemont, 2018).

1.2.3 Egocentric space

Standard accounts of perception acknowledge the importance of the spatial relation between the subject and the perceived object: objects are seen on the left or right, as up or down, and even as close or far from the subject's body. These egocentric coordinates are especially important for action planning. One might then wonder what is so special about peripersonal perception. For some, there is actually no difference between egocentric space and peripersonal space. They are both body-centred and have tight links to action (Briscoe, 2009; Ferretti, 2016). However, it is important to clearly distinguish the frame of reference that is exploited by peripersonal perception from the egocentric frame of perception in general. Egocentric location of objects is encoded in external space, and not in bodily space. Although body parts, such as the eyes, the head, and the torso, are used to anchor the axes on which the egocentric location is computed, the egocentric location is not on those body parts themselves. By contrast, the perceptual system anticipates objects seen or heard in peripersonal space to be in contact with one part or another of the body. Hence, although the objects are still located in external space, they are also anticipated to be in bodily space. The reference frame of peripersonal perception is thus similar to the somatotopic frame used by touch and pain (also called 'skin-based' or 'bodily' frame). To illustrate the difference from the egocentric frame, consider the following example:

> There is a rock next to my right foot (somatotopic coordinates) on my right (egocentric coordinates). I then cross my legs. In environmental space, the egocentric coordinates do not change (I still see the rock on my right) but the rock is now close to my left foot, and the somatotopic coordinates thus change.

An interesting hypothesis is that the multisensory-motor mechanism underlying peripersonal processing contributes to egocentric processing by linking multisensory processing about the relationship between bodily cues and environmental cues (computed by posterior parietal and premotor areas) with visual and vestibular information about the orientation of oneself in the environment (computed in the temporo-parietal junction). This proposal might provide insight into how egocentric space processing is computed, which is less studied and less well understood compared with allocentric space processing, which is computed by place and grid cells in medio-temporal regions (Moser et al., 2008).

1.2.4 Probing peripersonal space

We saw that one of the main challenges in the field is to offer a satisfactory definition of peripersonal space that is specific enough to account for its peculiar spatial, multisensory, and motor properties. There have been a multitude of proposals but they remain largely controversial and have given rise to much confusion. Another source of confusion can be found in the multiplicity of methods used to experimentally investigate peripersonal space. Using diverse experimental tasks, different studies have highlighted specific measures and functions of peripersonal space (see Table 1.1).

Emphasis can be put either on perception or on action, on impact prediction or on defence preparation. A prominent category of studies employed tactile detection or discrimination tasks, using multisensory stimulation to probe perceptual processes. One of these approaches (the CCE and its variants) consists in the presentation of looming audio-visual stimuli combined with tactile ones. It has consistently revealed the fairly limited extension of the proximal space for which approaching stimuli can facilitate tactile detection. It has also been used to show the plasticity of the representation of peripersonal space after manipulations such as the rubber hand illusion, full body illusion, and tool use. Other paradigms focus on bodily capacities. For instance, the reachability judgement task probes the effect of perceived distance on the participants' estimate of their potential for action. The

Table 1.1 Experimental measures of peripersonal space

Stimulus Modality	Stimulus Position	Response	Peripersonal Effect
Visual & Tactile (V-T) Auditory & Tactile (A-T) *V & A task irrelevant* *T target modality*	Visual and auditory at various distances or looming Touch on the hand, head, or trunk	Tactile detection	Better accuracy and/ or faster reaction times with V closer to the body
Visual & Tactile (V-T) *V task irrelevant* *T target modality*	Visual at various distances Touch on index or thumb	Tactile discrimination (index or thumb)	Slower reaction times with V closer to the body
Visual *V target modality*	Visual at various distances	Visual object detection or discrimination	Faster reaction times with V closer to the body
Visual *V target modality*	Visual at various distances	Judgement if reachable by the hand or not (without moving)	Slower reaction times and accuracy at chance level (50%, i.e. threshold) when V presented at around arm length
Strong tactile or nociceptive *No task*	Touch at the hand, kept at various distances from the face	Hand-blink reflex (HBR; motor-evoked response recorded from the face)	HBR increases at shorter hand–face distances
Visual & Tactile (V-T) *V task irrelevant* *T target modality*	Visual object far from hand Touch on index or thumb of the grasping hand	Tactile discrimination (index or thumb)	Slower reaction times when planning to move or moving toward object

extent of peripersonal space and its dynamic features have also been tackled by joining perceptual multisensory and action tasks concurrently. Finally, in probing the defensive function of peripersonal space, the relative proximity to the head of nociceptive or startling tactile stimuli has been shown to modulate physiological responses (e.g. the hand blink reflex).

As can be seen from inspection of the (non-exhaustive) Table 1.1, a variety of methods have been proposed to study the representation of peripersonal space, which likely tap into diverse physiological and psychological mechanisms. There is a risk of losing the homogeneity of the notion of peripersonal space within this multiplicity of methods. Ideally, one should design experimental paradigms for them to best investigate the notion to study. However, in the absence of a robust definition, the tasks may come first and the notion is constantly redefined on the basis of their results. If there is more than one method, this may give rise to a multiplicity of notions. We believe that it will be beneficial to use multiple tasks to address the same question, especially for the purposes of determining whether there are different notions of peripersonal representation. Beyond the methodological value of such an effort, this route holds the potential to provide more operationally defined and testable definitions of peripersonal space, with the promise of strengthening the theoretical foundations of the notion.

In addition to the central issue of 'which test for which peripersonal space?', we want to introduce some of the many questions for future research to consider from both an empirical and a theoretical perspective.

Outstanding questions:

1. Can we distinguish purely peripersonal space processes from attentional ones? Do we even need the concept 'attention' for near stimuli, if we characterize peripersonal space as a peculiar way of processing stimuli occurring near the body? And, conversely, do we need the concept 'peripersonal space' if it is only a matter of attention?

2. Is peripersonal space a matter of temporal immediacy in addition to spatial immediacy? How do the spatial and temporal factors interact?

3. Can there be cognitive penetration of peripersonal perception? At what stage does threat evaluation intervene?

4. Can there be a global whole-body peripersonal representation in addition to local body part-based peripersonal representations? If so, how is it built? What is its relation to egocentric space?

5. To what extent does peripersonal space representation, as a minimal form of the representation of the self in space for interaction, relate to allocentric representation of space for navigation?

6. Did peripersonal space evolve as a tool for survival? What was peripersonal space for? What is it for today? What will it be for tomorrow? Do we still need peripersonal space in a future in which brain–machine interfaces feature heavily?

7. Are there selective deficits of peripersonal space or are they always associated with other spatial deficits? What impact do they have on other abilities?

8. What are the effects of environmental and social factors, known to be relevant to personal space, on peripersonal space?

9. What are the effects of peripersonal space on social interactions? For instance, is emotional contagion or joint action facilitated when the other is within one's peripersonal space?

10. How are fear and pain related to the defensive function of peripersonal perception?

1.2 Philosophical implications

The computational specificities of the processing of peripersonal space are such that one can legitimately wonder whether peripersonal processing constitutes a sui generis psychological kind of perception. Whether it is the case or not, one can ask whether the general laws of perception, as characterized by our philosophical theories, apply to the special case of the perception of peripersonal space. More specifically, one needs to reassess the relationship between perception, action, emotion, and self-awareness in the highly special context of the immediate surroundings of one's body. Here we briefly describe the overall directions that some of these discussions might take.

1.2.1 Self-location and body ownership

Where do we locate ourselves? Intuitively, we locate ourselves where our bodies are, and this is also the place that anchors our egocentric space and our peripersonal space. Discussions on egocentric experiences have emphasized the importance of self-location for perspectival experiences. One might even argue that one cannot be aware of an object as being on the left or the right if one is not aware of one's own location. Being aware of the object in egocentric terms involves as much information about the perceived object as about the subject that perceives it. It has thus been suggested that the two terms of the relation (the perceived object and the self) are both experienced: when I see the door on the right, my visual content represents that it is on *my* right (Schwenkler, 2014; Peacocke, 2000). Many, however, have rejected this view (Evans, 1982; Campbell, 2002; Perry, 1993; Schnellenberg, 2007). They argue that there is no need to represent the self, which can remain implicit or unarticulated. A similar question can be raised concerning peripersonal space. Spatial proximity, which plays a key role in the definition of peripersonal space, is a *relational* property. One may then ask questions about the terms of the relation: are objects perceived close to the self, or only to the body? And is the self, or the body, represented in our peripersonal experiences? Another way to think about this is to ask: does peripersonal perception require some components of self-awareness? Or is it rather the other way around: can peripersonal perception ground some components of self-awareness?

Interestingly, it has been proposed that multisensory integration within peripersonal space is at the basis of body ownership and that this can be altered by manipulating specific features of sensory inputs (Makin et al., 2008; Serino, 2019). For instance, in the rubber hand illusion, tactile stimulation on the participant's hand synchronously coupled with stimulation of a rubber hand induces an illusionary feeling that the rubber hand is one's own (see the enfacement illusion, the full body illusion, or the body swap illusion for analogous manipulations related to the face or the whole body). The hypothesis here is that the body that feels as being one's own is the body that is surrounded by the space processed as being peripersonal. On this view, peripersonal space representation plays a key role in self-consciousness. A crucial question is how such low-level multisensory-motor mechanisms relate to the subjective experience of body ownership. A further question is how self-location, body location and the 'zero point' from which we perceive the world normally converge. A fruitful path to investigate their mutual relationship is to examine 'autoscopic phenomena', such as out-of-body experiences and heautoscopic experiences in patients (Blanke and Arzy, 2005), and in full body illusions (Lenggenhager et al., 2007; Ehrsson, 2007). In these cases, subjects tend to locate themselves toward the location of the body that they experience as owning. Likewise, the extent and shape of peripersonal space are altered along with the induction of illusory ownership toward the location of the illusory body (Serino et al., 2015). This suggests that peripersonal space is tied to self-location and body ownership—that peripersonal space goes with self-location, and that the latter goes with body ownership. This raises interesting questions about cases of heautoscopic patients who have illusions of a second body, which they own, and who report feeling located in both bodies. What does their peripersonal space look like? Measuring peripersonal space appears to provide us with an empirical and conceptual tool for thinking about self-location.

1.2.2 Sensorimotor theories of perception

In what manner does perceptual experience contribute to action? And, conversely, in what manner does action contribute to perceptual experience? The relationship between perceptual experience and action has given rise to many philosophical and empirical debates, which have become more complex since the discovery of dissociations between perception and action (Milner and Goodale, 1995). For instance, in the 'hollow face' illusion, a concave (or hollow) mask of a face appears as a normal convex (or protruding) face, but, if experimental subjects are asked to quickly flick a magnet off the nose (as if it were a small insect), they direct their finger movements to the actual location of the nose in the hollow face. In other words, the content of the illusory visual experience of the face does not correspond to the visually guided movements directed toward the face (Króliczak et al., 2006). Although not without controversy, results such as these have been taken as evidence against sensorimotor theories of perception that claim that action is constitutive of perceptual experience (Noë, 2004; O'Regan, 2011). They can also be taken as an argument against what Clark (2001) calls the assumption of experience-based control, according to which perceptual experience is what guides action. Here we claim that empirical findings on peripersonal space can shed a new light on this debate and, more specifically, that they can challenge a strict functional distinction between perception and action in one's immediate surroundings.

No matter how complex actions can be and how many sub-goals they can have, all bodily movements unfold in peripersonal space. Typically, while walking, the step that I make is made in the peripersonal space of my foot and, while I move forward, my peripersonal space follows—or, better, it anticipates my body's future position. Action guidance thus depends on the constant fine-grained monitoring and remapping of peripersonal space while the movement is planned and performed. It is no surprise, then, that peripersonal space is mainly represented in brain regions that are dedicated to action guidance. In addition, the practical knowledge of one's motor abilities determines whether objects and events are processed as being peripersonal or not. Consider first the case of tool use. One can act on farther objects with a tool than without. This increased motor ability leads to a modification of perceptual processing of the objects that are next to the tool. After tool use, these objects are processed as being peripersonal (e.g. Iriki et al., 1996; Farnè and Làdavas, 2000). Consider now cases in which the motor abilities are reduced. It has been shown that, after ten hours of right arm immobilization, there is a contraction of peripersonal space such that the distance at which an auditory stimulus is able to affect the processing of a tactile stimulus is closer to the body than before (Bassolino et al., 2015). It may thus seem that the relationship between perception and action goes both ways in peripersonal space: peripersonal perception is needed for motor control whereas motor capacities influence peripersonal perception. Can one then defend a peripersonal version of the sensorimotor theory of perception? And, if so, what shape should it take? Although all researchers working on peripersonal space agree on the central role of action, little has been done to precisely articulate this role (for discussion, see de Vignemont, forthcoming).

A first version is to simply describe peripersonal perception exclusively in terms of unconscious sensorimotor processing subserved by the dorsal visual stream. Interestingly, in the hollow face illusion, the face was displayed at fewer than 30 cm from the participants. Hence, even in the space close to us, the content of visual experiences (e.g. convex face) does not guide fine-grained control of action-oriented vision (e.g. hollow face). On this view, there is nothing really special about peripersonal space. We know that there is visuomotor processing, and this is true whether what one sees is close or far. This does not challenge the standard functional distinction between perception and action.

However, this account does not seem to cover the purely perceptual dimension of peripersonal processing. As described earlier, most multisensory tasks, which have become classic measures of peripersonal space, do not involve action at all. Participants are simply asked to either judge or detect a tactile stimulus, while seeing or hearing a stimulus closer or farther from the body. In addition, visual shape recognition is improved in peripersonal space (Blini et al., 2018). This seems to indicate that there is something special about *perceptual experiences* in peripersonal space. And because motor abilities influence these perceptual effects, as in the case of tool use, these peripersonal experiences must bear a close relationship to action. One may then propose to apply the classic perception-action model to the perception of peripersonal space. At the unconscious sensorimotor level, peripersonal processing provides the exact parameters about the immediate surroundings required for performing the planned movements. But at the perceptual level, peripersonal experiences also play a role. One may propose, for instance, that peripersonal experiences directly contribute in selecting at the motor level the type of movement to perform, such as arm withdrawal when an obstacle appears next to us.

Here one may reply that there is still nothing unique to peripersonal space. Since Gibson (1979), many have defended the view that we can perceive what have been called 'affordances'—that is, dispositions or invitations to act, and we can perceive them everywhere (e.g. Chemero, 2003). The notion of affordance, however, is famously ambiguous. One may then use instead Koffka's (1935) notion of *demand character*, which actually inspired Gibson. For instance, Koffka describes that you feel that you *have to* insert the letter in the letterbox when you encounter one. One may assume that you do not experience the pull of such an attractive force when the letterbox is far. In brief, the perception of close space normally presents the subject with actions that she needs to respond to. But again, is this true only of peripersonal space? When I see my friend entering the bar, I can feel that I have to go to greet her. The difference between peripersonal space and far space in their respective relation to action may then be just a matter of degree. One may claim, for instance, that one is more likely to experience demand characters in peripersonal space than in far space. This gradient hypothesis has probably some truth in it, but the direction of the gradient remains to be understood: why is the force more powerful when the object is seen as close? Bufacchi and Iannetti (2018) have recently proposed an interesting conceptualization of peripersonal space as an 'action field' or, more properly, as a 'value field'—that is, as a map of valence for potential actions. Why and how such a field is referenced and develops around the body, and how it is related to attention, remain to be explained.

1.2.3 Affective perception

If there is a clear case in which the objects of one's experiences have demand characters, then seeing a snake next to one's foot is such a case. One feels that one has to withdraw one's foot. But then it raises the question of the relationship between peripersonal perception and evaluation, and about the admissible contents of perception. Can *danger* be perceptually represented? More specifically, does one visually experience the snake close to one's foot as being dangerous?

A conservative approach to perception would most probably reply that danger is represented only at a post-perceptual stage. On this view, one visually experiences the shape and the colour of the snake, on the basis of which one forms the belief that there is a snake, which then induces fear, which then motivates protective behaviour. Danger awareness is manifested only later, in cognitive and conative attitudes that are formed on the basis of visual experiences. However, a more liberal approach to perception has recently challenged the conservative approach and argued that the content of visual experiences can also include higher-level properties (Siegel, 2011). On this view, for example, one can visually experience causation. Interestingly, discussions on the admissible contents of perception have been even more recently generalized to evaluative properties, but so far they have focused on aesthetic and moral properties (Bergqvist and Cowan, 2018). Can one see the gracefulness of a ballet? Can one see the wrongness of a murder? The difficulty in these examples is that assessing such aesthetic and moral properties requires sophisticated evaluative capacities that are grounded in other more or less complex abilities and knowledge. Danger, on the other hand, is a more basic evaluative property, possibly the most basic one from an evolutionary standpoint. To be able to detect danger is the first—although not the only—objective to meet for an organism to survive. Because of the primacy of danger detection, one can easily conceive that it needs to occur very early on, that it does not require many conceptual resources, and that it needs to be in direct connection with action. This provides some intuitive plausibility to the hypothesis that danger can be visually experienced.

This hypothesis seems to find empirical support in the existence of peripersonal perception. For many indeed, peripersonal space is conceived of as a margin of safety, which is encoded in a specific way to elicit protective behaviours as quickly as possible if necessary. Graziano (2009), for instance, claims that prolonged stimulation of regions containing peripersonal neurons in monkeys triggers a range of defensive responses such as eye closure, facial grimacing, head withdrawal, elevation of the shoulder, and movements of the hand to the space beside the head or shoulder. Disinhibition of these same regions (by bicuculline) leads the monkeys to react vividly even for non-threatening stimuli (when seeing a finger gently moving toward the face, for instance), whereas their temporary inhibition (by muscimol) has the opposite effect: monkeys no longer blink or flinch when their body is under real threat. In humans, just presenting stimuli close by (Makin et al., 2009; Serino et al., 2009) or as approaching (Finisguerra et al., 2015) results in the hand modulating the excitability of the cortico-spinal tract so as to implicitly prepare a potential reaction. When fake hands, for which subjects have an illusory sense of ownership, are similarly approached, the corticospinal modulation occurs as rapidly as within 70 milliseconds from vision of the approaching stimulus. What is more interesting is that this time delay is sufficient for the peripersonal processing to distinguish between right and left hands (Makin et al., 2015). This is precisely the kind of fast processing one would need to decide which hand of one's body is in danger and needs to be withdrawn. In short, having a dedicated sensory mechanism that is specifically tuned to the immediate surroundings of the body, and directly related to the motor system, is a good solution for detecting close threats and for self-defence.

There are, however, many ways that one can interpret how one perceives danger in peripersonal space. We shall here briefly sketch three possible accounts:

1. Sensory account: the sensory content includes the property of danger.
2. Attentional account: attentional priority is given to some low-level properties of the sensory content.
3. Affective account: the sensory content is combined with an affective content.

The first interpretation is simply that the perceptual content includes the property of danger along the properties of shape, colour, movement, and so forth. One way to make sense of this liberal hypothesis is that danger is conceived here as a natural kind. On this view, when you see a snake next to your foot, you visually experience not only the snake but also the danger. The difficulty with such an account, however, is that too many things can be dangerous. Although there are many different kinds of snakes, there is still something like a prototypical sensory look of snakes. By contrast, there is no obvious sensory look of danger.

In brief, there is a multidimensional combination of perceptual features (e.g. shapes, colours, sounds...) characterizing threats and it is impossible to find a prototype. Furthermore, what is perceived as dangerous depends on one's appraisal of the context. Unlike natural kinds, danger is an evaluative property.

One may then be tempted by a more deflationary interpretation in attentional terms. It has been suggested that we have an attentional control system that is similar in many respects to bottom-up attention but that is specifically sensitive to stimuli with affective valence (Vuilleumier, 2015). It occurs at an early sensory stage, as early as 50–80 milliseconds after stimulus presentation. Consequently, it can shape perceptual content by highlighting specific information. This attentional account is fully compatible with a conservative approach to perception. Indeed, giving attentional priority to some elements of the perceptual content is not the same as having perceptual content representing danger as such. However, one may wonder whether this deflationary account suffices to account for our phenomenology. There is a sense indeed in which we can *experience* danger. In addition, attention can possibly help in prioritizing tasks, but would attention help in interpreting the affective valence within such a short delay?

A possibility then is to propose that we experience danger not at the level of the perceptual content, but at the level of an associated affective content. On this last interpretation, perception can have a dual content, both sensory and affective. One may then talk of affective colouring of perception (Fulkerson, 2020) but this remains relatively metaphorical. A better way to approach this hypothesis might be by considering the case of pain. It is generally accepted that pain has two components, a sensory component that represents the location and the intensity of the noxious event, and an affective component that expresses its unpleasantness. One way to account for this duality in representational terms is known as 'evaluativism' (Bain, 2013): pain consists in a somatosensory content that represents bodily damage and that content also represents the damage as being bad. One may then propose a similar model of understanding for the visual experience of the snake next to one's foot: danger perception consists in visual content that represents a snake and that content also represents the snake as being bad.

We believe that a careful analysis of the results on peripersonal space and its link to self-defence should shed light on this debate, and, more generally, raise new questions about the admissible contents of perception.

The series of chapters included in the present volume directly or indirectly tap into these (and other) open questions about the nature of our interactions with the environment, for which peripersonal space is a key interface. Without aiming to be exhaustive, we hope that the contributions collected here will provide an updated sketch of the state of the art and provide new insights for future research.

References

Bain, D. (2013). What makes pains unpleasant? *Philosophical Studies, 166*, S69–S89.

Bassolino, M., Finisguerra, A., Canzoneri, E., Serino, A., & Pozzo, T. (2015). Dissociating effect of upper limb non-use and overuse on space and body representations. *Neuropsychologia, 70,* 385–392.

Bergqvist, A., & Cowan, R. (eds) (2018). *Evaluative perception.* Oxford: Oxford University Press.

Blanke, O., & Arzy, S. (2005). The out-of-body experience: disturbed self-processing at the temporo-parietal junction. *Neuroscientist, 11*(1), 16–24.

Blini, E., Desoche, C., Salemme, R., Kabil, A., Hadj-Bouziane, F., & Farnè, A. (2018). Mind the depth: visual perception of shapes is better in peripersonal space. *Psychological Science, 29*(11), 1868–1877.

Briscoe, R. (2009). Egocentric spatial representation in action and perception. *Philosophy and Phenomenological Research, 79*, 423–460.

Bufacchi, R. J., & Iannetti, G. D. (2018). An action field theory of peripersonal space. *Trends in Cognitive Sciences, 22*(12), 1076–1090.

Campbell, J. (2002). *Reference and consciousness.* New York: Oxford University Press.

Chemero, A. (2003). An outline of a theory of affordances. *Ecological Psychology, 15*(2), 181–195.

Clark, A. (2001). Visual experience and motor action: are the bonds too tight? *Philosophical Review, 110*(4), 495–519.

de Vignemont, F. (2018). *Mind the body: an exploration of bodily self-awareness.* Oxford: Oxford University Press.

de Vignemont, F. (forthcoming). Peripersonal perception in action. *Synthese.*

di Pellegrino, G., Làdavas, E., & Farnè, A. (1997). Seeing where your hands are. *Nature, 388*, 730.

Ehrsson, H. H. (2007). The experimental induction of out-of-body experiences. *Science, 317*(5841), 1048.

Evans, G. (1982). *The varieties of reference.* Edited by John McDowell. Oxford: Oxford University Press.

Farnè, A., & Làdavas, E. (2000). Dynamic size-change of hand peripersonal space following tool use. *Neuroreport, 11*(8), 1645–1649.

Ferretti, G. (2016). Visual feeling of presence. *Pacific Philosophical Quarterly, 99*, 112–136.

Finisguerra, A., Canzoneri, E., Serino, A., Pozzo, T., & Bassolino, M. (2015). Moving sounds within the peripersonal space modulate the motor system. *Neuropsychologia, 70*, 421–428.

Fulkerson, M. (2020). Emotional perception. *Australasian Journal of Philosophy, 98*(1), 16–30.

Gibson, J. J. (1979). *The ecological approach to visual perception.* Boston: Boston Mifflin.

Graziano, M. (2009). *The intelligent movement machine: an ethological perspective on the primate motor system.* New York: Oxford University Press.

Graziano, M. S., & Gross, C. G. (1993). A bimodal map of space: somatosensory receptive fields in the macaque putamen with corresponding visual receptive fields. *Experimental Brain Research, 97*(1), 96–109.

Hall, E. T. (1966). *The hidden dimension.* New York: Doubleday & Co.

Hediger, H. (1950). *Wild animals in captivity.* London: Butterworths Scientific Publications.

Iriki, A., Tanaka, M., & Iwamura, Y. (1996). Coding of modified body schema during tool use by macaque postcentral neurones. *Neuroreport, 7*(14), 2325–2330.

Koffka, K. (1935). *Principles of Gestalt psychology.* London: Kegan Paul, Trench, Trubner & Co.

Króliczak, G., Heard, P., Goodale, M. A., & Gregory, R. L. (2006). Dissociation of perception and action unmasked by the hollow-face illusion. *Brain Research, 1080*(1), 9–16.

Leinonen L., & Nyman G. (1979). II. Functional properties of cells in anterolateral part of area 7 associative face area of awake monkeys. *Experimental Brain Research, 34*, 321–333.

Lenggenhager, B., Tadi, T., Metzinger, T., & Blanke, O. (2007). Video ergo sum: manipulating bodily self-consciousness. *Science, 317*(5841), 1096–1099.

Makin, T. R., Brozzoli, C., Cardinali, L., Holmes, N. P., & Farnè, A. (2015). Left or right? Rapid visuo-motor coding of hand laterality during motor decisions. *Cortex, 64*, 289–292.

Makin, T. R., Holmes, N. P., Brozzoli, C., Rossetti, Y., & Farnè, A. (2009). Coding of visual space during motor preparation: approaching objects rapidly modulate corticospinal excitability in hand-centered coordinates. *Journal of Neuroscience*, *29*(38), 11841–11851.

Makin, T. R., Holmes, N. P., & Ehrsson, H. H. (2008). On the other hand: dummy hands and peripersonal space. *Behavioural Brain Research*, *191*(1), 1–10.

Maselli, A., & Slater, M. (2014). Sliding perspectives: dissociating ownership from self-location during full body illusions in virtual reality. *Frontiers in Human Neuroscience*, *8*, 693.

Milner, A. D., & Goodale, M. A. (1995). *The visual brain in action*. Oxford: Oxford University Press.

Moser, E. I., Kropff, E., & Moser, M.-B. (2008). Place cells, grid cells, and the brain's spatial representation system. *Annual Review of Neuroscience*, *31*, 69–89.

Noë, A. (2004). *Action in perception*. Cambridge, MA: The MIT Press.

O'Regan, J. K. (2011). *Why red doesn't sound like a bell: understanding the feel of consciousness*. New York: Oxford University Press.

Patané, I., Farnè, A., & Frassinetti, F. (2017). Cooperative tool-use reveals peripersonal and interpersonal spaces are dissociable. *Cognition*, *166*, 13–22.

Peacocke, C. (2000). *Being known*. New York: Oxford University Press.

Perry, J. (1993). Thought without representation. In his *The problem of the essential indexical and other essays*, 205–226. New York: Oxford University Press.

Rizzolatti, G., Scandolara, C., Matelli, M., & Gentilucci, M. (1981). Afferent properties of periarcuate neurons in macaque monkeys. II. Visual responses. *Behavioural Brain Research*, *2*(2), 147–163.

Schnellenberg, S. (2007). Action and self-location in perception. *Mind*, *116*, 603–631.

Schwenkler, J. (2014). Vision, self-location, and the phenomenology of the 'point of view'. *Noûs*, *48*(1), 137–155.

Serino, A. (2019). Peripersonal space (PPS) as a multisensory interface between the individual and the environment, defining the space of the self. *Neuroscience & Biobehavioral Reviews*, *99*, 138–159.

Serino, A., Annella, L., & Avenanti, A. (2009). Motor properties of peripersonal space in humans. *PloS ONE*, *4*(8), e6582.

Serino, A., Noel, J. P., Galli, G., Canzoneri, E., Marmaroli, P., Lissek, H., & Blanke, O. (2015). Body part-centered and full body-centered peripersonal space representations. *Scientific Reports*, *5*, 18603.

Siegel, S. (2011). *The contents of visual experience*. New York: Oxford University Press.

Spence, C., Pavani, F., & Driver, J. (2004). Spatial constraints on visual-tactile cross-modal distractor congruency effects. *Cognitive, Affective, and Behavioral Neuroscience*, *4*(2), 148–169.

Vuilleumier, P. (2015). Affective and motivational control of vision. *Current opinion in neurology*, *28*(1), 29–35.

2

Peri-personal space as an interface
for self-environment interaction

A critical evaluation and look ahead

Jean-Paul Noel, Tommaso Bertoni, and Andrea Serino

2.1 Introduction

In this chapter, we present the empirical evidence that has led researchers to conceive of the space near one's body, the peri-personal (PPS) space (or spaces; Brain, 1941; Hyvärinen & Poranen, 1974; Mountcastle, 1976; Leinonen & Nyman, 1979; Leinonen et al., 1979; Rizzolatti et al., 1981a, Rizzolatti et al., 1981b) as a multisensory-motor interface between the individual and the environment: as a spatial extension of the body, playing a key role in defensive and approaching behaviour. We emphasize that an array of fronto-parietal networks demonstrate distinctions between the near and far space, but not all of these neural networks are necessarily strictly encoding PPS. In the present chapter, our definition of PPS refers to a specific spatial representation resulting from the processing of a defined pool of multisensory neurons that possess receptive fields responding both to touch and to visual and/or auditory stimuli when these are presented near versus far from the body (Colby et al., 1993; Graziano et al., 1997; Duhamel et al., 1998). A certain portion of PPS neurons will evoke movements if electrically stimulated strongly enough (Graziano et al., 2002; Graziano & Cooke, 2006). Thus, PPS representation is considered multisensory and motor. However, it is also true that veritable integration (Sugihara et al., 2006; Bizley et al., 2007; Meijer et al., 2017) has seldom (if ever) been observed in PPS neurons. In fact, multisensory integration, at least as it was defined historically (Stein & Stanford, 2008)—in terms of non-linear additivity in firing rates—is seldom observed in the cortex generally. Similarly with the motor aspect, while micro-stimulation of PPS neurons in the pre-motor cortex arguably evokes stereotyped defensive-like behaviours that abate upon stopping stimulation, similar micro-stimulation of parietal PPS regions only evokes motor output upon high enough intensity so that this stimulation could potentially drive a whole array of regions to output motor responses (e.g. Mitz & Wise, 1987; Luppino et al., 1991). Further, this parietal stimulation evokes movements well beyond the duration of stimulation, which could putatively indicate that the parietal neurons themselves are not evoking movement, but solely catalyse the circuit that is causally involved in movement initiation and execution (Clery et al., 2015b; see Zhang & Britten, 2011, for evidence that VIP micro-stimulation is causally involved in heading perception, and not movement generation).

The first sections of this chapter will focus on the debate as to whether PPS is truly multisensory or motor. We will then present emerging areas of study within PPS, highlighting that the networks encoding for the near space may serve a larger, more abstract, more computationally-oriented purpose. To this regard, we review novel evidence in the human

Jean-Paul Noel, Tommaso Bertoni, and Andrea Serino, *Peri-personal space as an interface for self-environment interaction* In: *The World at Our Fingertips*. Edited by Frédérique de Vignemont, Andrea Serino, Hong Yu Wong, and Alessandro Farnè, Oxford University Press (2021). © Oxford University Press. DOI: 10.1093/oso/9780198851738.003.0002.

psychophysical literature suggesting that PPS is modulated by social context (Teneggi et al., 2013; Pellencin et al., 2017), and inclusively plays a role in scaffolding a primitive sense of self-awareness (Graziano et al., 2000; Noel et al., 2015a; Salomon et al., 2017). Lastly, we highlight recent computational models of PPS as a promising path forward for further research.

It is important to highlight that we do indeed conceive of PPS as an interface between self and environment (Serino et al., 2017). Our goal here is to stress where gaps still exist in an attempt to prompt further research into this endlessly fascinating topic of inquiry.

2.2 Back to the origins: what is peri-personal space and what neural networks are involved?

A number of fronto-parietal areas exist, both anatomically and functionally very close to one another. This proximity in brain regions, but most importantly in functional specialization, has at times lead researchers to use quick heuristics, and hence at times to some confusion (see Noel et al., 2018a, for our own example of a study confounding PPS with reaching space).

In the medial parietal area V6A (Luppino et al., 2005), for instance, a significant proportion of neurons are modulated by gaze position in three-dimensional space: azimuth, elevation and depth. These neurons are also modulated by vergence signals—that is, the simultaneous movement of the pupils of the eyes toward or away from one another while foveating on an object moving in depth. Interestingly, at a population level, these V6A neurons prefer fixation distances at about 30 cm from the monkey's body. A similar preference for eye fixations relatively near the body, in this case about 50 cm, has been reported for area 7a (Sakata et al., 1980).

Not far from V6A or area 7a, on the lateral bank of the intraparietal sulcus, lies LIP—the lateral intraparietal sulcus. In this area approximately three-quarters of neurons have discharge rates that are higher when visual stimuli are presented between the observer's body and their location of fixation (Genovesio & Ferraina, 2004). That is, they seemingly prefer visual disparity cues that indicate objects occurring closer than the fixation point. However, fixation depth may vary, and thus the absolute distance between the object and the observer cannot in principle be decoded from LIP. Interestingly, however, La Chioma et al. (2019) recently showed disparity selectivity for the near space in mice (higher-order visual rostrolateral area [RL]). In the case of the mouse, RL may indeed encode for PPS generally—absolute distance—as these animals do not saccade much.

Lastly, also in the parietal cortex, a set of neurons collectively form the so-called 'parietal reach region' (PRR; Andersen & Buneo, 2002). This region is situated medially from the ventral intraparietal cortex (VIP), facing LIP on the opposite bank of the intraparietal sulcus. These neurons have receptive fields that are multiplicatively stimulated through the combination of gaze and hand position—an operation known as 'gain fields' (see Blohm & Crawford, 2009) —which arguably allows them to perform a reference frame transformation, from eye-centered to hand-centered, and to encode the amount of hand displacement needed in order to reach a visual object. Thus, this area is widely considered to be a critical node in reaching behaviour (Andersen, 1997; Cohen & Andersen, 2002).

Now, while all these areas—namely, V6A, 7a, LIP and PRR/medial intraparietal area (MIP)—show some special response for objects or events occurring in the near space (defined in various formats: i.e. relative to reach, absolute, or relative fixation), in our opinion these areas do not strictly speaking encode PPS. This is because their depth preference is not absolute with respect to distance from the body, or, more regularly, because they do not respond to tactile stimulation on the body and their reference frame is most commonly eye-centered and not body-part centered. Contrarily, PPS representation is defined—at least to us—by the receptive fields of bimodal neurons that are anchored on specific body parts, and thus respond both to tactile stimulation on the body and to visual and/or auditory stimulation when this latter one occurs near the body. Neurons in VIP (Colby et al., 1993; Duhamel et al., 1997, 1998; Guipponi et al., 2013) and in area 7b (Hyvärinen, 1981; Leinonen et al., 1979; Leinonen & Nyman, 1979) show these characteristics. Beyond the parietal cortex, these features are also found in the pre-motor cortex, specifically in areas F4 and partially F5 (Gentilucci et al., 1988; Rizzolatti et al., 1988; Graziano et al., 1994; Gross & Graziano, 1995; Fogassi et al., 1996). Lastly, a report by Graziano & Gross (1993) suggests that PPS neurons are also present in subcortical areas, specifically in the putamen.

Using functional magnetic resonance imaging (fMRI) in non-human primates—and thus relying on hemodynamic as opposed to neurophysiological indices—Cléry et al. (2018) recently mapped all areas of the macaque brain that i) show a preference, and ii) are selective, for the near space. The same group has chartered the areas of the brain that respond both to visual and tactile information (Guipponi et al., 2015). The conjunction of these papers (Cléry et al., 2018; Guipponi et al., 2015), therefore, highlights all areas that according to fMRI are strictly speaking PPS: multisensory, including the tactile modality, and selective for the near space. These fMRI papers pointed to many of the same areas physiological measures had emphasized, while perhaps accentuating relatively early sensory areas in the visual and somatosensory hierarchy (see Cléry et al., 2018; Guipponi et al., 2015, for details).

Interestingly, the location of these putative PPS areas in non-human primates largely correspond to regions of the human brain that neuroimaging studies described as differently responding to stimuli near the body versus stimuli far apart (that is potentially reflecting global responses of PPS neurons whose multisensory receptive fields are activated by near-body stimulation). A recent quantitative meta-analysis on these findings reported consistent PPS-specific activations in humans in the inferior parietal lobule, intraparietal sulcus, superior parietal lobule, primary somatosensory cortex, and ventral/dorsal premotor cortex (Grivaz et al., 2017). In summary, there are a whole host of fronto-parietal regions that broadly can be construed as involved in translating perceptual representation (mainly from vision and in eye-centered reference frames) into formats appropriate for motor output. Many of these regions show certain selectivity for given regions of space, but this is not sufficient to represent PPS. Instead, encoding of PPS is implemented by bimodal responses (from a neurons population found in areas VIP/7b and F4/F5): a tactile response on the body, and an exteroceptive one (vision or audition) occurring in a limited space near the body. The exteroceptive component of the receptive fields is depth-limited, with the body as reference. Reaching, for instance, has strong similarities to PPS, in that it can only occur near the body, yet the neural underpinning of reaching involves a somewhat distinct neural network and representation.

In the next section, we first note idiosyncrasies within and between the circuits encoding for PPS. Subsequently, we narrow in to describe particularities of individual PPS cells, which are illustrative of the potential role, or roles, of PPS.

2.3 Properties of peri-personal space neurons and circuits: clues as to utility

Given that the space near one's body is encoded by bimodal neurons in VIP, 7b, F4, and F5, and neurons do not work in isolation, we can start delineating PPS networks (see Cléry et al.,

2018; Guipponi et al., 2015). PPS networks have been defined in terms of their supported function (i.e. defensive vs approaching) or in terms of their mapped body part/portion of PPS. In terms of function, a first circuit is that constituted by projections from parietal area VIP to prefrontal area F4 (Rizzolatti & Luppino, 2001; Matelli & Luppino, 2001). These areas are strongly interconnected anatomically yet show diverging properties (Matelli & Luppino, 2001). While visual information in F4 is completely mapped with respect to limbs, thus possessing congruent visuo-tactile receptive fields, in VIP this is seemingly solely true for tactile receptive fields on the head and face. For the rest (e.g. tactile receptive fields on the arm), the visual receptive fields are either in an eye-centered reference frame, or show some sort of mixed selectivity (Duhamel et al., 1997; Avillac et al., 2005). As a result of the seeming specialization of F4 in limb representation, and of VIP in head/face representation, researchers (e.g. Cléry et al., 2015b) have hypothesized that, within this VIP-F4 network, VIP functions more in a sensory/perceptual capacity than in the role of informing F4 about reaching or hand-centered action possibilities. This conjecture is supported by electrical stimulation studies wherein electrical perturbation of VIP produces characteristically defensive-like movements such as eye squinting and blinking, ear folding, shoulder shrugging, lifting of the upper lip in grimace, and retraction of the face as well as lifting of the arm contralateral to stimulation (Thier & Andersen, 1998; Cooke & Graziano, 2003; Stepniewska et al., 2005). Likewise, micro-stimulation of F4 produces either a motor repertoire similar to that described above for VIP, or results in fast withdrawal of the hand to a protective posture behind the back (Cooke & Graziano, 2004; Graziano et al., 2002; Graziano & Cooke, 2006). Thus, the VIP-F4 network seems to most closely correspond to a defensive network.

The second network possessing bimodal neurons that respond both to touch and to audio and/or vision when presented near, is that composed of parietal area 7b and prefrontal region F5. As the reader can probably guess, in contrast to the above-mentioned VIP-F4 network, this second network has been most closely associated with approaching behaviour (e.g. Cléry et al., 2015a). It has also been linked to select higher-order cognitive tasks (although the connection here is more tenuous) given the fact that some of these neurons seemingly respond not only when an exteroceptive stimuli is near one's own body, but also the body of others (Ishida et al., 2010). In more detail, a considerable number of area 7b neurons—but not all—respond to visual and tactile stimulation. Further, they also discharge to motor activity, be it simple grasping movements (Hyvärinen & Poranen, 1974; Hyvärinen & Shelepin, 1979; Leinonen et al., 1979; Leinonen & Nyman, 1979; Robinson et al., 1978) or more complex 'grasp to bring to the mouth' actions (Fogassi et al., 2005; Fogassi & Luppino, 2005). The ventral pre-motor F5 neurons are typically divided into 'canonical' and 'mirror' neuron categories. The former category of neurons discharge to visual stimuli, and to action generated toward the mentioned objects. Hence, they are suggested to play a role in visuo-motor encoding for grasping behaviour (Murata et al., 1997; Raos et al., 2006). The latter category—'mirror' neurons—respond both to the execution of a given action and to the observation of the same action performed by another agent (Gallese et al., 1996; Rizzolatti et al., 1996). Now, the neurons present in area F5 are distinct from 'mirror' neurons that may be described in other cortical areas by the fact that, for these neurons to respond, graspable objects ought to be presented close to the body (Caggiano et al., 2009; Bonini et al., 2014; see Brozzoli et al., 2013, for a similar finding in humans). Namely, the performed and observed actions have to be spatially or proximity-wise 'feasible'.

Taken together, analyses of the general properties of the different PPS networks seem to support the conclusion that there are at least two different forms of PPS, one for approaching behaviour—which may also support a certain social role, given the presence of mirror neurons—and one for defensive or avoidance behaviour (see de Vignemont & Iannetti, 2014, for a similar bifurcation of PPS functions). These conclusions are further bolstered and perhaps most clearly exemplified by two landmark studies of PPS published in 1996 that we have not yet discussed. The first is by Fogassi and colleagues (1996), who recorded from PPS neurons in F4 (a putative 'PPS for defence' network) and showed that the size of visual receptive fields increased with the velocity of the incoming stimuli. Noel et al. (2018b) recently demonstrated that this dynamic velocity-dependent resizing of PPS receptive fields could be accounted for by neural adaptation within the PPS network itself. That is, the resizing of PPS as a function of incoming stimuli velocity does not depend on the PPS network coordinating with any other sort of neural network—for example, one attributing a value to the rapidly approaching stimuli. Instead, the resizing is intrinsic to the PPS network. The second now classical study of PPS is by Iriki and colleagues (1996). These researchers recorded from the anterior bank of the IPS (AIPS/7b, part of the putative 'PPS for action' network) and showed that, after monkeys utilized a rake in order to drag objects of interest closer to themselves, the size of their PPS receptive fields increased so as to incorporate the rake and the far space. Thus, the size of PPS neurons' receptive fields seem to change both dynamically (within trials: Fogassi et al., 1996; Noel et al., 2018b) and plastically (between trials: Iriki et al., 1996), and further reinforce the notion that PPS neurons serve as a multisensory-motor interface between the body and the environment, both for approaching and defensive behaviour. Of course, these putative networks (for action and for defence) are highly interconnected and likely convergent at points, for example in VIP, an area that responds when objects loom toward the face (Cléry et al., 2015a), as well as possessing self-generated heading signals (Chen et al., 2013; Shao et el., 2018).

In terms of represented body parts, another distinction—or gradient—can be proposed. Most VIP neurons show tactile receptive fields centered on the head/face, with visual receptive fields extending proportionally over the surrounding space. They are usually not sensitive to the position of the arms in space, while they process information related to head and gaze direction (Duhamel et al., 1997). Conversely, neurons in the premotor cortex show arm-centered multisensory receptive fields, extending a few centimeters from the upper limbs, and in particular the distal parts of them (Graziano et al., 1997). Finally, neurons in area 7 have tactile receptive fields covering the trunk and the proximal part of the upper limb. They are not sensitive to hand position or head/eye direction (Graziano & Gross, 1995; Leinonen, 1980). Thus, there is no single PPS representation, but there are multiple representations of PPS around specific body parts. Note that describing the PPS network in terms of (mainly) defensive (e.g. VIP neurons) versus approaching (e.g. F5 neurons) function, or body parts (head: VIP; hand: F5) is not contradictory because it is the case that the face region is probably the most critical to be protected, whereas approaching behaviours are in the first place implemented by arm movements.

In the close to 30 years since the original description of PPS neurons, much of the scientific inquiry in this area of investigation has moved into the human realm. First, di Pellegrino, Ladavas, Farnè, and colleagues (di Pellegrino et al., 1997; di Pellegrino & Làdavas, 2015; Farnè et al., 2005a, 2005b; Farnè & Ladavas, 2000, 2002; Làdavas et al., 1998; Làdavas & Farnè, 2004; Làdavas & Serino, 2008), showed that upon brain injury, certain patients' ability to detect tactile stimuli was more selectively impaired for presentations of a contralateral stimuli when this stimulus was near rather than far from the body. That is, these patients not only demonstrated extinction (the competition between sensory representations that results in a lack of detection of one of the competing stimuli) and not only cross-modal extinction (extinction across sensory modalities, such as a visual stimulus suppressing a tactile presentation), but inclusively demonstrated a proximity-dependent cross-modal extinction (see Bisiach et al., 1986; Berti & Frassinetti, 2000; Halligan & Marshall, 1991, for related reports of hemi-spatial neglect that are restricted to the near space). Next, in early psychophysical studies (Spence et al., 2000, 2004a, 2004b; Macaluso & Maravita, 2010), cross-modal congruency effects involving the tactile modality were shown to be dependent on observer-stimuli distance. More recently, a number of neuroimaging projects have revealed that a largely analogous PPS network exists in humans to that described in non-human primates (Bremmer et al., 2001; Brozzoli et al., 2011, 2012b, 2013; Gentile et al., 2013; see Grivaz et al., 2017, for review), and finally our group has developed a multisensory facilitation reaction time task (Canzoneri et al., 2012; Serino et al., 2017), which has been employed under a wide variety of experimental designs in order to both replicate and extend much of the monkey work in humans. This task was inspired by the fact that PPS neurons are most responsive to dynamically approaching stimuli (Fogassi et al., 1996). Thus, participants are asked to respond as fast as possible to tactile stimulation while 'task-irrelevant' auditory (Serino et al., 2015; Pfeiffer et al., 2018), visual (Noel et al., 2018c), or audio-visual (Serino et al., 2017) stimuli approach a given body part. Results have repetitively shown that participants become faster as the external stimulus approaches. Measuring the location in space where exteroceptive sensory signals significantly facilitate tactile detection (with respect to unimodal tactile stimulation) provides a proxy for the size of PPS that several studies have used to quantify changes in PPS representation following different manipulations or in different contexts (see e.g. Canzoneri et al., 2013a, 2013b; Teneggi et al., 2013). In addition, measuring the sharpness of the transition function between no-tactile-facilitation to tactile-facilitation provides a proxy for the sharpness of the PPS 'boundary', or the degree at which PPS is distinguished from the extra-personal space (see e.g. Noel et al., 2018c).

Importantly, the bulk of the human studies regarding PPS have fallen in line with the early non-human primate physiology work, concurring that PPS is a multisensory-motor interface for human–environment interaction and having a feet both in approaching and defensive behaviours. Thus, the human psychophysical work has demonstrated that PPS is modular and distinct for different body parts (Farnè et al., 2005b; Serino et al., 2015; Noel et al., 2018d; Stone et al., 2018), and resizes upon changes in actions or action possibility ('PPS for action'; e.g. goal-directed reaching: Brozzoli et al., 2009, 2010; walking: Noel et al., 2015b; using a wheelchair: Galli et al., 2015; immobilization: Bassolino et al., 2015; amputation: Canzoneri et al., 2013), or upon changes in the valence of contextual cues ('PPS for defence'— e.g. spider vs butterfly: de Hann et al., 2016; cynophobia: Taffou & Viaud-Delmon, 2014; claustrophobia: Lourenco et al., 2011).

In summary, there is strong evidence for the existence of bimodal neurons with tactile receptive fields on the body, and depth-restricted auditory or visual receptive fields anchored on the body. These neurons are distributed among at least two specific fronto-parietal networks (VIP-F4 and 7b-F5), respectively thought to be involved in defensive and approaching behaviour. Research with human participants has suggested the existence of analogous PPS systems in the human brain and has led researchers to conclude that PPS is a multisensory-motor interface for human–environment interactions. Positioning ourselves as devil's advocates and in the interest of highlighting existing gaps in knowledge, in the next section we question how strong is the evidence for the fact that PPS is truly i) multisensory and ii) motor.

2.4 Is peri-personal space truly multisensory?

The PPS literature is fraught with a plethora of claims that the encoding of the near space is 'multisensory' (see Ursino et al., 2007; Makin et al., 2007; Legrand et al., 2007; Holmes et al., 2007; Spence et al., 2004b; Serino et al., 2011; Salomon et al., 2017; Pellencin et al., 2017; Brozzoli et al., 2012b; Guterstam et al., 2013; Gentile et al., 2013; Grivaz et al., 2017, for a few examples of reports claiming a 'multisensory PPS' in their titles). This denomination, contrary to the 'bimodal' or 'cross-modal' nomenclatures, implies that some sort of sensory integration has occurred. And, critically, integrating information across the senses is well known to result in behavioural or encoding benefits (e.g. Nozawa et al., 1994; Frens et al., 1995; Wilkinson et al., 1996; Frassinetti et al., 2002; Lovelace et al., 2003; Diederich & Colonius, 2004; Noel & Wallace, 2016) above and beyond benefits derived from the presentation of stimuli in isolation. Thus, questioning whether PPS is truly multisensory amounts to questioning what is the evidence for an integrative process in PPS. And this is important, because the answer to this question dictates what class of behavioural and/or encoding benefits is seemingly present within the PPS system.

Determining whether PPS is multisensory depends on the particular type of definition for multisensory integration utilized (which in itself is problematic). From a psychophysical perspective, the differentiation between multisensory versus unisensory responses has been highlighted via two commonly used approaches: beating the race-model (i.e. responses in multisensory conditions are better—faster—than the statistical sum of unisensory responses; Miller, 1982, 1986; Raab, 1962; Ulrich et al., 2007) and demonstrating optimal cue combination (Ernst & Banks, 2002; Fetsch et al., 2013; although see Rahnev & Denison, 2018, and Noel, 2018, for critiques in utilizing this last definition). From the neural standpoint, according to a classic view focusing on single-cell responses (see Stein & Stanford, 2008), multisensory integration is present when neural response in a condition of stimulation in more than one modality is different (i.e. supra- or sub-additive) than response to the sum of the response to each of the single modalities in isolation. A more recent perspective focuses on (optimal) integration of signals in multisensory neurons, given the responses from upstream multisensory neurons. Here the question is: do behavioural or neurophysiological data on PPS meet those criteria for multisensory integration?

2.4.1 Definitions of multisensory integration in behaviour

Researchers early on noted that, if two sensory signals pertaining to the same object/event are given (e.g. auditory and visual) in contrast to a single one (e.g. only visual), detection of the object/event would be facilitated by simple statistical facilitation (Raab, 1962; Miller, 1982). Namely, if we imagine that object detection occurs upon a signal eliciting a given neural state, then when two signals are given, and even if they do not interact, on particular trials one signal (e.g. auditory) could induce this state space before the other (e.g. vision), and vice versa for other trials. If the detection can be triggered by either signal (the 'winner' of the race), then object detection would occur on average quicker for double cue presentation that for a single cue. The amount of facilitation can be described statistically, and this predicted facilitation is what we call the 'race-model' (Miller, 1982, 1986). Now, a key assumption of the race-model is that the sensory signals race independently: there is no need for interaction between the cues to produce statistical facilitation. This property makes the race-model a very useful benchmark in multisensory integration, because evidence for facilitation above and beyond statistical facilitation must be interpreted as the race between signals not being independent. Evidence of the race-model being 'violated' is evidence for a 'co-activation' model.

Within the cadre of PPS studies, reaction time studies are common, in fact potentially a majority (see Canzoneri et al., 2012, 2013; Serino et al., 2015; Noel et al., 2015a, 2015b, 2018b, 2018c; Galli et al., 2015; Pfeiffer et al., 2018; Stone et al., 2018, etc.). And thus, in principle, race-model violation analyses could be commonplace within PPS research. However, in practice these are seldom utilized, mostly because reactions are measured for the key modality for PPS study—i.e. touch—whereas the processing of the exteroceptive stimulus in the other modality (vision or audition) is considered only for its modulatory effect on tactile reactions. In very informative exceptions, Teramoto & Kayuka (2015) first indexed the race-model under a multisensory task involving the tactile system and proximity. In this study, they used two distances (one near and one far), and indeed demonstrated that race-model violations occurred near but not far, implying multisensory integration in PPS. In a second step (Teramoto et al., 2017), this same group expanded their purview in order to include more distances: close, middle, and far. They also tested PPS in different age groups (Teramoto et al., 2017). Unfortunately, here their results were arguably fairly inconclusive. At opposition to their first report, race-model violations were encountered at all distances, and thus did not categorically differentiate between near and far space. Further, the particular percentiles over which race-model violations were observed were somewhat inconsistent and at times not continuous. Most vexingly, while mean reaction times showed a visuo-tactile facilitation with respect to visual reaction times, when incorporating the distance parameter, there was no significant interaction. Namely, distance did not play a greater role in the multisensory than unisensory case. Thus, no multisensory *interaction* was observed, which questions the validity of querying multisensory *integration* (a more stringent bar to clear vs interaction). Lastly, visuo-tactile facilitation (vs unisensory reaction times) was observed at all distance, and thus there was no true demonstration of a space-dependent cross-modal facilitation, which makes it hard to conclude that PPS was indexed.

Thus, while these reports are definitely pushing the field forward by questioning whether PPS is truly multisensory—operationalized as beating the race-model—this topic remains an open question for further research. Much of the existent literature demarking PPS via reaction paradigms—many of them our own—are likely best categorized under the umbrella of cross-modal facilitation. It is true, however, that these auditory or visually driven speeding reaction times to touch are well in line with the spatial principle of multisensory integration—the observation that signals are most likely to be bound when presented in close spatial proximity (Calvert et al., 2004; Murray & Wallace, 2012)—hence supporting the possibility that a veritable multisensory process is being indexed. Anecdotally, Van der Stoep and colleagues (Van der Stoep et al., 2015a, 2015b; Noel et al., 2018e) have demonstrated over a number of observations that, when pairing auditory and visual signals, race-model violations occur more frequently when signals are presented far, as opposed to near. The study of race-model violations as a function of distance and their particular sensory pairings (e.g. audio-visual vs visuo-tactile vs audio-tactile) is an understudied and interesting avenue for further research (see Noel et al., 2018a, for a similar argument).

The second commonly utilized definition for observing multisensory integration equates this process with optimal cue combination. Ernst and Banks (2002) took inspiration from the computer vision field (e.g. Knill & Richards, 1996) to highlight to the multisensory community that there is an optimal way of integrating two signals.

In essence, our best estimate for, say, the location of an object in the external world—a ground truth our central nervous system does not have access to—is a weighted sum where each of the unisensory signals is weighted according to its reliability. This mechanism produces estimates for multisensory properties (e.g. location of an audio-visual object) that are usually intermediate between the unisensory signals and, most importantly, result in a reduction of the uncertainty associated with the multisensory object. We are more certain of the location of an object given two cues than a single one. The exact degree of uncertainty reduction can be specified mathematically (see Ernst & Banks, 2002, for details), and thus experimentalists can compare the computationally optimal fashion of combining cues with the manner in which humans empirically combine cues. Over the years, researchers have demonstrated that audio-visual (Alais & Burr, 2004), visuo-tactile (Wozny & Shams, 2006), visuo-vestibular (Morgan et al., 2008; Fetsch et al., 2011, 2013), and visuo-proprioceptive (van Beers et al., 1998, 1999, 2002) pairings, among others, integrate in a seemingly (near) optimal fashion in humans.

Returning to the question of whether PPS is truly multisensory, therefore, one may question whether optimal cue combination occurs in PPS. Surprisingly, however, to the best of our knowledge, only a single study has posed this question (see Samad et al., 2015, for a closely related undertaking). Noel and colleagues (2018d) asked participants to reach toward visual or proprioceptive targets. Then, based on the variance associated with these unimodal reaches, they predicted what ought to be the variance of multisensory (visuo-proprioceptive) reaches if they were optimal. Lastly, they had participants reach toward visual targets that were spatially aligned with the proprioceptive cue—and thus within PPS—or misaligned by different magnitudes. The findings suggested that, when visual targets were very close (~25 mm) to the proprioceptive location of their index finger, these were combined optimally. Because the visual targets were further away from the proprioceptive target, the degree of disparity between the optimal prediction and the variance and bias measured experimentally grew. In other words, seemingly, optimal cue combination occurs when visual targets are presented very close to proprioceptive signals. When presented far, the two cues are *de facto* not integrated. Overall, the results seem to be compatible with a slightly more complex model of cue combination than classical optimal integration. In the most commonly described forms of optimal integration, the two cues are assumed to be originating from the same physical cause, and then combined following a Bayesian model based on such an assumption. In the proposed interpretation, the probability that they have the same cause is not set to one but allowed to vary based on distance, and this determines the weight given to integration versus segregation of the two cues. Following this line of reasoning, PPS could be interpreted as a spatial prior on multisensory binding of egocentric and allocentric information. Unfortunately, this is a single demonstration of optimal cue combination within PPS, and in this report there was no independent (and more classical) measure of PPS. Hence, it remains an open question whether one can directly relate the spatial disparity at which optimal cue combination occurs with traditional conceptions of PPS. Also, while finger-specific PPS has not been delineated in other studies (which is arguably what was measured in Noel et al., 2018d), 25 mm (or 2.5 cm) is much smaller than the standard reports of about 30 cm for peri-hand space, 45 cm for peri-face space, and 70+ cm for peri-trunk space (Serino et al., 2015). Whether this discrepancy is consequent to the different body parts indexed (a finger being physically smaller than any other body part where PPS has been measured in the past), due to the different approach (measuring localization error as opposed to reaction times), or both, remains to be determined. More generally, whether signals are integrated optimally in PPS is still very much an open question for further inquiry.

2.4.2 Definitions of multisensory integration in neurophysiology

Regarding the neural underpinnings of multisensory integration, as for behaviour, there are largely two camps. Directly inspired by the behavioural and theoretical work on optimal cue combination (Ernst & Banks, 2002), theoreticians (Ma et al., 2006) have proposed that given the noise properties of population of neurons, optimal cue integration (e.g. audio-visual) should follow from the simple sum of neural activity of two upstream populations (e.g. auditory and visual). That is, the weighted sum observed behaviourally should follow from noise properties within the unisensory populations and should not be a result of a weighted sum, but instead of a linear sum, at the neural level. However, when researchers probed this probabilistic population coding (Ma et al., 2006) scheme in the brain, they found that linear sums were the exception rather than the rule (Morgan et al., 2008). Thus, a number of researchers are actively querying how cue combination occurs in the brain (see Fetsch et al., 2013, for review; see Rohe & Noppeney, 2015, 2016; Rohe et al., 2019, for neuroimaging work relevant to this topic). Regardless, it is widely assumed that the solution to this problem lies at the level of populations of neurons, as opposed to single cells. Unfortunately, the vast majority of PPS-related neural recordings (e.g. Duhamel et al., 1997, 1998; Graziano et al., 1997, 1999) occurred at a time when neurotechnology only allowed measuring a single cell at a time, and thus there is no single unit data recorded simultaneously (e.g. V1, S1, and VIP) that is pertinent to our question at hand. Clearly, this is an area wide open for investigation, whereby more advanced multi-site recording could provide important insights (see, e.g. Jun et al., 2017, for state of the art large-scale electrophysiology).

The second approach in indexing multisensory integration at the neural level relates to quantifying supra- and sub-additivity in single neurons. Namely, Stein and colleagues (Meredith & Stein, 1983, 1986, 1996; Stein & Meredith, 1993) observed in the late 1980s and early 1990s that when certain neurons—most notoriously in the superior colliculus—were presented with multisensory stimuli (e.g. AV), their response was not a simple sum of their unisensory responses (e.g. A + V), but sometimes higher (supra-additive) or lower (sub-additive). Thus, these neurons ought to be performing some sort of interesting input–output transformation. This manner of indexing multisensory integration has been a dominating approach in neurophysiology, and seemingly translates to other neural recording methodologies, such as EEG and fMRI (Cappe et al., 2012; Quinn et al., 2014, but see Stanford & Stein, 2007, for critiques regarding both supra-additivity in multisensory integration and the different 'principles' of multisensory integration). Nonetheless, the usage of this approach to study multisensory integration in PPS has been fairly limited. In the seminal neural recordings defining PPS, while visual and tactile stimuli were given, there was never a formal analysis regarding whether multisensory presentations resulted in supra- or sub-additivity. A prominent exception is a study by Avillac and colleagues (2007) who recorded from VIP neurons—an area known to possess PPS neurons, see section 2.3 above—while presenting non-human primates with visual, tactile, or visuo-tactile stimuli. They reported that approximately 70% of neurons in this area showed non-linear additivity, and thus demonstrated multisensory integration, as classically defined by Meredith and colleagues (Meredith & Stein, 1983, 1986, 1996; Stein & Meredith, 1993). Unfortunately, this report was not centered on questions regarding PPS, and thus there was no exploration of the depth-related characteristics of these neurons, nor whether the visual receptive fields of the neurons were anchored on the body. Interestingly, Cléry et al. (2017) did recently provide imaging data pertinent to this question. They presented monkeys with looming visual stimuli and at discrete times gave tactile stimulation. In addition to improvements in detection (measured in d'), when a predictive visual cue is added to the tactile target (Cléry et al., 2015b), these authors demonstrated that a number of brain areas showed a multisensory non-linearity—namely, V1-V3, MST, FST, VIP, and F4 (Cléry et al., 2017). In summary, therefore, whether PPS neurons show a multisensory non-linearity still remains somewhat of an open question, particularly from the single neuron perspective. Note, however, that this seems a strong possibility, given Cléry et al. (2017), and a recent report by Bernasconi and colleagues (2018), who demonstrated that, in electrocorticographical recording approximately 30% of electrodes that demonstrated multisensory supra-additive or sub-additive responses also demonstrated PPS processing (also see Gentile et al., 2011, for a similar approach in fMRI with humans). Further, the electrodes that showed PPS processing were located in regions well in line with the original neurophysiological descriptions (Bernasconi et al., 2018).

Taken together, we wish to reinforce that we do consider PPS to be multisensory. We have seen this in our own data (Bernasconi et al., 2018) and a number of emerging studies indexing the race-model in PPS (Teramoto & Kayuka, 2015; Teramoto et al., 2017; personal communication with Dr. Dijkerman) seem to support this notion. Similarly, on numerous occasions (Noel et al., 2015a, 2015b, 2018b, 2018c, etc.), we have reported cross-modal facilitations that are in line with the spatial principle of multisensory integration (Murray & Wallace, 2012) and hence likely indicative of multisensory processing. In this section, we have simply aimed at emphasizing that this is an open area of investigation given that the evidence for true integration in PPS is weaker than one might expect given the common mentions of the 'multisensory PPS' in the literature.

2.5 How strong is the link between peri-personal space and the motor system?

The second commonly accepted notion we wish to examine is that PPS is part of, or very closely related to, the motor system. Again, we do not wish to overturn this conception here. Indeed, a number of researchers have provided convincing evidence that visual or auditory stimuli presented near the body result in enhanced corticospinal excitability (Makin et al., 2009), and inclusively our own group (Serino et al., 2009; Avenanti et al., 2012; Finisguerra et al., 2015) has observed this strong connection. Even more convincingly, the premotor cortex has projections both to the spinal cord (Martino & Strick, 1987) and primary motor cortex (Matsumura & Kubota, 1979; Pandya & Vignolo, 1971), and PPS neurons respond during movement (Rozzi et al., 2008). A full line of work on PPS processing even assesses a likely brainstem mediated reflex—i.e. the hand-blink reflex and its modulation as a function of the position of the hand in space—to study PPS representation (Sambo et al., 2012; see Bufacchi & Iannetti, 2018). Thus, unequivocally, there is an intimate relation between the coding of the space near one's body and the motor system, at its different levels. However, in the interest of sparking future research in this area, we wish to highlight a few oddities that point toward the problem being more complex than simply stating that PPS 'is motor'.

As noted earlier, electrical stimulation of the VIP or ventral pre-motor PPS neurons can result in stereotyped defensive behaviours, such as ducking or deflecting incoming stimuli (Graziano & Cooke, 2006; see Dijkerman et al., 2009, for evidence in a human subject). This feature is routinely quoted as the centerpiece in any argument linking PPS with defensive behaviour (Graziano & Cooke, 2006). However, it must be acknowledged that electrical stimulation of a wide array of neural areas can evoke movements. For instance, electrical stimulation of the supplementary motor area (SMA; Goldberg, 1985)—an area not known to possess PPS neurons—can either evoke the urge to move (Penfield & Welch, 1951; Fried et al., 1991; Chauvel et al., 1996) or directly move the face, neck, and upper trunk (Mitz & Wise, 1987; Luppino et al., 1991). Thus, evoking complex movements upon electrical stimulation is not a property solely of PPS neurons. Once this is taken into account, the burden of proof that electrical stimulation of VIP and/or ventral pre-motor neurons yields specifically defensive behaviours grows enormously. That is, the demonstration that VIP/F4 neurons drive defensive-like behaviours is strong evidence that this network is involved in avoidance behaviour. However, if one supposes that (strong enough) electrical stimulation of any area would evoke a motor response, then it's not convincing enough to qualitatively state that movements engendered by VIP/F4 seem 'defensive-like': this must be demonstrated quantitatively.

In any case, accepting that electrical stimulation of VIP/F4 drives defensive behaviours, it must be emphasized that these neural areas possess strong idiosyncrasies in their motoric properties. In the pre-motor cortex, the defensive motor repertoire can be evoked with stimulation intensities of approximately 20 mA, in both the awake and anesthetized monkey. Contrarily, the intensity of the current administered has to be approximately five-fold larger in VIP than in ventral pre-motor in order to elicit motor responses. Further, in the posterior parietal cortex, at opposition from the ventral pre-motor, anesthesia depth has a strong impact on the motor repertoire that is outputted (Graziano & Cooke, 2006; Cléry et al., 2015b). In a similar vein, while electrical stimulation in both areas F4 and VIP disrupts the ongoing behaviour of monkeys—and replaces it by defensive-like movements—upon secession of the electrical stimulation to F4, monkeys resume ongoing behaviour while this secession of stimulation in VIP does not always imply stoppage of defensive behaviour. Hence, one could interpret that in VIP but not in F4, even after discontinuation of electrical stimulation, a potential percept that originated defensive behaviours remains present (see Graziano & Cooke, 2006, for further discussion). Lastly, in VIP, the induced defensive repertoire often diminishes over repeated trials, suggesting adaptation to the stimulation. This feature has not been reported in ventral pre-motor. Taken together, seemingly F4 is more directly linked to motor output than VIP is, the former area truly being a driving force in motor output, while the latter one could potentially only result in motor output upon electrical stimulation due to off-targets effects: eliciting a percept of danger, as opposed to eliciting a movement in the absence of a percept. The distinctions in intensity needed to evoke movements, and the fact that VIP adapts but ventral pre-motor does not, also suggest a different activation threshold among these areas: VIP is difficult to drive and silence, while F4 functions more as a direct, all-or-none switch. This proposal fits with neurophysiological evidence in humans (Avenanti et al., 2012) showing that inhibitory non-invasive stimulation (i.e. cathodal transcranial electrical stimulation) over the ventral premotor cortex (including the human homologues of F4) abolishes the modulation of corticospinal excitability induced by exteroceptive stimuli within (as compared to outside) PPS. No effect is induced by the same stimulation over the PPS-related areas in the posterior parietal cortex (including the human homologues of VIP), reinforcing the more direct link between PPS premotor areas and motor responses, rather than parietal areas. This, however, does not imply a simple model whereby VIP processes PPS-related perceptual information and transfers it to premotor areas, mediating a motor response. Indeed, the latency of evoked movements by VIP stimulation can be significantly faster than those evoked by F4 stimulation (e.g. 10 vs 30 ms for blink response; see Graziano & Cooke, 2006, for comments), making a simplistic parietal-to-premotor serial model not compatible with data. Thus, a more complex interaction between multisensory parietal and premotor regions, and cortical and subcortical motor structures needs to be conceived.

While neurophysiological data mainly showed PPS-related modulation of motor responses, psychophysical studies have demonstrated the symmetric side of the relationship—that is, how motor acts remap PPS (Brozzoli et al., 2009; Noel et al., 2015b) in an action-specific manner (Brozzoli et al., 2010). Most strikingly, this remapping occurs even before action initiation, as if it were involved in the planning phase of action execution (Patane et al., 2018). These findings have widely been taken to demonstrate the strong association between PPS and the motor system, which they do. However, an understated conclusion that must also be drawn from these findings is that they reveal a potential for temporal disparity between action execution and PPS remapping. PPS neurons may fire during movement, but this system can also predict movement (Patane et al., 2018). Thus, the association between PPS and the motor system is not an absolute one, but one that can toggle flexibly back and forth in time. To us, an interesting possibility suggested by Patane and colleagues' (2018) results is that PPS remapping may co-occur with intention. Although this is only a speculation, a shift of PPS neurons' receptive fields towards the location of an intended movement, before the movement onset, has been described in the case of saccades in parietal neurons (Duhamel et al., 1992). Libet et al. (1983) famously reported that, while subjects were able to report a will to move that preceded self-generated movements, electrical activity in the brain indicative of movement onset preceded the will by up to 1,000 ms. This neural marker preceding report of will has been denominated the 'readiness potential'. In the future, it may be interesting to measure the receptive fields of PPS neurons (via direct or indirect measures), in a self-generated motor paradigm, while recording neural activity. This would allow placing on a timeline the onset of PPS remapping, motor execution, and motor intention.

In summary, the association between PPS processing and the motor system is strong. Nonetheless, we are far from understanding exactly what the specific role of PPS parietal and frontal neurons is in modulating motor responses, at the different cortical and subcortical levels of motor processing. Similarly, we are far from understanding the temporal relation between PPS encoding and motor output, and how this temporal relation seemingly recalibrates flexibly.

2.6 A new wave of psychophysical peri-personal space research

In the previous two sections, we have highlighted that, while it is clear that PPS processing is at an interesting junction between the multisensory and the motor systems, there are still a whole host of questions that remain to be answered within those sensorimotor domains: are there really two parallel fronto-parietal systems (VIP-F4 and 7b-F5), with preferential defensive and approaching functions (and how do these interact)? Perhaps PPS is frankly multisensory from the classical 'non-linear additivity' framework (Stein & Stanford, 2008)—but this framework seemingly is most applicable to the subcortex and does not necessarily apply to the neocortical mantle (e.g. Meijer et al., 2017), where the majority of PPS neurons are found. Perhaps a better question, thus, is whether we find evidence of optimal cue combination—arguably a better suited definition for multisensory integration in the cortex—in PPS? How are the motor properties of VIP transformed along the fronto-parietal network to engender the very distinct motor features in F4? If PPS is involved in impact prediction (see Kandula et al., 2015; Clèry & Hamed, 2018, for this argument), why do these neurons respond to tactile offset and receding exteroceptive stimuli (Duhamel et al., 1997, 1998)? Given the time course of PPS remapping (Patane et al., 2018), is it potentially associated with motor intentions?

Notwithstanding these open questions within sensory and motor domains, much of the effort—and progress—within the study of PPS for the past decade has lied in exploring novel domains. Here, we will highlight two of these domains: social cognition and bodily self-consciousness. We consider that it is valuable to highlight two exemplars and distinct arenas where PPS appears to play a role beyond sensorimotor transformation because this demonstration suggests that a full account of PPS will necessitate an abstract and computational framework. Finally, in section 2.7 of this chapter, we will emphasize how computational models of PPS are starting to mechanistically explain certain aspects of the multisensory-motor PPS, and by doing so they are putatively also accounting for findings within the novel domains of study within PPS (i.e. PPS and social cognition). Thus, conversely, by moving forward in studying PPS under more abstract and computationally oriented conditions, we may equally account for long-standing multisensory-motor questions within the study of PPS.

2.6.1 Peri-personal space in social cognition

A decade ago, Ishida and colleagues (2010) described 'body-matching neurons' in VIP that responded not only to visual stimuli near a specific body part, but also to visual stimuli presented near a corresponding body part of the human experimenter. That is, these neurons respond just as the classically defined PPS neurons (Fogassi et al., 1996; Duhamel et al., 1998; Graziano et al., 1997, 1999), yet additionally they seemingly also respond when visual stimuli are presented within the PPS of other individuals (inclusively of a non-conspecific). A similar neural response has been observed in humans via fMRI (Brozzoli et al., 2013; Cardini et al., 2011). Furthermore, it has been shown that reaction times to tactile stimulation (Teramoto, 2018; see also Heed et al., 2010) or the hand-blink reflex (Fossataro et al., 2016) are influenced not only when visual stimuli are close to the self, but also when they approach another person. Together, these findings suggest two main interesting conclusions. First, primates (human and other) possess a dedicated neural network to encode for the space near their own body, and part of this network (e.g. 48 out of 441 neurons reported by Ishida et al., 2010) also encodes for the space near the body of others. Second, one's own PPS representation is affected by social factors. Indeed, utilizing the audio-tactile reaction time paradigm described earlier (see Canzoneri et al., 2012; Serino et al., 2017), Teneggi et al. (2013) showed that not only is PPS affected by the presence of another (vs a mannequin, or another person, Heed et al., 2010, and Fossataro et al., 2016; Teramoto, 2018) but, critically, that PPS reshapes as a function of the quality of the social interaction with others. In more detail, initially when a confederate experimenter was placed facing the participants (vs. a mannequin being placed), the subjects' PPS shrank, as if to allow space for the other individual. Then, upon cooperative interactions, the subjects' PPS grew, so as to encompass the confederate. This last observation was only true when the cooperation was positive, as opposed to negative (Teneggi et al., 2013). In a follow-up experiment, Pellencin et al. (2017) suggested that PPS is more generally sensitive to the social perception of others by demonstrating that PPS is extended when participants face another person they perceive as being moral, as opposed to immoral. Interestingly, this extension of PPS, observed both in Teneggi et al. (2013) and Pellencin et al. (2017), does not apply to all social contexts. In fact, when the social manipulation between self and other is in terms of shared sensory experiences, instead of social perception, PPS does not extend as to include the other, but remaps. That is, Maister et al. (2015) had participants experience the 'enfacement illusion'—an illusion whereby touch given to one's own face is seen on another person's face (Tsakiris, 2008), eliciting ownership over the other person's face—and observed that tactile reaction times were facilitated when exteroceptive signals were given either near the self or the other, but not in between (Maister et al., 2015). Thus, PPS did not expand (tactile facilitation would have been observed throughout the space mapped) but remapped. Together, these empirical findings suggest that response properties of the PPS system, likely a subpopulation of PPS neurons, modulate the representation of the space between the self and the others in a rather flexible way, depending of the nature of social interactions.

2.6.2 Peri-personal space in bodily self-consciousness

A second novel avenue of research for the PPS field is that of bodily self-consciousness. Blanke & Metzinger (2009) described that the minimal self ought to possess at least a sense of body ownership (i.e. a body that is one's own), a sense of self-location (i.e. a body that is located somewhere in the world), and a sense of first-person perspective (i.e. a heading direction). Approximately a decade before, Botvinick & Cohen (1998) supplied researchers interested in bodily self-consciousness with the necessary tools to study this topic, or at least body-part ownership, from an experimental point of view. That is, these researchers demonstrated that, by synchronously applying touch on a fake hand a subject observed, and touch on a participant's real hand which they could feel but not see, they could elicit the illusion of ownership over a fake hand. Further, in this rubber hand illusion (Botvinick & Cohen, 1998), subjects would report feeling touch on the rubber hand and, when asked to localize their real hand, they would show a bias toward the fake hand. These effects are absent or weaker in the condition of asynchronous visuo-tactile stimulation. A few years following Botvinick & Cohen's (1998) report, Graziano and colleagues (2000) recorded from PPS neurons while providing the same synchronous (vs asynchronous) stimulation used to induce the rubber hand illusion in non-human primates. These recordings demonstrate that before synchronous visuo-tactile stroking, PPS neurons encoded the position of the real hand and were indifferent to the location of the fake hand. Astonishingly, however, after synchronous visuo-tactile stroking (arguably inducing the illusion—a fact that of course could not be confirmed from the subjective point of view of the monkey), these same neurons started mapping the position of the fake arm (Graziano et al., 2000). Given the homology with the pattern of visuo-tactile stimulation administered to induce the rubber hand illusion in humans, these data suggest that the multisensory mechanism implemented by PPS neurons are involved in body ownership. Further data support this suggestion. Brozzoli et al. (2012a) showed that ventral premotor and parietal regions, coding for the PPS around one's own hand process visual information close to an artificial hand after visuo-tactile stimulation inducing the illusory ownership of the artificial hand. Recently, Grivaz et al. (2017) demonstrated a more general link between PPS and body ownership/self-identification. Results from the quantitative meta-analysis showed a clear overlap (but also a dissociation) and, importantly, redundant connections, between the brain regions reported by fMRI studies as involved in PPS processing and those modulated by changes in body ownership/self-identification during multisensory bodily illusions.

A further strong hypothesis, originally proposed by Blanke (2012), is that under a full-body version of the rubber hand illusion, it would not only be the hand that is displaced but also the self. Thus, a full-body version of the rubber hand illusion was developed, the full-body illusion (Lenggenhager et al., 2007), whereby synchronous stimulation is applied tactilely to the participant's torso and visually to an avatar seen at a distance. This manipulation extended Botvinick & Cohen's (1998) paradigm by not only inducing change in body-part ownership but also in full-body ownership (or self-identification) and, most importantly, by altering self-location. That is, when one's arm is perceived to be located at an erroneous location, this body part may still be described from an egocentric perspective and in relation to the self. Contrary, under the full-body illusion (Lenggenhager et al., 2007), the body is shifted in allocentric coordinates, because it is the self that is translated. With this paradigm in hand, therefore, we (Noel et al., 2015a) tested whether PPS would encode the location of the self and corroborated Blanke's (2012) hypothesis by demonstrating that when tactile stimuli on participant's back is given synchronously to touch seen on the back of a virtual avatar in front of the participant, the participant's PPS expands in the front-space and retracts in the back-space, as if shifting to encode the location of the perceived bodily-self and not the physical body. A couple years later we reinforced the connection between PPS and self-location by replicating Noel et al. (2015b), while administering subliminally either the stimuli inducing the full-body illusion or the stimuli used to delineate PPS (Salomon et al., 2017). This manipulation not only demonstrated that mapping of PPS was uncontaminated from attentional confounds but, more importantly, these findings provided the first empirical evidence for the fact that self-location, a core component of bodily self-consciousness, is processed pre-reflexive—i.e. without requiring explicit awareness of the stimuli processed to generate it (see Legrand et al., 2007).

Finally, we have recently demonstrated that—within an impoverished sensory environment and in absence of actions, reducing the amount of bodily and exteroceptive information, the boundary of PPS becomes 'ill-defined'—i.e. there is a shallow gradient between the far and near space. Further, participants reported feeling 'lost in space' (Noel et al., 2018c). These findings show a direct link between PPS representation and self-location, which is one of the components of bodily self-consciousness. There might be other components of bodily self-consciousness, or of body awareness in general, which are not directly related to PPS processing. Currently, no link has yet been established between PPS and the direction of the first person perspective. Also, body awareness from the inside—i.e. interoception—which is considered a key element for self-consciousness (Damasio, 2003; Seth & Tsakiris, 2018) seems less related to the specific multisensory mechanism underlying PPS representation as presented in this chapter.

Nevertheless, seemingly PPS encodes the perceived location of the bodily self, rather than of the physical body and, putatively, inclusively the sharpness with which the near and far spaces are separated may be related to the degree with which participants are confident in where they perceived them to be located. Ongoing studies within our group are aimed at developing a neural marker of PPS (see Bernasconi et al., 2018; Noel et al., 2018a, for a first approximation) and utilizing these measures to index PPS in patients with distinct disorders of consciousness (Noel et al., 2019).

2.7 A look ahead for peri-personal space research

Throughout this chapter, we have argued that, while PPS is clearly multisensory-motor, there are still a vast number of open questions within these domains that we have not been able to explain. More strikingly, in recent years, a large number of studies have demonstrated that PPS is altered by experimental conditions seemingly far from the sensorimotor domain (e.g. social cognition and bodily self-consciousness). Arguably, therefore, conceptions of PPS may need to be abstracted; a fuller account of PPS is likely to be expressed within a computational perspective (see Klein, this volume, for a similar argument). In turn, in this last section, we first briefly review existing models of PPS. Then, in closing, we leverage one of the existing models of PPS (Noel et al., 2018b), a model that was developed to account for the velocity-dependent resizing of PPS (Fogassi et al., 1996), to sketch a speculative answer to the seeming conundrum as to why anxiety differentially resizes PPS (e.g. enlarges or shrinks) as a function of context. We provide this speculative answer as a demonstration that attempting to understand PPS within its larger computational demands (i.e. goal-directed and avoidance behaviour, reference frame transformation, optimal multisensory integration, etc.), and as a single computational node within the nervous system—whose goals are survival and reproduction—can resolve apparent mysteries.

A surprisingly large number of models accounting for different aspects of PPS exist. These models take on a number of different forms, from simple mathematical and hypothesis-driven models (Bufacchi et al., 2016), to complex multi-component and robotically instantiated models (Roncone et al., 2016). In more detail, Bufacchi et al. (2016) start from potential shapes they consider the peri-face could potentially have (e.g. circular, ellipsoidal) and then perform model fitting to determine the fine-grain topographical structure of peri-face space based on data from the hand-blink reflex. Roncone et al. (2016), in contrast demonstrate that given the statistics of the natural environment (e.g. giving touch on the body) and with relatively simple engineering, a humanoid robot can learn from scratch to perform time-to-contact estimates (hence fulfilling what is widely considered one of the main functions of PPS: impact prediction, Cléry & Ben Hamed, 2018; Noel et al., 2018f), and to avoid potential threats (Roncone et al., 2016).

Further inspired by biology and attempting to provide mechanistic insight, Magosso and colleagues have built a series of neural network models to account for an array of features of PPS. These models count with tactile neurons with receptive fields on the body, and with audio (Serino et al., 2015b) or visual (Magosso et al., 2010a, 2010b; Serino et al., 2015a) neurons with receptive fields encompassing the body, the near space, and the far space. Both unisensory areas (tactile and visual/or auditory) project to a multisensory area. Postulating that the strength of connections between the exteroceptive area and the multisensory layer is stronger in the near space than the far space naturally leads to simulated reaction times that are facilitated in the near space versus the far space—i.e. PPS encoding. Perhaps more interestingly, postulating a Hebbian learning mechanism (Hebb, 1949) within this framework leads to the prediction that, after frequent congruent tactile stimulation on the body with visual/auditory stimulation at a far location, PPS should enlarge. Serino and colleagues (2015) tested and corroborated this hypothesis. Similarly, Noel et al. (2018b) recently demonstrated that, when adding a neural adaptation mechanism to Magosso and colleagues' model, these networks could mimic Fogassi et al.'s (1996) observation that PPS enlarges with the speed of incoming stimuli. Thus, there is no need for another sub-system to communicate with the PPS network to inform it about the value of the incoming stimuli (e.g. 'fast and hence dangerous'), but the resizing of PPS with velocity of incoming stimuli occurs naturally given the biological constraints of the PPS network.

The latest wave of PPS models live within the machine learning space. Both Straka & Hoffmann (2017) and ongoing work within our group (Bertoni et al., 2020) have applied restricted boltzmann machines (an artificial neural network that can learn probability distributions over its set of inputs in an unsupervised manner; see Hinton, 2010) to the study of PPS. Straka & Hoffmann (2017) show that, if given a velocity vector and allowed to learn in an unsupervised manner, a neural network will reconfigure its connections so as to perform time to contact estimates. Recent work from our group (Bertoni et al., 2020) suggests that an analogous architecture can engender a PPS field that is anchored to the body—i.e. it is capable of performing reference frame transformations. Namely, a PPS representation naturally arises in this network as the encoding of external (visual) stimuli in a body-centered (tactile) space. In this framework, PPS representation and reference frame transformations spontaneously arise as complementary epiphenomena of a unitary neural process and cannot be disentangled. From a more abstract perspective, the veritable point of notice here is that any network that is built to learn in an unsupervised fashion probability distributions can create a PPS representation, provided it gets exposed to the appropriate set of training stimuli (i.e. sensory representations of external stimuli interacting with the body). Previously, Makin et al. (2013) showed that this same network architecture could perform reference frame transformation, Bayes optimal integration, and causal inference (Körding et al., 2007). It is therefore tempting to conclude that PPS may arise naturally from the constraints of the nature environment and the need to perform a whole host of operations (not only time to contact, but also reference frame transformation and causal inference, for instance.)

In closing, we wish to provide a more concrete (but speculative) example as to how a computational perspective may in the future facilitate the study of PPS. Studies show that anxiety seems to increase the size of one's PPS (e.g. Sambo & Iannetti, 2013), yet on occasions anxiety can also shrink PPS (Graydon et al., 2012). Hunley & Lourenco (2018) postulated that these differential effects are driven by the fact that in the former example the defensive PPS is indexed, and in the latter case the approaching PPS is indexed. We do not entirely disagree but wish to provide a computational perspective. Namely, in Noel et al. (2018d), we built a network with auditory and tactile neurons projecting to a common multisensory neuron. As for the rest of Magosso and colleagues' models, the auditory connections are stronger the closer they are to the tactile receptive field. Further, in this model we include a neural adaptation mechanism, and this mechanism we show can account for the enlargement of PPS when sounds approach more rapidly. Intuitively, when sounds are slow, they are located in the same position in space for a longer period and thus, if there is neural adaptation, these neurons will not drive the multisensory neuron as strongly as they could. Interestingly, in performing 'sensitivity analyses' (e.g. determining whether the effect was driven by particular parameters or the architecture and mechanism themselves; Noel et al., 2018b) we observed that, if we increased the intensity of incoming sounds, PPS enlarged. Similarly, if we increased the standard deviation (i.e. the uncertainty) associated with the representation of the auditory stimuli, PPS enlarged. Given these observations, one could hypothesize that, in a defensive framework, what is important to determine is whether something is near or not. It is not important to compute the fine-grain characteristics of the stimuli, because we do not wish to interact with it. Potentially, therefore, sensitivity to intensity is increased under a defensive framework. Conversely, under a goal-directed framework, one in which we may potentially want to reach out and grasp the object, it is important to localize it well. Hence, putatively, the reliability of the sensory representation is prioritized—that is, the variance decreases. Under these hypotheses, and given the neural network described in Noel et al. (2018b), PPS would enlarge in defensive contexts and shrink in approaching contexts. Importantly, the network is always the same: it is simply that upstream regions have emphasized either the intensity or spatial reliability of the stimuli. There is no need for a large array of different PPSs (Bufacchi & Iannetti, 2018). This is an example of how potentially complex modulations of multisensory PPS processing by cognitive or affective factors (e.g. anxiety), apparently very far from one another in an information-processing perspective, can be explained by the computational properties of the PPS system. The open question in the field is how the different nodes of the PPS networks interact with other brain systems (see Bufachi & Iannetti, 2018), mediating different behaviours (e.g. social interactions) or experience (e.g. bodily self-consciousness).

2.8 Conclusion

The scientific study of the space near one's body, the PPS, is at an incredibly exciting junction. It is clear that the neurons and networks described by Graziano and colleagues are multisensory-motor. However, exactly *how* multisensory and motor is not fully understood (contrary to what a large number of scientific publications would suggest). Further, in recent years, we have seen PPS be associated with a large array of tasks—for example, in social cognition (Teneggi et al., 2013; Pellencin et al., 2017), visual perception (Blini et al., 2018), or conceptual processing (Canzoneri et al., 2016). This promiscuity of the PPS system has led some to essentially question whether there is a PPS, if proximity to the body is a defining characteristic of a subset of neurons (Bufachi & Iannetti, 2018). We consider that encoding the space near one's body is an essential stage in the larger computational goal of agents in surviving and reproducing. As such, proximity is a key feature for a number of neurons and neural circuits, yet precisely understanding their operation will require a computational lens that goes beyond simple sensorimotor contingencies (Noel & Serino, 2019). Thus, we welcome the new wave of psychophysical studies indicating that PPS may play a role in disparate domains, far from approaching or avoidant behaviour. In fact, we predict that in the near future PPS will become an important topic of study for those interested in even more far-fetched neuroscientific areas, such as navigation and path integration, or the transformation from allocentric spatial representations (e.g. grid and place cells; Moser et al., 2008) to egocentric representations (e.g. head direction cells, PPS encoding; Taube et al., 1996), memory, and decision making.

References

Alais, D., & Burr, D. (2004). The ventriloquist effect results from near-optimal bimodal integration. *Curr Biol, 14*, 257–262. doi:org/10.1016/j.cub.2004.01.029

Andersen, R. A. (1997). Multimodal integration for the representation of space in the posterior parietal cortex. *Philos Trans R Soc Lond B Biol Sci, 352*(1360), 1421–1428.

Andersen, R. A., & Buneo, C. A. (2002). Intentional maps in posterior parietal cortex. *Annu Rev Neurosci, 25*, 189–220.

Avenanti, A., Annella, L., & Serino, A. (2012). Suppression of premotor cortex disrupts motor coding of peripersonal space. *Neuroimage, 63*, 281–288.

Avillac, M., Ben Hamed, S., & Duhamel, J.-R., (2007). Multisensory integration in the ventral intraparietal area of the macaque monkey. *J. Neurosci, 27*, 1922–1932. doi:10.1523/JNEUROSCI.2646-06.2007

Avillac, M., Denève, S., Olivier, E., Pouget, A., & Duhamel, J.-R. (2005). Reference frames for representing visual and tactile locations in parietal cortex. *Nat Neurosci, 8*, 941–949. doi:10.1038/nn1480

Bassolino, M., Finisguerra, A., Canzoneri, E., Serino, A., & Pozzo, T. (2015). Dissociating effect of upper limb non-use and overuse on space and body representations. *Neuropsychologia, 20*, 385–392. doi:org/10.1016/j.neuropsychologia/2014.11.028

Bernasconi, F., Noel, J.-P., Park, H. D., Faivre, N., Seeck, M., Spinelli, L., Schaller, K., Blanke, O., & Serino, A. (2018). Spatio-temporal processing of multisensory peripersonal space in human parietal and temporal cortex: an intracranial EEG study. *Cereb Cortex.* doi:org/10.1093/cercor/bhy156

Berti, A., & Frassinetti, F. (2000). When far becomes near: remapping of space by tool use. *J. Cogn Neurosci, 12*, 415–420.

Bertoni, T., Magosso, E., & Serino, A. (2020). From statistical regularities in multisensory inputs to peripersonal space representation and body ownership: insights from a neural network model. *European Journal of Neuroscience,* https://doi.org/10.1111/ejn.14981

Bisiach, E., Perani, D., Vallar, G., & Berti, A. (1986). Unilateral neglect: personal and extra-personal. *Neuropsychologia, 24*(6), 759–767.

Bizley, J. K., Nodal, F. R., Bajo, V. M., Nelken, I., & King, A. J. (2007). Physiological and anatomical evidence for multisensory interactions in auditory cortex. *Cereb Cortex, 17*, 2172–2189.

Blanke, O. (2012). Multisensory brain mechanisms of bodily self-consciousness. *Nat Rev Neurosci, 13*(8), 556–571.

Blanke, O., & Metzinger, T. (2009). Full-body illusions and minimal phenomenal selfhood. *Trends Cogn Sci, 13*(1), 7–13.

Blini, E., Desoche, C., Salemme, R., Kabil, A., Hadj-Bouziane, F., & Farnè, A. (2018). Mind the depth: visual perception of shapes is better in peripersonal space. *Psychol. Sci, 29*(11), 1868–1877.

Blohm, G., & Crawford, J. D. (2009). Fields of gain in the brain. *Neuron, 64*(5), 598–600.

Bonini, L., Maranesi, M., Livi, A., Fogassi, L., & Rizzolatti, G. (2014). Space-dependent representation of objects and other's action in monkey ventral premotor grasping neurons. *J. Neurosci, 34*, 4108–4119. doi:10.1523/JNEUROSCI.4187- 13.2014

Botvinick, M., & Cohen, J. D. (1998). Rubber hand 'feels' what eyes see. *Nature, 391*, 756.

Brain, W. R. (1941). Visual disorientation with special reference to lesions of the right cerebral hemisphere. *Brain, 64*, 244–272.

Bremmer, F., Schlack, A., Shah, N. J., Zafiris, O., Kubischik, M., Hoffmann, K., Zilles, K., Fink, G. R. (2001). Polymodal motion processing in posterior parietal and premotor cortex: a human fMRI study strongly implies equivalencies between humans and monkeys. *Neuron, 29*, 287–296.

Brozzoli, C., Cardinali, L., Pavani, F., & Farnè, A. (2010). Action-specific remapping of peripersonal space. *Neuropsychologia, 48*(3), 796–802.

Brozzoli, C., Gentile, G., Bergouignan, L., & Ehrsson, H. H. (2013). A shared representation of the space near oneself and others in the human premotor cortex. *Curr Biol, 23*, 1764–1768.

Brozzoli, C., Gentile, G., & Ehrsson, H. H. (2012a). That's near my hand! Parietal and premotor coding of hand-centered space contributes to localization and self-attribution of the hand. *J. Neurosci, 32*(42), 14573–14582.

Brozzoli, C., Gentile, G., Petkova, V. I., & Ehrsson, H. H. (2011). fMRI adaptation reveals a cortical mechanism for the coding of space near the hand. *J. Neurosci, 31*, 9023–9031.

Brozzoli, C., Makin, T., Cardinali, L., Holmes, N., & Farnè, A. (2012b). Peripersonal space: a multisensory interface for body-object interaction. In M. M. Micah, & M. T. Wallace (eds), The neural bases of multisensory processes (pp. 449–466). Boca Raton (FL): CRC Press/Taylor & Francis.

Brozzoli, C., Pavani, F., Urquizar, C., Cardinali, L., & Farnè, A. (2009). Grasping actions remap peripersonal space. *NeuroReport, 20*(10), 913–917.

Bufacchi, R., & Iannetti, G. (2018). An action field theory of peripersonal space. *Trends Cogn Sci*, *22*(12), 1076–1090. doi:10.1016/j.tics.2018.09.004

Bufacchi, R. J., Liang, M., Griffin, L. D., & Iannetti, G. D. (2016). A geometric model of defensive peripersonal space. *J. Neurophysiol*, *115*, 218–225.

Caggiano, V., Fogassi, L., Rizzolatti, G., Thier, P., & Casile, A. (2009). Mirror neurons differentially encode the peripersonal and extrapersonal space of monkeys. *Science*, *324*, 403–406. doi: 10.1126/science.1166818

Calvert, G. A., Spence, C., & Stein, B. E. (2004). The handbook of multisensory processes. Cambridge, MA: The MIT Press.

Canzoneri, E., di Pellegrino, G., Herbelin, B., Blanke, O., & Serino, A. (2016). Conceptual processing is referenced to the experienced location of the self, not to the location of the physical body. *Cognition*, *154*, 182–192.

Canzoneri, E., Magosso, E., & Serino, A. (2012). Dynamic sounds capture the boundaries of peripersonal space representation in humans. *PLoS ONE*, *7*:e44306.

Canzoneri, E., Marzolla, M., Amoresano, A., Verni, G., & Serino, A. (2013a). Amputation and prosthesis implantation shape body and peripersonal space representations. *Sci Rep*, *3*, 2844.

Canzoneri, E., Ubaldi, S., Rastelli, V., Finisguerra, A., Bassolino, M., & Serino, A. (2013b). Tool-use reshapes the boundaries of body and peripersonal space representations. *Exp Brain Res*, *228*(1), 25–42.

Cappe, C., Thelen, A., Romei, V., Thut, G., & Murray, M. M. (2012). Looming signals reveal synergistic principles of multisensory interactions. *J. Neurosci*, *32*, 1171–1182.

Cardini, F., Costantini, M., Galati, G., Romani, G. L., Làdavas, E., & Serino, A. (2011). Viewing one's own face being touched modulates tactile perception: an fMRI study. *J. Cogn Neurosci*, *23*(3), 503–513.

Chauvel, P. Y., Rey, M., Buser, P., & Bancaud, J. (1996). What stimulation of the supplementary motor area in humans tells about its functional organization. *Adv Neurol*, *70*(199e209), 199–209.

Chen, A., DeAngelis, G. C., & Angelaki, D. E. (2013). Functional specialization of the ventral intraparietal area for multisensory heading discrimination. *J. Neurosci*, *33*, 3567–3581.

Cléry, J., & Ben Hamed, S. (2018). Frontier of self and impact prediction. *Psychol*, *9*. doi:org/10.3389/fpsyg.2018.01073

Cléry, J., Guipponi, O., Odouard, S., Pinede, S., Wardak, C., & Ben Hamed, S. (2017). The prediction of impact of a looming stimulus onto the body is subserved by multisensory integration mechanisms. *J. Neurosci*, *37*(44), 10656–10670.

Cléry, J., Guipponi, O., Odouard, S., Wardak, C., & Ben Hamed, S. (2015a). Impact prediction by looming visual stimuli enhances tactile detection. *J. Neurosci*, *35*, 4179–4189.

Cléry, J., Guipponi, O., Odouard, S., Wardak, C., & Ben Hamed, S. (2018). Cortical networks for encoding near and far space in the non-human primate. *Neuroimage*, *176*, 164–178. doi:org/10.1016/j.neuroimage.2018.04.036

Cléry, J., Guipponi, O., Wardak, C., & Ben Hamed, S. (2015b). Neuronal bases of peripersonal and extrapersonal spaces, their plasticity and their dynamics: knowns and unknowns. *Neuropsychologia*, *70*, 313–326.

Cohen, Y. E., & Andersen, R. A. (2002). A common reference frame for movement plans in the posterior parietal cortex. *Nat Rev Neurosci*, *3*(7), 553–562.

Colby, C. L., Duhamel, J. R., & Goldberg, M. E. (1993). Ventral intraparietal area of the macaque: anatomic location and visual response properties. *J. Neurophysiol*, *69*, 902–914.

Cooke, D. F., & Graziano, M. S. A. (2003). Defensive movements evoked by air puff in monkeys. *J. Neurophysiol, 90*, 3317–3329. doi:10.1152/jn.00513.2003

Cooke, D. F., & Graziano, M. S. A. (2004). Sensorimotor integration in the precentral gyrus: polysensory neurons and defensive movements. *J. Neurophysiol, 91*, 1648–1660. doi:10.1152/jn.00955.2003

de Haan, A. M., Smit, M., Van der Stigchel, S., & Dijkerman, H. C. (2016). Approaching threat modulates visuotactile interactions in peripersonal space. *Exp Brain Res, 234*, 1875–1884.

de Vignemont, F., & Iannetti, G. D. (2014). How many peripersonal spaces? *Neuropsychologia, 70*, 327–334.

di Pellegrino, G., & Làdavas, E. (2015). Peripersonal space in the brain. *Neuropsychologia, 66*, 126–133.

di Pellegrino, G., Làdavas, E., & Farnè, A. (1997). Seeing where your hands are. *Nature, 388*(6644), 730.

Damasio A. (2003). Feelings of emotion and the self. *Ann N. Y. Acad. Sci, 1001*, 253–261.

Diederich, A., & Colonius, H. (2004). Bimodal and trimodal multisensory enhancement: effects of stimulus onset and intensity on reaction time. *Percept Psychophys, 66*, 1388–1404.

Dijkerman, H. C., Meekes, J., Ter Horst, A., Spetgens, W. P. J., de Haan, E. H. F., & Leijten, F. S. S. (2009). Stimulation of the parietal cortex affects reaching in a patient with epilepsy. *Neurology, 73*(24), 2130.

Duhamel, J. R., Bremmer, F., Ben Hamed, S., & Graf, W. (1997). Spatial invariance of visual receptive fields in parietal cortex neurons. *Nature, 389*, 845–848. doi:10.1038/39865

Duhamel, J. R., Colby, C. L., & Goldberg, M. E. (1992). The updating of the representation of visual space in parietal cortex by intended eye movements. *Science, 255*, 90–92. PMID: 1553535.

Duhamel, J. R., Colby, C. L., & Goldberg, M. E. (1998). Ventral intraparietal area of the macaque: congruent visual and somatic response properties. *J. Neurophysiol, 79*, 126–136.

Ernst, M. O., & Banks, M. S. (2002). Humans integrate visual and haptic information in a statistically optimal fashion. *Nature, 415*, 429–433.

Farnè, A., Demattè, M. L., & Làdavas, E. (2005a). Neuropsychological evidence of modular organization of the near peripersonal space. *Neurology, 65*, 1754–1758.

Farnè, A., Dematte, M. L., & Làdavas, E. (2005b). Neuropsychological evidence of modular organization of the near peripersonal space. *Neurology, 65*(11), 1754–1758.

Farnè, A., & Làdavas, E. (2000). Dynamic size-change of hand peripersonal space following tool use. *NeuroReport, 11*(8), 1645–1649.

Farnè, A., & Làdavas, E. (2002). Auditory peripersonal space in humans. *J. Cogn Neurosci, 14*(7), 1030–1043.

Fetsch, C. R., DeAngelis, G. C., & Angelaki, D. E. (2013). Bridging the gap between theories of sensory cue integration and the physiology of multisensory neurons. *Nat Rev Neurosci, 14*, 429–442. PMID: 23686172, doi:10.1038/nrn3503

Fetsch, C. R., Pouget, A., DeAngelis, G. C., & Angelaki, D. E. (2011). Neural correlates of reliability-based cue weighting during multisensory integration. *Nat Neurosci, 15*, 146–154. PMID: 22101645, doi:10.1038/nn.2983

Finisguerra, A., Canzoneri, E., Serino, A., Pozzo, T., & Bassolino, M. (2015). Moving sounds within the peripersonal space modulate the motor system. *Neuropsychologia, 70*, 421–428. doi: 10.1016/j.neuropsychologia.2014.09.043

Fogassi, L., Ferrari, P. F., Gesierich, B., Rozzi, S., Chersi, F., & Rizzolatti, G. (2005). Parietal lobe: from action organization to intention understanding. *Science, 308*, 662–667. doi:10.1126/science.1106138

Fogassi, L., Gallese, V., Fadiga, L., Luppino, G., Matelli, M., & Rizzolatti, G. (1996). Coding of peripersonal space in inferior premotor cortex (area F4). *J. Neurophysiol, 76*, 141–157.

Fogassi, L., & Luppino, G. (2005). Motor functions of the parietal lobe. *Curr Opin Neurobiol, 15*, 626–631. doi:10.1016/j.conb.2005.10.015

Fossataro, C., Sambo, C. F., Garbarini, F., & Iannetti, G. D. (2016). Interpersonal interactions and empathy modulate perception of threat and defensive responses. *Sci Rep, 6*, 19353.

Frassinetti, F., Bolognini, N., & Ladavas, E. (2002). Enhancement of visual perception by crossmodal visuo-auditory interaction. *Exp Brain Res, 147*, 332–343.

Frens, M. A., Van Opstal, A. J., & Van der Willigen, R. F. (1995). Spatial and temporal factors determine auditory-visual interactions in human saccadic eye movements. *Percept Psychophys, 57*, 802–816.

Fried, I., Katz, A., McCarthy, G., Sass, K. J., Williamson, P., Spencer, S. S, & Spencer, D. D. (1991). Functional organization of human supplementary motor cortex studied by electrical stimulation. *J. Neurosci, 11*(11), 3656e3666.

Gallese, V., Fadiga, L., Fogassi, L., & Rizzolatti, G., (1996). Action recognition in the premotor cortex. *Brain, 119*, 593–609. doi:10.1093/brain/119.2.593

Galli, G., Noel, J.-P., Canzoneri, E., Blanke, O., & Serino, A. (2015). The wheelchair as a full-body tool extending the peripersonal space. *Front. Psychol, 6*, 639. doi:10.3389/fpsyg.2015.00639

Genovesio, A., & Ferraina, S., 2004. Integration of retinal disparity and fixation-distance related signals toward an egocentric coding of distance in the posterior parietal cortex of primates. *J. Neurophysiol, 91*, 2670–2684. doi:10.1152/jn.00712.2003

Gentile, G., Guterstam, A., Brozzoli, C., & Ehrsson, H. H. (2013). Disintegration of multisensory signals from the real hand reduces default limb self-attribution: an fMRI study. *J. Neurosci, 33*(33), 13350–13366.

Gentile, G., Petkova, V. I., & Ehrsson, H. H. (2011). Integration of visual and tactile signals from the hand in the human brain: an fMRI study. *J. Neurophysiol, 105*, 910–922.

Gentilucci, M., Fogassi, L., Luppino, G., Matelli, M., Camarda, R., & Rizzolatti, G. (1988). Functional organization of inferior area 6 in the macaque monkey. I. Somatotopy and the control of proximal movements. *Exp Brain Res, 71*, 475–490.

Goldberg, G. (1985). Supplementary motor area structure and function – review and hypotheses. *Behav. Brain Sci, 8*(4), 567–588.

Graydon, M. M., Linkenauger, S. A., Teachman, B. A., & Proffitt, D. R. (2012). Scared stiff: the influence of anxiety on the perception of action capabilities. *Cogn Emot, 26*, 1301–1315. doi:10.1080/02699931.2012.667391

Graziano, M. S., & Cooke, D. F. (2006). Parieto-frontal interactions, personal space, and defensive behavior. *Neuropsychologia, 44*, 845–859.

Graziano, M. S., Cooke, D. F. & Taylor, C. S. (2000). Coding the location of the arm by sight. *Science, 290*, 1782–1786.

Graziano, M. S., & Gross, C. G. (1993). A bimodal map of space: somatosensory receptive fields in the macaque putamen with corresponding visual receptive fields. *Exp Brain Res, 97*, 96–109.

Graziano, M. S., & Gross, C. G. (1995). The representation of extrapersonal space: a possible role for bimodal, visual-tactile neurons. *Cogn Neurosci*, 1021–1034.

Graziano, M. S., Hu, X. T., & Gross, C. G. (1997). Visuospatial properties of ventral premotor cortex. *J. Neurophysiol, 77*, 2268–2292.

Graziano, M. S., Reiss, L. A., & Gross, C. G. (1999). A neuronal representation of the location of nearby sounds. *Nature, 397*, 428–430.

Graziano, M. S. A., Taylor, C. S. R, & Moore, T. (2002). Complex movements evoked by microstimulation of precentral cortex. *Neuron, 34*, 841–851.

Graziano, M. S., Yap, G. S., & Gross, C. G. (1994). Coding of visual space by premotor neurons. *Science, 266*, 1054–1057.

Grivaz, P., Blanke, O., & Serino, A. (2017). Common and distinct brain regions processing multisensory bodily signals for peripersonal space and body ownership. *Neuroimage, 147*, 602–618.

Gross, C. G., & Graziano, M. S. A. (1995). Multiple representations of space in the brain. *Neuroscientist, 1*, 43–50. doi:10.1177/107385849500100107

Guipponi, O., Wardak, C., Ibarrola, D., Comte, J.-C., Sappey-Marinier, D., Pinède, S., & Ben Hamed, S. (2013). Multimodal convergence within the intraparietal sulcus of the macaque monkey. *J. Neurosci, 33*, 4128–4139. doi:10.1523/JNEUROSCI.1421- 12.2013

Guipponi, O., Cléry, J., Odouard, S., Wardak, C., & Ben Hamed, S. (2015). Whole brain mapping of visual and tactile convergence in the macaque monkey. *Neuroimage, 117*, 93–102. doi: 10.1016/ j.neuroimage.2015.05.022

Guterstam, A., Gentile, G., & Ehrsson, H. H. (2013). The invisible hand illusion: multisensory integration leads to the embodiment of a discrete volume of empty space. *J. Cogn Neurosci, 25*(7), 1078–1099.

Halligan, P. W., & Marshall, J. C. (1991). Left neglect for near but not far space in man. *Nature, 350*, 498–500. doi:10.1038/350498a0

Hebb, D. O. (1949). The organization of behavior. New York: Wiley.

Heed, T., Habets, B., Sebanz, N., & Knoblich, G. (2010). Others' actions reduce crossmodal integration in peripersonal space. *Curr Biol, 20*, 1345–1349.

Hinton, G. (2010). A practical guide to training restricted Boltzmann machines. *Momentum, 9*(1), 926.

Holmes, N. P., Sanabria, D., Calvert, G. A., & Spence, C. (2007). Tool-use: capturing multisensory spatial attention or extending multisensory peripersonal space? *Cortex, 43*(3), 469–489.

Hunley, S. B., & Lourenco, S. F. (2018). What is peripersonal space? An examination of unresolved empirical issues and emerging findings. *Wiley Interdiscip. Rev. Cogn. Sci*, e1472. doi:org/10.1002/ wcs.1472

Hyvärinen, J. (1981). Regional distribution of functions in parietal association area 7 of the monkey. *Brain Res, 206*(2), 287–303.

Hyvärinen J., & Poranen A. (1974). Function of the parietal associative area 7 as revealed from cellular discharges in alert monkeys. *Brain, 97*, 673–692.

Hyvärinen, J., & Shelepin, Y. (1979). Distribution of visual and somatic functions in the parietal associative area 7 of the monkey. *Brain Res, 169*, 561–564.

Iriki, A., Tanaka, M., & Iwamura, Y. (1996). Coding of modified body schema during tool use by macaque postcentral neurones. *NeuroReport, 7*, 2325–2330.

Ishida, H., Nakajima, K., Inase, M., & Murata, A. (2010). Shared mapping of own and others' bodies in visuotactile bimodal area of monkey parietal cortex. *J. Cogn Neurosci, 22*, 83–96. doi:10.1162/ jocn.2009.21185

Jun, J. J., Steinmetz, N. A., Siegle, J. H., Denman, D. J., Bauza, M., Barbarits, B., et al. (2017). Fully integrated silicon probes for high-density recording of neural activity. *Nature, 551*(7679), 232. PMID: 29120427

Kandula, M., Hofman, D., & Dijkerman, H. C. (2015). Visuo-tactile interactions are dependent on the predictive value of the visual stimulus. *Neuropsychologia, 70,* 358–366.

Knill, D. C., & Richards, W. (1996). Perception as Bayesian inference. New York: Cambridge University Press.

Körding, K. P., Beierholm, U., Ma, W. J., Quartz, S., Tenenbaum, J. B., & Shams L. (2007). Causal inference in multisensory perception. *PloS ONE, 2*(9): e943. PMID: 17895984, doi:10.1371/journal.pone.0000943

La Chioma, A., Bonhoeffer, T., & Huberner, M. (2019). Area-specific mapping of binocular disparity across mouse visual cortex. *BioRxiv.* doi:org/10.1101/591412

Làdavas, E., di Pellegrino, G., Farnè, A., & Zeloni, G. (1998). Neuropsychological evidence of an integrated visuotactile representation of peripersonal space in humans. *J. Cogn Neurosci, 10*(5), 581–589.

Làdavas, E., & Farnè, A. (2004). Visuo-tactile representation of near-the-body space. *J. Physiol, Paris, 98*(1–3), 161–170.

Làdavas, E., & Serino, A. (2008). Action-dependent plasticity in peripersonal space representations. *Cogn Neuropsychol, 25*(7–8), 1099–1113.

Legrand, D., Brozzoli, C., Rossetti, Y., & Farnè, A. (2007). Close to me: multisensory space representations for action and pre-reflexive consciousness of oneself-in-the-world. *Conscious Cogn, 16*(3), 687–699.

Leinonen, L. (1980). Functional properties of neurones in the posterior part of area 7 in awake monkey. *Acta Physiol. Scand, 108*(3), 301–308.

Leinonen, L., & Nyman, G. (1979). II. Functional properties of cells in anterolateral part of area 7 associative face area of awake monkeys. *Exp Brain Res, 34,* 321–333.

Leinonen, L., Hyvärinen, J., Nyman, G., & Linnankoski, I. (1979). Functional properties of neurons in lateral part of associative area 7 in awake monkeys. *Exp Brain Res, 34,* 299–320.

Lenggenhager, B., Tadi, T., Metzinger, T., & Blanke, O. (2007). Video ergo sum: manipulating bodily self-consciousness. *Science, 317*(5841), 1096–1099.

Libet, B., Gleason, C. A., Wright, E. W., & Pearl, D. K. (1983). Time of conscious intention to act in relation to onset of cerebral activity (readiness-potential): the unconscious initiation of a freely voluntary act. *Brain, 106,* 623–642. PMID: 6640273, doi:10.1093/brain/106.3.623

Lourenco, S. F., Longo, M. R., & Pathman, T. (2011). Near space and its relation to claustrophobic fear. *Cognition, 119,* 448–453.

Lovelace, C. T., Stein, B. E., & Wallace, M. T. (2003). An irrelevant light enhances auditory detection in humans: a psychophysical analysis of multisensory integration in stimulus detection. *Cogn Brain Res, 17,* 447–453.

Luppino, G., Ben Hamed, S., Gamberini, M., Matelli, M., & Galletti, C. (2005). Occipital (V6) and parietal (V6A) areas in the anterior wall of the parieto-occipital sulcus of the macaque: a cytoarchitectonic study. *Eur. J. Neurosci, 21,* 3056–3076. doi:10.1111/j.1460- 9568.2005.04149.x

Luppino, G., Matelli, M., Camarda, V., Gallese, V., & Rizzolatti, G. (1991). Multiple representations of body movement in mesial area 6 and the adjacent cingulate cortex: an intracortical microstimulation study. *J. Comp Neurol, 311,* 463–482.

Ma, W. J., Beck, J. M., Latham, P. E., & Pouget, A. (2006). Bayesian inference with probabilistic population codes. *Nat Neurosci, 9,* 1432–1438.

Macaluso, E., & Maravita, A. (2010). The representation of space near the body through touch and vision. *Neuropsychologia, 48*(3), 782–795.

Magosso, E., Ursino, M., di Pellegrino, G., Làdavas, E., & Serino, A. (2010a). Neural bases of peri-hand space plasticity through tool-use: insights from a combined computational-experimental approach. *Neuropsychologia*, *48*, 812–830. doi:10.1016/j.neuropsychologia.2009.09.037

Magosso, E., Zavaglia, M., Serino, A., di Pellegrino, G., & Ursino, M. (2010b). Visuotactile representation of peripersonal space: a neural network study. *Neural Comput*, *22*, 190–243. doi:10.1162/neco.2009.01-08-694

Maister, L., Cardini, F., Zamariola, G., Serino, A., & Tsakiris, M. (2015). Your place or mine: shared sensory experiences elicit a remapping of peripersonal space. *Neuropsychologia*, *70*, 455–461.

Makin, J. G., Fellows, M. R., & Sabes, P. N. (2013). Learning multisensory integration and coordinate transformation via density estimation. *PLoS Comput Biol*, *9*(4), e1003035.

Makin, T. R., Holmes, N. P., Brozzoli, C., Rossetti, Y., & Farnè, A. (2009). Coding of visual space during motor preparation: approaching objects rapidly modulate corticospinal excitability in hand-centered coordinates. *J. Neurosci*, *29*(38), 11841–11851.

Makin, T. R., Holmes, N. P., & Zohary, E. (2007). Is that near my hand? multisensory representation of peripersonal space in human intraparietal sulcus. *J. Neurosci*, *27*(4), 731–740. doi:10.1523/JNEUROSCI.3653-06.2007

Martino, A. M., & Strick, P. L. (1987). Corticospinal projections originate from the arcuate premotor area. *Brain Res*, *404*, 307–312.

Matelli, M., & Luppino, G., (2001). Parietofrontal circuits for action and space perception in the macaque monkey. *Neuroimage*, *14*, S27–S232. doi:10.1006/nimg.2001.0835

Matsumura, M., & Kubota, K. (1979). Cortical projection of hand-arm motor area from postarcuate area in macaque monkey: a histological study of retrograde transport of horseradish peroxidase. *Neurosci Lett*, *11*, 241–246.

Meijer, G. T., Montijn, J. S., Pennartz, C. M. A., & Lansink, C. S. (2017). Audiovisual modulation in mouse primary visual cortex depends on crossmodal stimulus configuration and congruency. *J. Neurosci*, *37*, 8783–8796.

Meredith, M. A., & Stein, B. E. (1983). Interactions among converging sensory inputs in the superior colliculus. *Science*, *221*, 389–391.

Meredith, M. A., & Stein, B. E. (1986). Spatial factors determine the activity of multisensory neurons in cat superior colliculus. *Brain Res*, *365*, 350–354.

Meredith, M. A., & Stein, B. E. (1996). Spatial determinants of multisensory integration in cat superior colliculus neurons. *J. Neurophysiol*, *75*, 1843–1857.

Miller, J. (1982). Divided attention: evidence for coactivation with redundant signals. *Cogn Psychol*, *14*, 247–279. doi: org/10.1016/0010-0285(82)90010-X

Miller, J. (1986). Timecourse of coactivation in bimodal divided attention. *Percept Psychophys*, *40*, 331–343.

Mitz, A. R., & Wise, S. P. (1987). The somatotopic organization of the supplementary motor area: intracortical microstimulation mapping. *J. Neurosci*, *7*, 1010–1021.

Morgan, M. L., DeAngelis, G. C., & Angelaki, D. E. (2008). Multisensory integration in macaque visual cortex depends on cue reliability. *Neuron*, *59*, 662–673. PMID 18760701, doi:10.1016/j.neuron.2008.06.024

Moser, E. I., Kropff, E., & Moser, M.-B. (2008). Place cells, grid cells, and the brain's spatial representation system. *Annu Rev. Neurosci*, *31*, 69–89.

Mountcastle, V. B. (1976). The world around us: neural command functions for selective attention. *Neurosci Res Progr Bull*, *14*, 1–47.

Murata, A., Fadiga, L., Fogassi, L., Gallese, V., Raos, V., & Rizzolatti, G. (1997). Object representation in the ventral premotor cortex (area F5) of the monkey. *J. Neurophysiol, 78*, 2226–2230.

Murray, M. M., & Wallace, M. T. (2012). *The neural bases of multisensory processes.* Boca Raton (FL): CRC Press.

Noel, J-P. (2018). Supra-optimality may emanate from suboptimality, and hence optimality is no benchmark in multisensory integration. *Behav. Brain Sci, 41*, e239. doi:10.1017/S0140525X18001280

Noel, J.-P., Blanke O., Magosso, E., & Serino, A. (2018b). Neural adaptation accounts for the resizing of peri-personal space representation: evidence from a psychophysical-computational approach. *J. Neurophysiol.* doi:org/10.1152/jn.00652.2017

Noel, J.-P., Blanke, O., & Serino, A. (2018f). From multisensory integration in peripersonal space to bodily self-consciousness: from statistical regularities to statistical inference. *Ann. N. Y. Acad. Sci.* doi:10.1111/nyas.13867

Noel, J.-P., Chatelle, C., Perdikis, S., Jöhr, J., Da Silva, M. L., Ryvlin, P., et al. (2019). Peri-personal space encoding in patients with disorders of consciousness and cognitive-motor dissociation. *Neuroimage Clin, 24*, 10190.

Noel, J.-P., Grivaz, P., Marmaroli, P., Lissek, H., Blanke, O., & Serino, A. (2015b). Full body action remapping of peripersonal space: the case of walking. *Neuropsychologia.* doi:10.1016/j.neuropsychologia.2014.08.030.

Noel, J.-P., Modi, K., Wallace, M. T., & Van der Stoep, N. (2018e). Audiovisual integration in depth: multisensory binding and gain as a function of distance. *Exp Brain Res.* doi:10.1007/s00221-018-5274-7

Noel, J.-P., Park, H., Pasqualini, I., Lissek, H., Wallace, M., Blanke, O., & Serino, A. (2018c). Audiovisual sensory deprivation degrades visuo-tactile peri-personal space. *Conscious Cogn, 61*, 61–75. doi:10.1016/j.concog.2018.04.001

Noel, J.-P., Pfeiffer, C., Blanke, O., & Serino, A. (2015a). Full body peripersonal space as the sphere of the bodily self? *Cognition, 144*(3), 49–57.

Noel, J.-P., Samad, M., Doxon, A., Clark, J., Keller, S., & Di Luca, M. (2018d). Peri-personal space as a prior in coupling visual and proprioceptive signals. *Sci Rep, 8*(1), 15819.

Noel, J-P., & Serino, A. (2019). High action values occur near our body. *Trends Cogn Sci, 23*(4), 269–270.

Noel, J-P., Serino, A., & Wallace, M. (2018a). Increased neural strength and reliability to audiovisual stimuli at the boundary of peri-personal space. *J. Cogn Neurosci.* doi.org/10.1162/jocn_a_01334

Noel, J.-P., & Wallace, M. (2016). Relative contributions of visual and auditory spatial representations to tactile localization. *Neuropsychologia, 82*, 84–90. doi:org/10.1016/j.neuropsychologia.2016.01

Nozawa, G., Reuter-Lorenz, P. A., & Hughes, H. C. (1994). Parallel and serial processes in the human oculomotor system: bimodal integration and express saccades. *Biol Cybern, 72*, 19–34.

Pandya, D. N., & Vignolo, L. A. (1971). Intra- and interhemispheric projections of the precentral premotor and arcuate areas in the rhesus monkey. *Brain Res, 26*, 217–233.

Patane, I., Cardinali, L., Salemme, R., Pavani, F., Farnè, A., & Brozzoli, C. (2018). Action planning modulates peri-personal space. *J. Cogn Neurosci.* doi:10.1162/jocn_a_01349

Pellencin, E., Paladino, M. P., Herbelin, B., & Serino, A. (2017). Social perception of others shapes one's own multisensory peripersonal space. *Cortex, 104*, 163–179. doi.org/10.1016/j.cortex.2017.08.033

Penfield, W., & Welch, K. (1951). Supplementary motor area of the cerebral cortex. *Arch Neurol Psychiat, 66*, 289–317.

Pfeiffer, C., Noel, J.-P., Blanke, O., & Serino, A. (2018). Vestibular modulation of peri-personal space boundaries. *Eur. J. Neurosci, 47,* 800–811. doi:10.1111/ejn.13872

Quinn, B. T., Carlson, C., Doyle, W., Cash, S. S., Devinsky, O., Spence, C., & Thesen T. (2014). Intracranial cortical responses during visual-tactile integration in humans. *J. Neurosci, 34*(1), 171–181.

Raab, D. H. (1962). Statistical facilitation of simple reaction times. *Trans N Y Acad Sci, 24,* 574–590.

Rahnev, D., & Denison, R. N. (2018). Suboptimality in perceptual decision making. *Behav. Brain Sci, 41.* doi:org/10.1017/S0140525X18000936

Raos, V., Umiltá, M.-A., Murata, A., Fogassi, L., & Gallese, V., (2006). Functional properties of grasping-related neurons in the ventral premotor area F5 of the macaque monkey. *J. Neurophysiol, 95,* 709–729. doi:10.1152/jn.00463.2005

Rizzolatti, G., Camarda, R., Fogassi, L., Gentilucci, M., Luppino, G., & Matelli, M., (1988). Functional organization of inferior area 6 in the macaque monkey. II. Area F5 and the control of distal movements. *Exp Brain Res, 71,* 491–507.

Rizzolatti, G., Fadiga, L., Gallese, V., & Fogassi, L. (1996. Premotor cortex and the recognition of motor actions. *Cogn Brain Res, 3,* 131–141.

Rizzolatti, G., & Luppino, G. (2001). The cortical motor system. *Neuron, 31,* 889–901.

Rizzolatti, G., Scandolara, C., Matelli, M., & Gentilucci, M. (1981a). Afferent properties of periarcuate neurons in macaque monkeys. I. Somatosensory responses. *Behav. Brain Res, 2*(2), 125–146.

Rizzolatti, G., Scandolara, C., Matelli, M., & Gentilucci, M. (1981b). Afferent properties of periarcuate neurons in macaque monkeys. II. Visual responses. *Behav. Brain Res, 2,* 147–163.

Robinson, D. L., Goldberg, M. E., & Stanton, G. B. (1978). Parietal association cortex in the primate: sensory mechanisms and behavioral modulations. *J. Neurophysiol, 41,* 910–932.

Rohe, T., Ehlis, A. C., & Noppeney, U. (2019). The neural dynamic of hierarchical Bayesian causal inference in multisensory perception. *Nat Comm, 10*(1), 1907. doi:10.1038/s41467-019-09664-2

Rohe, T., & Noppeney, U. (2015). Cortical hierarchies perform Bayesian causal inference in multisensory perception. Academic editor: Kayser, C. *PLoS Biol, 13*(2), e1002073. doi:10.1371/journal.pbio.1002073.s009

Rohe, T., & Noppeney, U. (2016). Distinct computational principles govern multisensory integration in primary sensory and association cortices. *Curr Biol, 26,* 509–514. PMID: 26853368, doi:10.1016/j.cub.2015.12.056

Roncone, A., Hoffmann, M., Pattacini, U., Fadiga, L., & Metta, G. (2016). Peripersonal space and margin of safety around the body: learning visuo-tactile associations in a humanoid robot with artificial skin. *PLoS ONE, 11*(10), e0163713.

Rozzi, S., Ferrari, P. F, Bonini, L., Rizzolatti, G., & Fogassi, L. (2008). Functional organization of inferior parietal lobule convexity in the macaque monkey: electrophysiological characterization of motor, sensory and mirror responses and their correlation with cytoarchitectonic areas. *Eur. J. Neurosci, 28*(8),1569–1588. doi:10.1111/j.1460-9568.2008.06395.x

Sakata, H., Shibutani, H., & Kawano, K. (1980). Spatial properties of visual fixation neurons in posterior parietal association cortex of the monkey. *J. Neurophysiol, 43,* 1654–1672.

Salomon, R., Noel, J.-P., Lukowska, M., Faivre, N., Metzinger, T., Serino, A., et al. (2017). Unconscious integration of multisensory bodily inputs in the peripersonal space shapes bodily self-consciousness. *Cognition, 166,* 174–183.

Samad, M., Chung, A. J., & Shams, L. (2015). Perception of body ownership is driven by Bayesian sensory inference. *PLoS ONE, 10*(2), e0117178.

Sambo, C. F., Forster, B., Williams, S. C., & Iannetti, G. D. (2012). To blink or not to blink: fine cognitive tuning of the defensive peripersonal space. *J. Neurosci*, *32*(37), 12921–12927.

Sambo, C. F., & Iannetti, G. D. (2013). Better safe than sorry? The safety margin surrounding the body is increased by anxiety. *J. Neurosci*, *33*(35), 14225–14230.

Serino, A., Annella, L., & Avenanti, A. (2009). Motor properties of peripersonal space in humans. PLoS ONE, *4*(8), e6582. doi:org/10.1371/journal.pone.0006582

Serino, A., Canzoneri, E., & Avenanti, A. (2011). Fronto-parietal areas necessary for a multisensory representation of peripersonal space in humans: an rTMS study. *J. Cogn Neurosci*, *23*, 2956–2967.

Serino, A., Canzoneri, E., Marzolla, M., di Pellegrino, G., & Magosso E. (2015a). Extending peripersonal space representation without tool-use: evidence from a combined behavioral-computational approach. *Front. Behav Neurosci*, *9*, (4). doi:10.3389/fnbeh.2015.00004

Serino, A., Noel, J.-P., Galli, G., Canzoneri, E., Marmaroli, P., Lissek, H., & Blanke, O. (2015b). Body part centered and full body-centered peripersonal space representations. *Sci Rep*, *5*, 18603. doi:10.1038/srep18603

Serino, A., Noel, J-P., Mange, R., Canzoneri, E., Pellencin, E., Bello-Ruiz, J., Bernasconi, F., Blanke, O., & Herbelin, B. (2017). Peri-personal space: an index of multisensory body-interaction in real, virtual, and mixed realities. *Front. ICT*, *4*, 31.

Seth, A. K., & Tsakiris, M. (2018). Being a beast machine: the somatic basis of selfhood. *Trends Cogn Sci*, *22*(11), 969–981.

Shao, M., DeAngelis, G. C., Angelaki, D. E., & Chen, A. (2018). Clustering of heading selectivity and perception-related activity in the ventral intraparietal area. *J. Neurophysiol*, *119*, 1113–1126.

Spence, C., Pavani, F., & Driver, J. (2000). Crossmodal links between vision and touch in covert endogenous spatial attention. *J. Exp Psychol Hum Percept Perform*, *26*(4), 1298–1319.

Spence, C., Pavani, F., & Driver, J. (2004a). Spatial constraints on visual-tactile crossmodal distractor congruency effects. *Cogn. Affect. Behav. Neurosci*, *4*(2), 148–169.

Spence, C., Pavani, F., Maravita, A., & Holmes, N. (2004b). Multisensory contributions to the 3-D representation of visuotactile peripersonal space in humans: evidence from the crossmodal congruency task. *J. Physiol, Paris*, *98*(1–3), 171–189.

Stanford, T. R., & Stein, B. E. (2007). Superadditivity in multisensory integration: putting the computation in context. *NeuroReport*, *18*, 787–792.

Stein, B. E., & Meredith, M. A. (1993). The merging of the senses (ed. Gazzaniga, M. S.) Cambridge, MA: The MIT Press.

Stein, B. E., & Stanford, T. R. (1993). Multisensory integration: current issues from the perspective of the single neuron. *Nat Rev Neurosci*, *9*, 255–266. doi:org/10.1038/nrn2331

Stepniewska, I., Fang, P.-C., & Kaas, J. H. (2005). Microstimulation reveals specialized subregions for different complex movements in posterior parietal cortex of prosimian galagos. *Proc. Natl. Acad. Sci. U.S.A.*, *102*, 4878–4883. doi:10.1073/pnas.0501048102

Stone, K. D., Kandula, M., Keizer, A., & Dijkerman, H. C. (2018). Peripersonal space boundaries around the lower limbs. *Exp Brain Res*, *236*(1), 161–173.

Straka, Z., & Hoffmann, M. (2017). Learning a peripersonal space representation as a visuo-tactile prediction task. In Proceedings of the 26th International Conference on Artificial Neural Networks (ICANN), pp. 101–109. Cham: Springer. doi:10.1007/978-3-319-68600-4_13

Sugihara, T., Diltz, M. D., Averbeck, B. B., & Romanski, L. M. (2006). Integration of auditory and visual communication information in the primate ventrolateral prefrontal cortex. *J. Neurosci*, *26*, 11138–11147.

Taffou, M., & Viaud-Delmon, I. (2014). Cynophobic fear adaptively extends peripersonal space. *Front. Psychiatry*, *5*, 122.

Taube, J. S., Goodridge, J. P., Golob, E. J., Dudchenko, P. A., & Stackman, R.W. (1996). Processing the head direction signal: a review and commentary. *Brain Res. Bull*, *40*, 477–486.

Teneggi, C., Canzoneri, E., di Pellegrino, G., & Serino, A. (2013). Social modulation of peripersonal space boundaries. *Curr Biol*, *23*, 406–411.

Teramoto, W. (2018). A behavioral approach to shared mapping of peripersonal space between oneself and others. *Sci Rep*, *8*(5432). doi:10.1038/s41598-018-23815-3

Teramoto, W., Honda, K., Furuta, K., & Sekiyama, K. (2017). Visuotactile interaction even in far sagittal space in older adults with decreased gait and balance functions. *Exp Brain Res*, 1–15. doi:10.1007/s00221-017-4975-7

Teramoto, W., & Kakuya T. (2015). Visuotactile peripersonal space in healthy humans: evidence from crossmodal congruency and redundant target effects. *Interdiscip Inf Sci*, *21*, 133–142. doi:10.4036/iis.2015.A.04

Thier, P., & Andersen, R. A. (1998). Electrical microstimulation distinguishes distinct saccade-related areas in the posterior parietal cortex. *J Neurophysiol*, *80*, 1713–1735.

Tsakiris M (2008). Looking for myself: current multisensory input alters self-face recognition. *PLoS ONE*, *3*(12), e4040.

Ulrich, R., Miller, J., Schröter, H. (2007). Testing the race model inequality: an algorithm and computer programs. *Behav. Res. Methods*, *39*, 291–302.

Ursino, M., Zavaglia, M., Magosso, E., Serino, A., & di Pellegrino, G. (2007). A neural network model of multisensory representation of peripersonal space: effect of tool use. *Conf. Proc. IEEE Eng. Med. Biol. Soc*, 2735–2739.

van Beers, R. J., Sittig, A. C., & Gon, J. J. (1998).The precision of proprioceptive position sense. *Exp Brain Res*, *122*, 367–377.

van Beers, R. J., Sittig, A. C., & Gon, J. J. (1999). Integration of proprioceptive and visual position-information: an experimentally supported model. *J. Neurophysiol*, *81*, 1355–1364. doi:org/10.1152/jn.1999.81.3.1355

van Beers, R. J., Wolpert, D. M., & Haggard, P. (2002). When feeling is more important than seeing in sensorimotor adaptation. *Curr Biol*, *12*, 834–837. doi:org/10.1016/S0960-9822(02)00836-9

Van der Stoep, N., Nijboer, T. C. W., Van der Stigchel, S., & Spence, C. (2015a). Multisensory interactions in the depth plane in front and rear space: a review. *Neuropsychologia*, *70*, 335–349. doi:10.1016/j.neuropsychologia.2014.12.007

Van der Stoep, N., Van der Stigchel, S., Nijboer, T. C. W., & Van der Smagt, M. J. (2015b). Audiovisual integration in near and far space: effects of changes in distance and stimulus effectiveness. *Exp Brain Res*, 1–14. doi:10.1007/s00221-015-4248-2

Wilkinson, L. K., Meredith, M. A., & Stein, B. E. (1996). The role of anterior ectosylvian cortex in cross-modality orientation and approach behavior. *Exp Brain Res*, *112*(1), 1–10.

Wozny, D. R., & Shams, L. (2006). Integration and segregation of visual-tactile-auditory information is Bayes optimal. *J. Vis*, *6*, 176.

Zhang, T., & Britten, K. H. (2011). Parietal area VIP causally influences heading perception during pursuit eye movements. *J. Neurosci*, *31*, 2569–2575.

3

Close is better

Visual perception in peripersonal space

Elvio Blini, Alessandro Farnè, Claudio Brozzoli, and Fadila Hadj-Bouziane

3.1 Peripersonal space: the multisensory origins

The intuition that the space closely surrounding our body is somehow special has a rather long history (Hall, 1966; Hediger, 1950). For example, Hall (1966) sketched several rough boundaries which, from a social perspective, are functional to different types of interaction (e.g. intimate, up to 45 cm; personal, up to 1.2 m; social, up to 3.6 m; or public; see de Vignemont, 2018). However, it is probably the seminal study of Rizzolatti et al. (1981) that shaped and gave impulse to the vast literature about the newly termed 'peripersonal space' (PPS; also see Hyvärinen & Poranen, 1974). Rizzolatti and colleagues recorded, in macaque monkeys, the responses of neurons located in the periarcuate cortex—part of the frontal lobe that receives input from associative sensory areas. They found a substantial proportion of neurons that were reliably coding for visual stimuli appearing in PPS, defined operationally as the space immediately surrounding the body (e.g. 10 to 30 cm). Strikingly, the vast majority of these visual neurons had bimodal (visual and somatosensory) receptive fields that were in register with the corresponding somatosensory area (e.g. neurons responding to visual stimuli close to the mouth had a tactile counterpart on the mouth), an observation that has been confirmed and extended afterwards to other brain regions (e.g. putamen, ventral intraparietal area, premotor cortex; Colby, Duhamel & Goldberg, 1993; Fogassi et al., 1996; Graziano & Cooke, 2006; Graziano & Gross, 1993). This seminal finding gave rise to a first neuroscientific approach to studying PPS, with a dominant multisensory perspective, as well as a framework to guide cognitive models.

The original account of Rizzolatti et al. (1981) advocated that the role of these bimodal neurons may essentially be action-oriented (also see Murata et al., 1997). PPS is the only region of space in which we can act on, reach, or manipulate objects directly: a tight link between visual and somatosensory input is particularly expected in PPS, because it may contribute to the efficiency of our goal-directed actions, on the one hand, and it may be shaped by the continuous experience of simultaneous multisensory stimulations, on the other hand. Arguably, studies exploring the effect of tool-use training (Iriki, Tanaka, & Iwamura, 1996; Maravita & Iriki, 2004), which can expand the region of space upon which we can purposefully act, were largely inspired by this formulation. In the seminal study of Iriki et al. (1996), macaque monkeys were trained to reach objects with a rake while the activity of multisensory neurons in the postcentral gyrus was recorded. The authors found that, during tool use, the visual receptive field of these neurons was enlarged up to covering the tip of the rake (Iriki et al., 1996). Tool use has been exploited ever since to probe the plasticity of PPS representation and body schema (Berti & Frassinetti, 2000; Cardinali et al., 2009; Farnè &

Elvio Blini, Alessandro Farnè, Claudio Brozzoli, and Fadila Hadj-Bouziane, *Close is better* In: *The World at Our Fingertips.* Edited by Frédérique de Vignemont, Andrea Serino, Hong Yu Wong, and Alessandro Farnè, Oxford University Press (2021). © Oxford University Press. DOI: 10.1093/oso/9780198851738.003.0003.

Làdavas, 2000; Maravita et al., 2002; Sposito et al., 2012), a closely related—yet different—representation of the body for actions (Cardinali, Brozzoli, & Farnè, 2009).

Defensive actions are particularly important responses that can be accomplished in PPS (Graziano & Cooke, 2006). Escape from a predator or a threatening stimulus has a much more pronounced motivational priority than the needs for food or mating, yet the sight of a threat is not generally sufficient to activate defensive responses: the threat must also violate some safety boundary around the body, termed 'flight distance' (Hediger, 1950), and be perceived as intrusive (Graziano & Cooke, 2006). PPS has also been conceptualized as a system evolved to be highly specialized in coding and maintaining this safety boundary (Graziano & Cooke, 2006), a task for which multisensory interactions are paramount. Think, for example, about looming objects: potentially harmful objects approaching enhance automatically the tactile sensitivity of the body part (and nearby body parts) upon which contact is expected to happen (Cléry et al., 2015; Colby, Duhamel, & Goldberg, 1993; Neppi-Mòdona et al., 2004). Stressing the defensive function of PPS does not contradict the first, action-related formulation. This account has provided instead fertile ground for studies extending the notion of PPS plasticity to include interpersonal distance regulation depending on emotion- or stress-related stimuli. For example, Ruggiero et al. (2017) presented to healthy participants visual avatars depicting different facial expressions with emotional valence (e.g. happy vs angry) approaching them in an immersive virtual reality environment. They asked participants to press a button as soon as the distance of the visual avatars was felt as uncomfortable, thus delineating a comfort distance or zone. Participants showed enlarged comfort zones for the avatars signalling anger (Ruggiero et al., 2017). The magnitude of this enlargement can be predicted on the basis of individuals' autonomic responses (Cartaud et al.,2018) and is modulated by personality traits such as the level of anxiety (Sambo & Iannetti, 2013). Overall, this picture is coherent with the notion that the presence of threatening cues within the close space surrounding us prompts avoidance behaviours (Ruggiero et al., 2017; also see Ferri et al., 2015; Teneggi et al., 2013). As such, PPS may help us move efficiently towards (to reach) or away from (to avoid) elements of our close environment. Separating the two functions of PPS into action- or defence-related is therefore probably artificial (de Vignemont & Iannetti, 2015). A more general take could assume a flexible organization of PPS that would depend on current task and environmental constraints (e.g. Bufacchi & Iannetti, 2018). It is worth stressing, however, that both formulations firmly build on, and were originally put forward to accommodate for, the multisensory and distance-tuned properties of neurons ascribed to PPS coding.

It would not be surprising, at this point, to note that most of the tasks devised to probe PPS-related processing involve bimodal sensory stimulations. One common scenario involves the presentation of one to-be-discriminated tactile stimulus coupled with an irrelevant visual or auditory one, delivered in either an overlapping, close, or distant position in space (Brozzoli et al., 2009; Canzoneri, Magosso, & Serino, 2012; Maravita, Spence, & Driver, 2003; Spence, Pavani, & Driver, 2004; Teneggi et al., 2013). These tasks have shown that, when both stimuli are overlapping in space and time, stronger multisensory interaction occurs: neural and behavioural responses are therefore enhanced, resulting in advantages (Makin et al., 2012). This effect is maximal near the body (i.e. where touch is delivered) and decreases as a function of the distance at which the irrelevant stimulus is presented. Psychophysical modelling of this decay allows one to estimate a rough point of 'indifference', in which multisensory interaction no longer occurs, and thus to functionally identify two seemingly different regions of space (Canzoneri et al., 2012; Noel et al., 2015; Teneggi et al., 2013). For example, Canzoneri et al. (2012) measured response times for tactile discrimination of stimuli presented with a concurrent dynamic sound (i.e. perceived to be looming or receding, and at different distances from the delivered touch). The results showed an audio-tactile interaction effect, stronger with approaching sounds, that was maximal close to the stimulated hand. Psychophysical modelling further suggested that a sigmoidal function could adequately capture this decay, and the inflection point of the curve (the aforementioned 'indifference point') was taken as a proxy for the putative limit and extension of PPS. This approach was proven fertile and capable of highlighting the peculiar plasticity of PPS. For example, it was exploited to show that sounds associated—by either physical or semantic properties—with negative emotions or contents are capable of pushing the PPS boundaries farther away (Ferri et al., 2015), in agreement with the PPS role in maintaining a safety zone around the body.

The wide use of multisensory tasks to probe PPS processing is paradigmatic of its conception as the region of space in which multisensory interactions occur. This is certainly the case for the integration of touch with other sensory stimulations, because touch clearly cannot happen in extrapersonal space. Indeed, PPS is inherently multisensory. However, this does not preclude the existence of unimodal advantages tied to PPS-specific processing. This chapter will review recent evidence for the existence of such unisensory (visual) advantages. However, before tackling this emerging field, we deem useful a brief overview of the literature concerning the distribution of spatial attention in depth.

3.2 Peripersonal space and attention: the inextricable link?

It is hard, perhaps impossible, to tease apart whether depth-specific neural and behavioural modulations result from enhanced attentional processing close to the body or rather to a dedicated system for PPS (perceptual) processing (but see Makin et al., 2009; Reed et al., 2013, for dissociations between effects depending on mere hand proximity and orienting of attention). This literature, however, allows us to access and appreciate several findings that, far from denying a privileged role of PPS in perception, may increase our understanding of PPS-specific unisensory advantages.

Spatial and hemispheric asymmetries hold a special place within the literature on human perceptual and attentional systems. General consensus has been reached on the notion that attention is not uniformly distributed along the three orthogonal axes (Gawryszewski et al., 1987; Shelton, Bowers, & Heilman, 1990). However, the majority of experiments has been carried exploiting two-dimensional screens, thereby neglecting the sagittal (near-to-far) plane (but see Couyoumdjian, Nocera, & Ferlazzo, 2003; Losier & Klein, 2004; Plewan & Rinkenauer, 2017).

As notable exceptions, few studies exploited cued detection tasks with stimuli appearing at different distances (Couyoumdjian et al., 2003; Gawryszewski et al., 1987; Losier & Klein, 2004). The first notion drawn from these studies is that spatial attention can be displaced along the sagittal plane just like it can be displaced along the horizontal and vertical planes (Couyoumdjian et al., 2003; Gawryszewski et al., 1987; Losier & Klein, 2004), as seen by cueing validity effects (i.e. better performance when cues and targets appear in the same region of space, and decreased performance when positions are incongruent). Second, participants are faster in responding to stimuli appearing close to their body, in PPS, suggesting that more attentional resources are allocated there (Gawryszewski et al., 1987; Plewan & Rinkenauer, 2017). A more specific manifestation of this phenomenon has been described for the space close to the hands (Reed, Grubb, & Steele, 2006). Reed et al. (2006) used a purely visual covert attention paradigm, a cued detection task like the ones described earlier in this paragraph, in which the only experimental manipulation was the position of the participants' hand (either close to left- or right-sided targets); thus, also in this case, the task had no explicit cross-modal component, being confined to the visual modality. Visual stimuli were detected faster when appearing closer to the perceived position of the hands (near-hand effect). This applied also when visual input was lacking—namely, when the hand was occluded, the proprioceptive input appeared sufficient for this effect to emerge (but see, for contrasting evidence on the role of hand proprioception, Blini et al., 2018; Di Pellegrino & Frassinetti, 2000; Làdavas et al., 2000). Hand position alone could, indeed, modulate both early and late attention-sensitive components of brain activity in a subsequent experiment exploiting event-related potentials (Reed et al., 2013). Behavioural results were later extended with different tasks (e.g. visual search, inhibition of return, attentional blink), all consistently showing that visual and attentional abilities are altered near the hands (Abrams et al. 2008).

It is interesting to notice that the accounts proposed to frame these results also call into cause the role of attention in maximizing action efficiency or in monitoring the nearby space for defensive purposes (Abrams et al., 2008). The maintenance of a defensive space, indeed, would necessarily involve monitoring of the nearby environment, a mental representation of it, and ultimately attentional resources to be constantly deployed (Graziano & Cooke, 2006). The question as to whether such a—constantly active—monitoring is biologically and evolutionarily plausible remains open, because this could come at very high costs for already limited resources. In addition, the monitoring of looming objects, with respect to receding ones, would be privileged because more likely to result in an impact with the body (Cléry et al., 2015; Neppi-Mòdona et al., 2004). Looming objects are indeed known to strongly capture visuospatial attention in human (Franconeri & Simons, 2003; Lin, Murray, & Boynton, 2009) and non-human primates (Ghazanfar, Neuhoff, & Logothetis, 2002; Schiff, Caviness, & Gibson, 1962), and cause increased multisensory interaction (Canzoneri et al., 2012; Maier et al., 2004), typically attributed to PPS-specific processing. Similarly, threatening stimuli or cues strongly capture and hold attention (Armony & Dolan, 2002; Koster et al., 2004). It could thus be argued that this specificity may explain, at least in part, the effects reported over PPS signatures (i.e. extended PPS limits, Ferri et al., 2015; but see Makin et al., 2009, for evidence of a dissociation). Another possibility would be that the domain-general mechanism of spatial attention, and cross-modal attention in particular (e.g. Eimer, Velzen, & Driver, 2002), may exploit the neural circuits specialized for PPS, as initially suggested (Làdavas et al., 1998). Disentangling these—potentially not mutually exclusive—alternatives may be difficult, especially with purely behavioural paradigms. Yet, better identifying the specific roles played by attention versus PPS perception is likely to provide valuable contributions to our understanding of the mechanisms that today are indistinguishably gathered within the PPS label. At any rate, phenomena like the near-hand effects typically represent instances in which advantages for stimuli presented close to the body occur in purely unisensory tasks.

3.3 Neural bases of peripersonal multi- (and uni-) sensory processing

The functional linkage between PPS and actions, supported by neurophysiological and anatomical evidence from primate work (see for review, Makin et al., 2012), prompted the idea that visual processing in PPS would mainly rely on the dorsal visual stream, optimized for action, whereas visual processing beyond it would mainly rely on the ventral stream, optimized for perception (Milner & Goodale, 2008; Previc, 1990). Because the dorsal stream recruits more extensively parietal networks and magnocellular neurons—that are highly specialized in responding to rapid changes in the visual scene in spite of their low resolution—this is also well fitting with the PPS role in monitoring a zone around the body. One could argue that, at least in some conditions, it would be better to ward off an insect close to us before knowing whether it is a wasp or a ladybird. This notion has been supported by behavioural studies showing faster detection times for stimuli occurring close to the body or the hands (Gawryszewski et al., 1987; Plewan & Rinkenauer, 2017; Reed et al., 2006; but see Makin et al., 2015). Furthermore, studies have shown that performance on tasks requiring speeded temporal-gap detection improves in the near-hand space (Goodhew et al., 2013; Gozli, West, & Pratt, 2012), whereas performance in spatial-gap tasks is hampered (Gozli et al., 2012). This has been discussed as coherent with a general magnocellular advantage for PPS processing (Bush & Vecera, 2014; Goodhew et al., 2015, for a review).

However, on the other side of the coin, on the bases of this account, performance benefits in fine-grained discrimination tasks could be predicted to occur beyond PPS, where the ventral pathway would play a more important role. Parvocellular neurons—with their small and contrast-sensitive receptive fields—appear indeed ideal matches to contribute to object identification, especially because, in everyday life, the size of an object (retinal size) scales with distance. In other words, we would be better at discriminating wasps from ladybirds when the insect was far; in this case, we could afford time to prepare an optimal response, shaped according to the significance of the threat (we would need a fast, automatic response when too late, the insect being already close). However, this view has recently been challenged, because visual discrimination appears to actually also improve in PPS (Blini et al., 2018). Although this classic account is well supported by neuropsychological and neurophysiological evidence, the dichotomy between ventral and dorsal pathways is not meant to be strict (Milner & Goodale, 2008). Accumulating evidence specifically points to the fact that the dorsal stream contains object representations that are to some extent independent of those in the ventral stream, and capable of contributing to human perception (Freud, Ganel et al., 2017; Freud, Plaut, & Behrmann, 2016; Quinlan & Culham, 2007; Wang et al., 2016). One recent study, for example, has shown that fundamental properties of shapes can be reliably decoded from posterior parietal regions, whose activation profile appears correlated with recognition performance (Freud, Culham et al., 2017), suggesting a functional role in shape identification. Candidate areas appear to be a set of subcortical (e.g. putamen, Graziano & Gross, 1993) and fronto-parietal cortical areas (i.e. inferior parietal and premotor, Brozzoli, Gentile, & Ehrsson, 2012; Brozzoli et al., 2011; di Pellegrino & Làdavas, 2015; Fogassi et al., 1996; Graziano & Cooke, 2006; Lloyd, Morrison, & Roberts, 2006) associated with PPS processing. For example, Brozzoli et al. (2011) presented, to healthy participants lying supine inside a magnetic resonance imaging (MRI) scanner, 3D objects either close (3 cm) or far (100 cm) from their outstretched hand (or in the same physical/visual position but while their hand was resting unseen on the torso). The authors capitalized on a robust property of neuronal responses measurable using functional MRI, which is neural adaptation: neural activity is reduced when a stimulus feature is repeated, but only for a subpopulation of neurons that is selective for the repeated feature itself (Grill-Spector, Henson, & Martin, 2006). The authors found evidence of neural adaptation only when visual stimuli appeared close to the outstretched hand, but not when the stimuli appeared in the same spatial position while the hand was placed on the torso (Brozzoli et al., 2011). Thus, in agreement with neurophysiological investigations in monkeys, they confirmed in humans that a set of interconnected premotor and posterior parietal regions specifically encodes the position of visual objects close to the body, in hand-centred coordinates. It is interesting to note that the set of areas described by Brozzoli et al. (2011) tightly overlapped with regions reported to respond to multisensory stimulations occurring in PPS (Brozzoli et al., 2012; Gentile, Petkova, & Ehrsson, 2010; Lloyd et al., 2006; Macaluso & Driver, 2005; Makin, Holmes, & Zohary, 2007), and yet were obtained via purely visual stimulation. This supports the idea that, although PPS is inherently multisensory, enhanced perceptual processing in it can be expected to occur for unimodal (e.g. visual) stimuli as well.

3.4 Visual discrimination advantages in peripersonal space

Purely visual advantages occurring in PPS, in contrast to multisensory ones, have been seldom reported (for a recent review, see de Vignemont, 2018), or have been framed in attentional terms. The aforementioned near-hand effects for target detection, for example, actually consist of purely visual advantages, which occur without direct multisensory stimulation and depend on a *static* proprioceptive feedback. Notwithstanding the difficulty in disentangling attentional and perceptual processing (but see Makin et al., 2009; Reed et al., 2013), a recent study attempted to investigate how shape perception—classically considered a function of the ventral visual pathway (Goodale & Milner, 1992)—is affected by proximity (Blini et al., 2018).

Blini et al. (2018) presented, to healthy participants, 3D shapes in the context of an immersive virtual reality environment (Figure 3.1). The task was adapted from that used by O'Connor et al. (2014), originally employed to test spatial sensitivity to reward. The geometrical shapes were presented either close (50 cm) or far (300 cm) from participants, thus within reach (in PPS) or not; the task consisted in a speeded discrimination of the presented shape (i.e. cube or sphere). As physical size scales, in everyday life, with depth (that is, farther shapes appear smaller), and this has arguably a profound impact on visual capacities (experiment 4), retinal size correction was applied. By equating the retinal size of close and far shapes, the latter appear illusorily bigger (because depth cues are accounted for by the visual system to estimate objects' size). Despite this striking visual illusion, participants were consistently faster in discriminating shapes appearing close to them (experiment 1, see Figure 3.1). Moreover, this effect could not be explained by upper/lower visual field confounds (i.e. in everyday life, close objects more commonly appear at the bottom of the visual field, which could therefore be privileged, Previc, 1990), or vergence eye-movements. First, the effect persisted when both shapes appeared at the same height—that of the fixation cross—to avoid any upper/lower visual field confound (experiment 3). Second, the effect persisted when the authors exploited a mere illusion of depth to avoid any vergence eye-movements confound (i.e. Ponzo illusion, experiment 2). In the illusion, perspectives cues (i.e. converging lines) were used as a background for two elements displayed at different heights, one of which therefore appeared illusorily farther away in space. Thus, this context stripped the task of many important depth indices, including vergence eye-movements, except for perspective cues. Interestingly, when Blini et al. (2018) probed the spatial distribution of this performance benefit, termed 'distance effect', by presenting shapes at six equi-spaced distances from the participants, they found that a sigmoid trend could adequately account for behavioural performances in terms of both accuracy and response times. As discussed in section 3.1, the sigmoidal trend has been considered a hallmark signature of PPS (multisensory) processing. Having described such a pattern for a purely unisensory (visual) task has one important

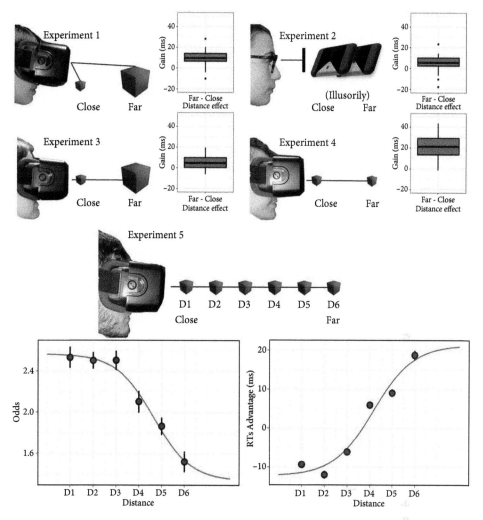

Figure 3.1 In the study of Blini et al. (2018), the authors sought to assess perceptual discrimination abilities across different depths. Geometrical shapes were presented close (50 cm) or far (300 cm) from participants, in a 3D virtual environment (VE) or in the context of a visual illusion of depth (experiment 2); in this context, participants saw 2D images depicting perspective cues that created an illusory perception of depth. In experiments 1 (3D VE) and 2 (2D Ponzo-like illusion), closer shapes appeared in the bottom part of the participants' visual field (below a fixation cross), and farther ones in the upper visual field; in experiments 3, 4, and 5, all shapes were presented at the same height of the fixation cross. In experiments 1, 2, and 3, retinal size was kept constant for close and far stimuli, whereas in experiments 4 and 5 it was naturally scaled with distance. The authors consistently found an advantage in discriminating shapes when these were presented close as compared to far—difference depicted in the boxplots. Furthermore, a sigmoid trend could capture the spatial distribution of this, purely unimodal, advantage (experiment 5).

theoretical implication: defining PPS as the region of space in which multisensory interaction occurs, and explaining PPS-related performance benefits in terms of multisensory
convergence, is probably limiting in not properly accounting for what appears to be a more
fundamental role of PPS circuitry in perception. Thus, researchers conducting multisensory
studies should not neglect the fact that unimodal stimulations alone (at least in the visual
modality) could capture behavioural signatures of PPS processing, and should be cautious
before ascribing them to multisensory convergence.

3.5 Close is better

Recent proposals have questioned the view of PPS as a unitary construct, but rather declined
several peripersonal space(s) according to their functional role (i.e. body protection vs goal-
directed action), and therefore sensory and motor requirements (de Vignemont & Iannetti,
2015). The lowest common denominators of these PPS constructs appear, however, to encompass two elements: the body—which, by definition, is involved in any action towards
an object or defensive behaviour—and the distance of a stimulus from it (i.e. proximity, in
its parametric and continuous meaning). Recent frameworks have also stressed the task-
dependent nature of PPS signatures, depending on 'the behavioural relevance of actions
aiming to create or avoid contact between objects and the body' (Bufacchi & Iannetti, 2018).
The latter definition can account for the manifold variables tapping onto PPS plasticity. The
magnitude of PPS functional measurements would additionally vary according to several
factors, listed under the umbrella concept of behavioural relevance—and, again, hardly dis-
tinguishable from enhanced attentional processing. However, proximity to the body by itself
attributes saliency to a cue (Spaccasassi, Romano, & Maravita, 2019). Interestingly, behavioural sensitivity to reward decreases in the far space (O'Connor et al., 2014) as if the intrinsic or learned value of stimuli presented close to the body is automatically increased and
gains in salience (Spaccasassi et al., 2019).

One could therefore put forward the general prediction that *everything* would be enhanced when close to the body, as long as the task at hand offers sufficient sensitivity and it is,
indeed, enhanced by increased attentional or PPS-specific processing. The latter two requirements are not trivial. First, behavioural effects can often reveal themselves as being very fragile and necessitating of well-powered and rigorous designs (Dosso & Kingstone, 2018, for
the near-hand effects). Second, one may debate whether the increased attentional salience of
body parts invariantly leads to improved behavioural performance on one task. There are,
indeed, instances in which hands proximity seems to *hamper* the task 'at hand'. Abrams et al.
(2008), for example, reported that people shifted their attention between items more slowly
when their hands were near the display, in comparison with when their hands were placed
farther apart. At odds with most previous near-hand effect studies, here both hands served
as spatial attentional wands, thus possibly increasing the cost of shifting attention by modulating either the engagement or disengagement attentional components. Leveraging a classic
inhibition of return paradigm, the same study indeed associated hands proximity to delayed
attentional disengagement for cued locations (also see Qi et al., 2019). This can actually reflect a more thorough visual assessment of the region of space around the hands (i.e. visual
enhancement, not hampering), which was supported by higher accuracy in target discrimination in the visual search task when hands were near to, as compared to far from, the display. There are, indeed, situations in which a more thorough assessment of the space around
the hands or the body is critical—namely, when goal-oriented actions (including reactive/

defensive ones) must be performed. In this case, sensory processing of objects presented in PPS may be effectively enhanced for the sake of guiding the motor system toward an optimal response (e.g. de Vignemont, 2018). Thus, this can be reconciled with views that stress the need of a purpose for PPS to serve in order to effectively enhance performance within PPS. In this chapter, we have focused on visual advantages, although such advantages may potentially extend to other modalities (e.g. audition, see Brungart, 1999; Brungart, Durlach, & Rabinowitz, 1999), or other dimensions of stimuli such as their perceived duration (Qi et al., 2019), provided the aforementioned conditions are met. More research is needed in this regard.

The system for PPS coding is ancient and subtended by a large neural network, already optimized for monitoring the space close to the body, as well as the distance of stimuli from it. Two scenarios are possible, and not mutually exclusive: the PPS-specific system may easily be exploited by domain-general mechanisms for saliency attribution (i.e. spatial attention) to promote the processing of relevant stimuli; the specialized PPS processing may bias spatial attention toward the region of space in which it excels. At any rate, as reviewed earlier, the role of PPS processing extends, as a consequence, from multisensory interaction to more basic features of (unisensory) perception, including visual shape discrimination. Its contribution appears, therefore, much broader than previously thought. In essence, while facing a potentially annoying insect, there may be no need to surrender to a speed/accuracy trade-off in visual discrimination as a function of depth: PPS-specialized processing could provide performance benefits for both processes (fast reaction and proper identification of the threat) concurrently. For a defensive system to work efficiently (i.e. by being quick without bugging constantly for stimuli not deserving protection from), such an extra perceptual boost appears indeed very convenient (Makin et al., 2015, 2009).

References

Abrams, R. A., Davoli, C. C., Du, F., Knapp III, W. H., & Paull, D. (2008). Altered vision near the hands. *Cognition*, *107*(3), 1035–1047. https://doi.org/10.1016/j.cognition.2007.09.006

Armony, J. L., & Dolan, R. J. (2002). Modulation of spatial attention by fear-conditioned stimuli: an event-related fMRI study. *Neuropsychologia*, *40*(7), 817–826. https://doi.org/10.1016/S0028-3932(01)00178-6

Berti, A., & Frassinetti, F. (2000). When far becomes near: remapping of space by tool use. *Journal of Cognitive Neuroscience*, *12*(3), 415–420.

Blini, E., Desoche, C., Salemme, R., Kabil, A., Hadj-Bouziane, F., & Farnè, A. (2018). Mind the depth: visual perception of shapes is better in peripersonal space. *Psychological Science*, *29*(11), 1868–1877. https://doi.org/10.1177/0956797618795679

Brozzoli, C., Gentile, G., & Ehrsson, H. H. (2012). That's near my hand! Parietal and premotor coding of hand-centered space contributes to localization and self-attribution of the hand. *Journal of Neuroscience*, *32*(42), 14573–14582. https://doi.org/10.1523/JNEUROSCI.2660-12.2012

Brozzoli, C., Gentile, G., Petkova, V. I., & Ehrsson, H. H. (2011). fMRI adaptation reveals a cortical mechanism for the coding of space near the hand. *Journal of Neuroscience*, *31*(24), 9023–9031. https://doi.org/10.1523/JNEUROSCI.1172-11.2011

Brozzoli, C., Pavani, F., Urquizar, C., Cardinali, L., & Farnè, A. (2009). Grasping actions remap peripersonal space. *NeuroReport*, *20*(10), 913–917. https://doi.org/10.1097/WNR.0b013e32832c0b9b

Brungart, D. S. (1999). Auditory localization of nearby sources. III. Stimulus effects. *Journal of the Acoustical Society of America*, *106*(6), 3589–3602. https://doi.org/10.1121/1.428212

Brungart, D. S., Durlach, N. I., & Rabinowitz, W. M. (1999). Auditory localization of nearby sources. II. Localization of a broadband source. *Journal of the Acoustical Society of America*, *106*(4), 1956–1968. https://doi.org/10.1121/1.427943

Bufacchi, R. J., & Iannetti, G. D. (2018). An action field theory of peripersonal space. *Trends in Cognitive Sciences*, *22*(12), 1076–1090. https://doi.org/10.1016/j.tics.2018.09.004

Bush, W. S., & Vecera, S. P. (2014). Differential effect of one versus two hands on visual processing. *Cognition*, *133*(1), 232–237. https://doi.org/10.1016/j.cognition.2014.06.014

Canzoneri, E., Magosso, E., & Serino, A. (2012). Dynamic sounds capture the boundaries of peripersonal space representation in humans. *PLoS ONE*, *7*(9), e44306. https://doi.org/10.1371/journal.pone.0044306

Cardinali, L., Brozzoli, C., & Farnè, A. (2009). Peripersonal space and body schema: two labels for the same concept? *Brain Topography*, *21*(3–4), 252–260. https://doi.org/10.1007/s10548-009-0092-7

Cardinali, L., Frassinetti, F., Brozzoli, C., Urquizar, C., Roy, A. C., & Farnè, A. (2009). Tool-use induces morphological updating of the body schema. *Current Biology*, *19*(12), R478–R479. https://doi.org/10.1016/j.cub.2009.05.009

Cartaud, A., Ruggiero, G., Ott, L., Iachini, T., & Coello, Y. (2018). Physiological response to facial expressions in peripersonal space determines interpersonal distance in a social interaction context. *Frontiers in Psychology*, *9*. https://doi.org/10.3389/fpsyg.2018.00657

Cléry, J., Guipponi, O., Odouard, S., Wardak, C., & Hamed, S. B. (2015). Impact prediction by looming visual stimuli enhances tactile detection. *Journal of Neuroscience*, *35*(10), 4179–4189. https://doi.org/10.1523/JNEUROSCI.3031-14.2015

Colby, C. L., Duhamel, J. R., & Goldberg, M. E. (1993). Ventral intraparietal area of the macaque: anatomic location and visual response properties. *Journal of Neurophysiology*, *69*(3), 902–914.

Couyoumdjian, A., Nocera, F. D., & Ferlazzo, F. (2003). Functional representation of 3d space in endogenous attention shifts. *Quarterly Journal of Experimental Psychology Section A*, *56*(1), 155–183. https://doi.org/10.1080/02724980244000215

de Vignemont, Frédérique. (2018). Peripersonal perception in action. *Synthese*. https://doi.org/10.1007/s11229-018-01962-4

de Vignemont, F., & Iannetti, G. D. (2015). How many peripersonal spaces? *Neuropsychologia*, *70*, 327–334. https://doi.org/10.1016/j.neuropsychologia.2014.11.018

di Pellegrino, G., & Frassinetti, F. (2000). Direct evidence from parietal extinction of enhancement of visual attention near a visible hand. *Current Biology*, *10*(22), 1475–1477. https://doi.org/10.1016/S0960-9822(00)00809-5

di Pellegrino, G., & Làdavas, E. (2015). Peripersonal space in the brain. *Neuropsychologia*, *66*(Supplement C), 126–133. https://doi.org/10.1016/j.neuropsychologia.2014.11.011

Dosso, J. A., & Kingstone, A. (2018). The fragility of the near-hand effect. *Collabra: Psychology*, *4*(1), 27. https://doi.org/10.1525/collabra.167

Eimer, M., Velzen, J. van, & Driver, J. (2002). Cross-modal interactions between audition, touch, and vision in endogenous spatial attention: ERP evidence on preparatory states and sensory modulations. *Journal of Cognitive Neuroscience*, *14*(2), 254–271. https://doi.org/10.1162/089892902317236885

Farnè, A., & Làdavas, E. (2000). Dynamic size-change of hand peripersonal space following tool use. *NeuroReport*, *11*(8), 1645.

Ferri, F., Panadura-Jiménez, A., Väljamäe, A., Vastano, R., & Costantini, M. (2015). Emotion-inducing approaching sounds shape the boundaries of multisensory peripersonal space. *Neuropsychologia*, *70*(Supplement C), 468–475. https://doi.org/10.1016/j.neuropsychologia.2015.03.001

Fogassi, L., Gallese, V., Fadiga, L., Luppino, G., Matelli, M., & Rizzolatti, G. (1996). Coding of peripersonal space in inferior premotor cortex (area F4). *Journal of Neurophysiology*, *76*(1), 141–157.

Franconeri, S. L., & Simons, D. J. (2003). Moving and looming stimuli capture attention. *Perception & Psychophysics*, *65*(7), 999–1010. https://doi.org/10.3758/BF03194829

Freud, E., Culham, J. C., Plaut, D. C., & Behrmann, M. (2017). The large-scale organization of shape processing in the ventral and dorsal pathways. *ELife*, *6*. https://doi.org/10.7554/eLife.27576

Freud, E., Ganel, T., Shelef, I., Hammer, M. D., Avidan, G., & Behrmann, M. (2017). Three-dimensional representations of objects in dorsal cortex are dissociable from those in ventral cortex. *Cerebral Cortex*, *27*(1), 422–434. https://doi.org/10.1093/cercor/bhv229

Freud, E., Plaut, D. C., & Behrmann, M. (2016). 'What' is happening in the dorsal visual pathway. *Trends in Cognitive Sciences*, *20*(10), 773–784. https://doi.org/10.1016/j.tics.2016.08.003

Gawryszewski, L. de G., Riggio, L., Rizzolatti, G., & Umiltá, C. (1987). Movements of attention in the three spatial dimensions and the meaning of 'neutral' cues. *Neuropsychologia*, *25*(1), 19–29. https://doi.org/10.1016/0028-3932(87)90040-6

Gentile, G., Petkova, V. I., & Ehrsson, H. H. (2010). Integration of visual and tactile signals from the hand in the human brain: an fMRI study. *Journal of Neurophysiology*, *105*(2), 910–922. https://doi.org/10.1152/jn.00840.2010

Ghazanfar, A. A., Neuhoff, J. G., & Logothetis, N. K. (2002). Auditory looming perception in rhesus monkeys. *Proceedings of the National Academy of Sciences*, *99*(24), 15755–15757. https://doi.org/10.1073/pnas.242469699

Goodale, M. A., & Milner, A. D. (1992). Separate visual pathways for perception and action. *Trends in Neurosciences*, *15*(1), 20–25. https://doi.org/10.1016/0166-2236(92)90344-8

Goodhew, S. C., Edwards, M., Ferber, S., & Pratt, J. (2015). Altered visual perception near the hands: a critical review of attentional and neurophysiological models. *Neuroscience & Biobehavioral Reviews*, *55*, 223–233. https://doi.org/10.1016/j.neubiorev.2015.05.006

Goodhew, S. C., Gozli, D. G., Ferber, S., & Pratt, J. (2013). Reduced temporal fusion in near-hand space. *Psychological Science*, *24*(6), 891–900. https://doi.org/10.1177/0956797612463402

Gozli, D. G., West, G. L., & Pratt, J. (2012). Hand position alters vision by biasing processing through different visual pathways. *Cognition*, *124*(2), 244–250. https://doi.org/10.1016/j.cognition.2012.04.008

Graziano, M. S. A., & Cooke, D. F. (2006). Parieto-frontal interactions, personal space, and defensive behavior. *Neuropsychologia*, *44*(6), 845–859. https://doi.org/10.1016/j.neuropsychologia.2005.09.009

Graziano, M. S. A., & Gross, C. G. (1993). A bimodal map of space: somatosensory receptive fields in the macaque putamen with corresponding visual receptive fields. *Experimental Brain Research*, *97*(1), 96–109. https://doi.org/10.1007/BF00228820

Grill-Spector, K., Henson, R., & Martin, A. (2006). Repetition and the brain: neural models of stimulus-specific effects. *Trends in Cognitive Sciences*, *10*(1), 14–23. https://doi.org/10.1016/j.tics.2005.11.006

Hall, E. T. (1966). *The hidden dimension.* New York: Doubleday & Co.

Hediger, H. (1950). *Wild animals in captivity.* London: Butterworths Scientific Publications.

Hyvärinen, J., & Poranen, A. (1974). Function of the parietal associative area 7 as revealed from cellular discharges in alert monkeys. *Brain*, *97*(4), 673–692. https://doi.org/10.1093/brain/97.4.673

Iriki, A., Tanaka, M., & Iwamura, Y. (1996). Coding of modified body schema during tool use by macaque postcentral neurones. *NeuroReport*, *7*(14), 2325–2330.

Koster, E. H. W., Crombez, G., Van Damme, S., Verschuere, B., & De Houwer, J. (2004). Does imminent threat capture and hold attention? *Emotion*, *4*(3), 312–317. https://doi.org/10.1037/1528-3542.4.3.312

Làdavas, E., Farnè, A., Zeloni, G., & di Pellegrino, G. P. (2000). Seeing or not seeing where your hands are. *Experimental Brain Research*, *131*(4), 458–467. https://doi.org/10.1007/s002219900264

Làdavas, Elisabetta, di Pellegrino, G., Farnè, A., & Zeloni, G. (1998). Neuropsychological evidence of an integrated visuotactile representation of peripersonal space in humans. *Journal of Cognitive Neuroscience*, *10*(5), 581–589. https://doi.org/10.1162/089892998562988

Lin, J. Y., Murray, S. O., & Boynton, G. M. (2009). Capture of attention to threatening stimuli without perceptual awareness. *Current Biology*, *19*(13), 1118–1122. https://doi.org/10.1016/j.cub.2009.05.021

Lloyd, D., Morrison, I., & Roberts, N. (2006). Role for human posterior parietal cortex in visual processing of aversive objects in peripersonal space. *Journal of Neurophysiology*, *95*(1), 205–214. https://doi.org/10.1152/jn.00614.2005

Losier, B. J., & Klein, R. M. (2004). Covert orienting within peripersonal and extrapersonal space: young adults. *Cognitive Brain Research*, *19*(3), 269–274. https://doi.org/10.1016/j.cogbrainres.2004.01.002

Macaluso, E., & Driver, J. (2005). Multisensory spatial interactions: a window onto functional integration in the human brain. *Trends in Neurosciences*, *28*(5), 264–271. https://doi.org/10.1016/j.tins.2005.03.008

Maier, J. X., Neuhoff, J. G., Logothetis, N. K., & Ghazanfar, A. A. (2004). Multisensory integration of looming signals by rhesus monkeys. *Neuron*, *43*(2), 177–181. https://doi.org/10.1016/j.neuron.2004.06.027

Makin, T. R., Brozzoli, C., Cardinali, L., Holmes, N. P., & Farnè, A. (2015). Left or right? Rapid visuomotor coding of hand laterality during motor decisions. *Cortex*, *64*, 289–292. https://doi.org/10.1016/j.cortex.2014.12.004

Makin, T. R., Holmes, N. P., Brozzoli, C., & Farnè, A. (2012). Keeping the world at hand: rapid visuomotor processing for hand–object interactions. *Experimental Brain Research*, *219*(4), 421–428. https://doi.org/10.1007/s00221-012-3089-5

Makin, T. R., Holmes, N. P., Brozzoli, C., Rossetti, Y., & Farnè, A. (2009). Coding of visual space during motor preparation: approaching objects rapidly modulate corticospinal excitability in hand-centered coordinates. *Journal of Neuroscience*, *29*(38), 11841–11851. https://doi.org/10.1523/JNEUROSCI.2955-09.2009

Makin, T. R., Holmes, N. P., & Zohary, E. (2007). Is that near my hand? Multisensory representation of peripersonal space in human intraparietal sulcus. *Journal of Neuroscience*, *27*(4), 731–740. https://doi.org/10.1523/JNEUROSCI.3653-06.2007

Maravita, A., & Iriki, A. (2004). Tools for the body (schema). *Trends in Cognitive Sciences*, *8*(2), 79–86. https://doi.org/10.1016/j.tics.2003.12.008

Maravita, A., Spence, C., & Driver, J. (2003). Multisensory integration and the body schema: close to hand and within reach. *Current Biology*, *13*(13), R531–R539. https://doi.org/10.1016/S0960-9822(03)00449-4

Maravita, A., Spence, C., Kennett, S., & Driver, J. (2002). Tool-use changes multimodal spatial inter-actions between vision and touch in normal humans. *Cognition, 83*(2), B25–B34. https://doi.org/10.1016/S0010-0277(02)00003-3

Milner, A. D., & Goodale, M. A. (2008). Two visual systems re-viewed. *Neuropsychologia, 46*(3), 774–785. https://doi.org/10.1016/j.neuropsychologia.2007.10.005

Murata, A., Fadiga, L., Fogassi, L., Gallese, V., Raos, V., & Rizzolatti, G. (1997). Object representa-tion in the ventral premotor cortex (area F5) of the monkey. *Journal of Neurophysiology, 78*(4), 2226–2230.

Neppi-Mòdona, M., Auclair, D., Sirigu, A., & Duhamel, J.-R. (2004). Spatial coding of the predicted impact location of a looming object. *Current Biology, 14*(13), 1174–1180. https://doi.org/10.1016/j.cub.2004.06.047

Noel, J.-P., Grivaz, P., Marmaroli, P., Lissek, H., Blanke, O., & Serino, A. (2015). Full body action re-mapping of peripersonal space: the case of walking. *Neuropsychologia, 70*, 375–384. https://doi.org/10.1016/j.neuropsychologia.2014.08.030

O'Connor, D. A., Meade, B., Carter, O., Rossiter, S., & Hester, R. (2014). Behavioral sensitivity to re-ward is reduced for far objects. *Psychological Science, 25*(1), 271–277. https://doi.org/10.1177/0956797613503663

Plewan, T., & Rinkenauer, G. (2017). Simple reaction time and size–distance integration in virtual 3D space. *Psychological Research, 81*(3), 653–663. https://doi.org/10.1007/s00426-016-0769-y

Previc, F. H. (1990). Functional specialization in the lower and upper visual fields in humans: its eco-logical origins and neurophysiological implications. *Behavioral and Brain Sciences, 13*(3), 519–542.

Qi, Y., Wang, X., He, X., & Du, F. (2019). Prolonged subjective duration near the hands: effects of hand proximity on temporal reproduction. *Psychonomic Bulletin & Review.* https://doi.org/10.3758/s13423-019-01614-9

Quinlan, D. J., & Culham, J. C. (2007). fMRI reveals a preference for near viewing in the human parieto-occipital cortex. *NeuroImage, 36*(1), 167–187. https://doi.org/10.1016/j.neuroimage.2007.02.029

Reed, C. L., Leland, D. S., Brekke, B., & Hartley, A. A. (2013). Attention's grasp: early and late hand proximity effects on visual evoked potentials. *Frontiers in Psychology, 4*, 420. https://doi.org/10.3389/fpsyg.2013.00420

Reed, Catherine L., Grubb, J. D., & Steele, C. (2006). Hands up: attentional prioritization of space near the hand. *Journal of Experimental Psychology: Human Perception and Performance, 32*(1), 166. https://doi.org/10.1037/0096-1523.32.1.166

Rizzolatti, G., Scandolara, C., Matelli, M., & Gentilucci, M. (1981). Afferent properties of periarcuate neurons in macaque monkeys. II. Visual responses. *Behavioural Brain Research, 2*(2), 147–163. https://doi.org/10.1016/0166-4328(81)90053-X

Ruggiero, G., Frassinetti, F., Coello, Y., Rapuano, M., di Cola, A. S., & Iachini, T. (2017). The effect of facial expressions on peripersonal and interpersonal spaces. *Psychological Research, 81*(6), 1232–1240. https://doi.org/10.1007/s00426-016-0806-x

Sambo, C. F., & Iannetti, G. D. (2013). Better safe than sorry? The safety margin surrounding the body is increased by anxiety. *Journal of Neuroscience, 33*(35), 14225–14230. https://doi.org/10.1523/JNEUROSCI.0706-13.2013

Schiff, W., Caviness, J. A., & Gibson, J. J. (1962). Persistent fear responses in rhesus monkeys to the optical stimulus of 'looming'. *Science, 136*(3520), 982–983. https://doi.org/10.1126/science.136.3520.982

Shelton, P. A., Bowers, D., & Heilman, K. M. (1990). Peripersonal and vertical neglect. *Brain, 113*(1), 191–205. https://doi.org/10.1093/brain/113.1.191

Spaccasassi, C., Romano, D., & Maravita, A. (2019). Everything is worth when it is close to my body: how spatial proximity and stimulus valence affect visuo-tactile integration. *Acta Psychologica, 192*, 42–51. https://doi.org/10.1016/j.actpsy.2018.10.013

Spence, C., Pavani, F., & Driver, J. (2004). Spatial constraints on visual-tactile cross-modal distractor congruency effects. *Cognitive, Affective, & Behavioral Neuroscience, 4*(2), 148–169. https://doi.org/10.3758/CABN.4.2.148

Sposito, A., Bolognini, N., Vallar, G., & Maravita, A. (2012). Extension of perceived arm length following tool-use: clues to plasticity of body metrics. *Neuropsychologia, 50*(9), 2187–2194. https://doi.org/10.1016/j.neuropsychologia.2012.05.022

Teneggi, C., Canzoneri, E., di Pellegrino, G., & Serino, A. (2013). Social modulation of peripersonal space boundaries. *Current Biology, 23*(5), 406–411. https://doi.org/10.1016/j.cub.2013.01.043

Wang, A., Li, Y., Zhang, M., & Chen, Q. (2016). The role of parieto-occipital junction in the interaction between dorsal and ventral streams in disparity-defined near and far space processing. *PLoS ONE, 11*(3), e0151838. https://doi.org/10.1371/journal.pone.0151838

4

Functional networks for peripersonal space coding and prediction of impact to the body

Justine Cléry and Suliann Ben Hamed

4.1 Introduction

In everyday life, space very often refers to what lies outside one's own body, or specifies a location in one's environment. It is, however, a more complex concept, both in psychological and neurophysiological terms. It can be divided into several sub-spaces depending, for example, on whether a goal-directed behaviour or action is required or not, or on whether social interactions are at play. In the past few years, the neurophysiological bases underlying these sub-space representations, their interactions, and their dynamics have been explored both in humans and in monkeys. This body of work includes psychological studies aiming at understanding the construction of such cognitive representations of space from birth into adulthood, characterizing their disruption in major psychiatric (e.g. schizophrenia, anorexia, anxiety, etc.) and neurological (e.g. neglect, apraxia, etc.) conditions and eventually contributing to the development of rehabilitation protocols. This body of work also includes neurophysiological studies aiming at identifying the cortical networks, the cortical areas, and the unitary computational processes that underlie these sub-space representations.

In the following review, we will first discuss the different spaces around the body and the functional networks involved in these processes. We will then present our perspective on the functional association between different body parts with different perceptual and action spaces. Lastly, we will propose an over-arching functional brain connectivity model accounting both for our current understanding of space representations and their interaction with major cognitive functions.

4.2 Multiple spaces from the skin to far away

Strictly speaking, the margin between oneself and the outside world is defined by the skin. The skin codes direct contact to the body, whether generated by the outside world (e.g. a kiss from a beloved, a bite from a mosquito) or by the body itself (e.g. my right hand touching my left cheek).

Just beyond this *stricto sensu* body margin lies a so-called 'peripersonal space' (PPS). This peripersonal space has been associated with the idea of a protective space, extending the *stricto sensu* body margin with a security space. PPS allows us to anticipate contact to the body and prepare goal-directed behaviour, be it an escape from a hunting lion or a hug from a loving child. PPS thus corresponds to the physical space surrounding the body (50 to 70 cm around the skin). From a sensory perspective, PPS serves to signal proximity to the body

Justine Cléry and Suliann Ben Hamed, *Functional networks for peripersonal space coding and prediction of impact to the body* In: *The World at Our Fingertips*. Edited by Frédérique de Vignemont, Andrea Serino, Hong Yu Wong, and Alessandro Farnè, Oxford University Press (2021). © Oxford University Press. DOI: 10.1093/oso/9780198851738.003.0004.

and is functionally linked to protective behaviour (Figure 4.1). It is constructed based on a combination of sensory cues: (1) static or dynamic visual cues that define a visual PPS constrained by gaze direction; (2) static or dynamic auditory cues that define an auditory PPS constrained by head position; (3) dynamic tactile cues that define an often under-considered tactile PPS, signalled by passive hair or whisker movement signaling air displacement within the PPS, or heat detectors signaling changes within the PPS even at a distance from the skin. These different sensory cues are integrated into functional representation of PPS. From a motor perspective, PPS serves to organize proximal goal-directed actions and is functionally linked to proactive behaviour (action-based PPS, see Figure 4.1).

Figure 4.1 illustrates these different body margins from the closest to the furthest away: the skin, *tactile-PPS* (coding proximity to the body based on tactile information), which includes the entire body; *auditory-PPS* (coding proximity to the body based on auditory information), mostly centered around the head; *visual-PPS* (coding proximity to the body based on visual information), defined by the visual field allowed by the eye-head geometry; and, last, *action-based PPS* (allowing for direct action of the subject onto his/her surrounding environment). While this schema clearly defines a PPS around the body, contrasting with the space far away from the body, it also highlights the functional heterogeneity of PPS, and raises the question of how these different PPSs are weighted and integrated in a unitary functional PPS, both from a psychological and a neurophysiological perspective. This is discussed in the next section.

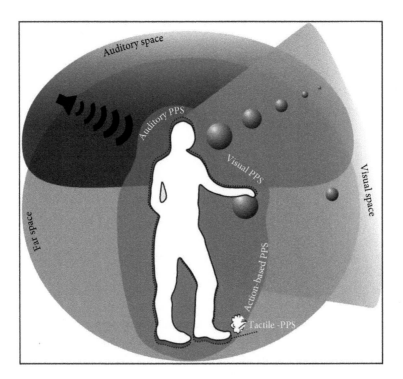

Figure 4.1 Peripersonal space arises from the combination of multiple sensory cues, and contributes to both proactive and protective behaviours.

Adapted from Cléry et al. Neuronal bases of peripersonal and extrapersonal spaces, their plasticity and their dynamics: knowns and unknowns. Neuropsychologia. 2015 Apr;70:313–26. doi: 10.1016/j.neuropsychologia.2014.10.022.

4.2.1 The functional cortical network for the coding
of *stricto sensu* body margin (the skin)

The *stricto sensu* body margin corresponds to the skin surrounding our body. As a result, the main associated sensory process is somatosensation, including touch and social touch, temperature, pain, and proprioceptive sensation. Tactile processing mainly activates somatosensory areas I (SI) and II (SII), respectively located in the central sulcus and its posterior convexity for SI, and within the lateral sulcus for SII (Wardak et al., 2016). Both these areas represent the body along well-organized somatotopic body maps (Kaas et al., 1979; Nelson et al., 1980; Pons et al., 1985; Krubitzer et al., 1995; Fitzgerald et al., 2004, 2006).

Somatosensory processes also recruit higher-order associative areas, as evidenced by single-cell recordings in non-human primates. For example, this is the case in premotor area 6 (Rizzolatti et al., 1981), parietal areas 5 and 7b (Murray and Mishkin, 1984; Dong et al., 1994; Rozzi et al., 2008), and the ventral intraparietal area VIP (Duhamel et al., 1998; Avillac et al., 2005, 2007; Guipponi et al., 2013, 2015a; Wardak et al., 2016; Cléry et al., 2017b), as well as prefrontal areas 45 and 46 (Gerbella et al., 2010, 2013). In most of these associative areas, somatosensory information is organized in coarse somatotopic maps. As a result, the stimulation of a given body part (e.g. face, hand, arm, trunk, leg, etc.) will co-activate specific functionally synergic subparts of these parieto-prefrontal areas. Very few studies have sought to characterize how different body parts are represented one with respect to the other at the whole brain level, to identify possible differences or similarities. A recent functional magnetic resonance imaging (fMRI) study captures the somatotopic organization of upper body tactile information at the whole brain level in non-human primates, covering the centre of the face, the periphery of the face, and the shoulders (macaque rhesus) (Wardak et al., 2016). This study used bilateral air puffs of neutral valence directed at different locations on the animal's skin.

4.2.1.1 Centre and periphery of the face touch network
Air puffs delivered to the centre of the face resulted in the largest activation patterns, recruiting numerous cortical regions and indicating an over-representation of this face fovea in our body representation (Figure 4.2[2]). As expected, peak activations are mainly found in somatosensory regions at the level of the central sulcus (areas 1, 2, 3a, and 3b) and of the lateral sulcus (area SII and parietal ventral area PV). These centre of the face tactile stimulations also elicited activations in the parietal cortex, specifically in the inferior parietal convexity (area 7b), in the medial parietal wall, and the fundus of the intraparietal sulcus (areas VIP and PIP), as well as in several specific anterior regions such as premotor cortex (premotor zone PMZ/F4/F5p, supplementary eye field SEF/F7), ventrolateral prefrontal cortex (area 9/46v) and orbitofrontal cortex (areas 11 and 13), insular cortex (Pi) and in the cingulate cortex (24c). Interestingly, activations were also reported in the visual striate (V1, V2v, V2d, V3/V3A) and extra-striate cortex (MST/MT), indicating direct cross-modal influences in early sensory cortices. While in most of these regions, the topographic granularity was not fine enough to allow us to distinguish between these two types of stimulations, most of the parietal, premotor, and cingulate regions showed stronger activations for centre than for periphery of the face tactile stimulations. Temporal activations showed stronger activations for the periphery of the face, possibly pointing towards distinct functional networks for processing face fovea versus face periphery.

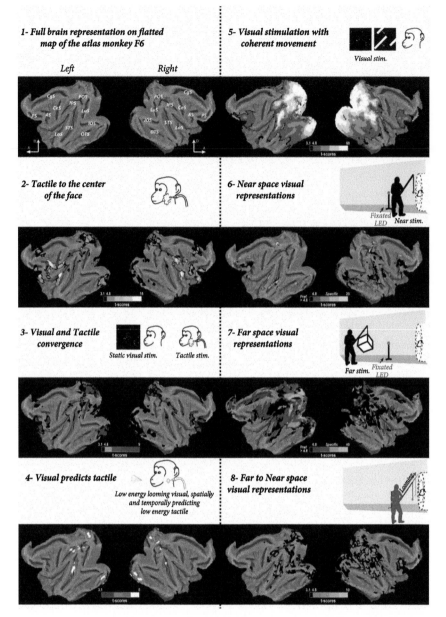

Figure 4.2 Functional cortical networks, represented in a flattened atlas monkey map for both hemispheres (1), involved in the coding of body margin (2), of body margin and vision (3), in impact prediction (4), in dynamic visual stimulation (5), in peripersonal visual space (6), in far space (7) and looming stimuli (8). Cortical sulci: AS, arcuate sulcus; CgS, cingulate sulcus; CeS, central sulcus; IOS, inferior occipital sulcus; IPS, intraparietal sulcus; LaS, lateral (Sylvian) sulcus; LuS, lunate sulcus; OTS, occipital temporal sulcus; POS, parieto-occipital sulcus; PS, principal sulcus; STS, superior temporal sulcus. Contrasts 2, 3, 5, and 8 are performed with levels of significance set at p<0.001 uncorrected level (t>3.1 black areas) and at p<0.05 corrected for multiple comparison (Family-wise error [FWE], t>4.8 coloured areas). Contrast 4 is performed with a level of significance set at p<0.001 uncorrected level (t>3.1 purple-coloured areas). Contrasts 6 and 7 are performed with a level of significance set at p<0.05 corrected for multiple comparison (FWE, t>4.8 coloured areas for specific near/far space encoding and black areas for preferred near/far space encoding).

Adapted from Cléry et al. Neuronal bases of peripersonal and extrapersonal spaces, their plasticity and their dynamics: knowns and unknowns. Neuropsychologia. 2015 Apr;70:313–26. doi: 10.1016/j.neuropsychologia.2014.10.022.

4.2.1.2 Shoulder touch network

Air puffs to the shoulders resulted in weaker and fewer activations than face stimulations (Wardak et al., 2016). These activations included the central and lateral sulci (area 2), the parietal cortex (the intraparietal sulcus, the medial wall, and the inferior parietal convexity), and the temporal cortex as well as more anterior regions (premotor: F4/PMZ/F5p, and ventrolateral prefrontal: area 9/46v, cingulate [24c], and insular cortex [Pi]).

This highlights the behavioural and functional relevance of the face relative to other upper body parts.

4.2.1.3 The functional specificity of the face touch
parieto-temporo-prefrontal network

This network includes areas 9/46v, F4/PMZ/F5p, 2, and SII/PV (Wardak et al., 2016). It demonstrates a clear preference for the face fovea to the degree that part of this network exclusively responds to centre of the face stimulations. This is corroborated by other studies describing face tactile neuronal responses in premotor areas F4 and F5 (Rizzolatti et al., 1981; Gentilucci et al., 1988; Fogassi et al., 1996; Guipponi et al., 2015a; Wardak et al., 2016) as well as in area 2 and SII/PV (Kaas et al., 1979; Nelson et al., 1980; Pons et al., 1985; Krubitzer et al., 1995; Fitzgerald et al., 2004, 2006; Wardak et al., 2016). Area 9/46v, an important node of this network, is directly connected to area SII and responds to complex visual and auditory stimulations, consistent with higher order representations of face and facial identity (Preuss and Goldman-Rakic, 1989; Ó Scalaidhe et al., 1997; Cohen et al., 2004; Romanski et al., 2005). A second node of this network, namely VIP, encodes space information relative to the head and shoulders (Bremmer et al., 2000, 2002a). This cortical region responds to visual information in a head-centred frame of reference (Duhamel et al., 1997; Avillac et al., 2004). It responds to both tactile and visual information, actively combining them (Duhamel et al., 1998; Avillac et al., 2005; Guipponi et al., 2013, 2015a). It contributes to the prediction of impact to the face (Cléry et al., 2017b) as well as to the coding of PPS (Cooke and Graziano, 2004; Cléry et al., 2018). Last but not least, this area represents observed touch on others' faces as well as the observation of others' actions (Ishida et al., 2009, 2015; Fiave et al., 2018). This area is strongly connected with the F4/PMZ/F5p complex (Matelli and Luppino, 2001; Rizzolatti and Luppino, 2001).

The increased functional relevance of the face relative to the shoulders thus extends well beyond the somatosensory homunculus described along the central sulcus and involves a large cortical network. It coincides with the evolutionary pressure to protect the head and face, which concentrate both the central nervous system, several sense organs as well as call/voice production. This predicts face PPS to be different from shoulder PPS (see section 4.2) A similar enhanced functional relevance also applies to the hand, whereby the somatosensory homunculus dedicated to hand processing extends along a large portion of the central sulcus. The hand-based tactile processing activates a large cortical network (for review, see Tessari et al., 2010), distinct from the above-described face network. Hand PPS has been extensively studied, in particular in relation to hand-directed actions (reaching, grasping) and tool use (see section 4.3). Importantly, while somatosensation is classically considered as defined by the skin, a recent study by Miller et al. (2018) demonstrates that tool manipulation in the context of sensing extends somatosensory processing beyond the hand. In other words, the tool gets incorporated into sensation very much like tools for action extend action PPS.

4.2.2 The functional cortical network coding for peri-personal space

The neural bases of PPS have been extensively studied in non-human primates at the single cell level (Hyvärinen and Poranen, 1974; Rizzolatti et al., 1983; Graziano et al., 1994; Fogassi et al., 1996; Bremmer et al., 2000, 2002a, 2002b, 2013) and more recently at the whole brain level, using fMRI (Cléry et al., 2018). Dosso and Kingstone (2018) found a difference in sensitivity to real objects compared to images presented on a screen, even when the latter shared the same affordance, reachability, or social significance with the real objects. Drawing on the specificity of real-world stimuli, Cléry et al. (2018) used a naturalistic and ecologically valid paradigm, and stimulated near and far space with either a small or a large 3D dynamical cube, while monkeys were required to fixate on an intermediate spatial position. This study reported a large cortical network involved in the coding of near space, activations including parts of the occipital striate and extrastriate areas, the temporal cortex (superior temporal sulcus), the parietal cortex, the prefrontal and premotor cortex (arcuate sulcus and posterior and anterior parts of the principal sulcus) as well as the orbitofrontal cortex. Some of these regions are also activated by far space stimulations (see section 4.2.3). However, a large portion of this functional network preferred near space (Figure 4.2[7], black activations) while other portions preferred far space. In other words, these cortical regions coded for both near and far space but showed stronger activations for near than for far stimuli. Areas preferring near space included *parietal areas*: posterior ventral (VIP) and medial (MIP) intraparietal areas as well as the anterior tip of IPS (possibly anterior intraparietal area AIP), medial parietal cortex (area PGm) and parietal opercular region area 7op; *temporal areas*: rostral temporoparietal occipital area TPOr in the medial mid-to-anterior bank of the superior temporal sulcus, the intraparietal sulcus associated area IPa, the inferior temporal area TEAa-m, the dorsal portion of the subdivision TE1-3; *insular regions*: the parainsular cortex PI; area SII within the medial bank of the lateral sulcus; *prefrontal and premotor regions*: dorsal premotor cortex F2, premotor area 4C or F4/F5 (including PMZ), SEF, FEF (area 8a as well as 8ac), prefrontal area 46p, prefrontal area 45B; *frontal regions*: the posterior orbitofrontal area 12. Importantly, some of these cortical regions exclusively coded for near space stimulations and not for far space stimulations (Figure 4.2[6], red-coloured scale activations). These near-specific activations were found bilaterally at the activation hotspot of most of the regions described above.

From a single neuron perspective, the areas of this near-specific network are expected to contain an extremely high concentration of neuronal populations that respond only to near stimuli. In contrast, the near preferring regions can either contain neuronal populations that respond to both near and far stimuli or a coexistence of two highly intermingled neuronal populations, one encoding near space and one encoding far space. Neurophysiological accounts of near space coding often describe a coding gradient between near and far space supporting the first view (see Cléry et al., 2015b for details). For example, Rozzi (2008) show that neurons in the inferior parietal lobule, at a cortical location identified by fMRI as a preferred near space region, respond to both far and near space stimulation.

This study supports previous research in PPS showing the involvement of two parietofrontal networks in the coding of this near space (see reviews: Cléry et al., 2015b; Cléry and Ben Hamed, 2018); a VIP-F4 network (Rizzolatti et al., 1981; Graziano et al., 1994, 1997, 1999; Fogassi et al., 1996; Duhamel et al., 1997, 1998; Matelli and Luppino, 2001; Avillac et al., 2005; Schlack et al., 2005; Bremmer et al., 2013; Guipponi et al., 2013; Cléry et al., 2018), which is dedicated to the definition of the body by the establishment of a safety margin, localizing objects around and with respect to the body in this space (Graziano and Cooke, 2006; Brozzoli et al., 2013, 2014; Chen et al., 2014); an AIP-F5 network, including associated areas, which is dedicated to goal-directed reaching or grasping actions within this PPS (Gallese et al., 1994; Iriki et al., 1996; Murata et al., 2000; Fogassi et al., 2001; Matelli and Luppino, 2001; Rizzolatti and Luppino, 2001; Rizzolatti and Matelli, 2003; Caprara et al., 2018). Importantly, the Cléry et al. (2018) study also provided the grounds for a more complex functional organization of PPS than previously described, with a sub-network selective for PPS embedded in a network coding both near and far space with a near space preference. Previous studies have focused on the selective PPS network. We argue that the broader PPS network might actually be crucial in subserving both the dynamic resizing of PPS (Cléry et al., 2015b; Bufacchi and Iannetti, 2018; Spaccasassi et al., 2019) and the functional and behavioural interactions between near and far space.

4.2.3 Functional cortical network involved in far space (away from the body)

Using naturalistic stimuli, Cléry et al. (2018) revealed a large widespread cortical network involved in the coding of far space with activations in the entire striate and extra-striate cortex, the temporal cortex (Vuilleumier et al., 1998; Weiss et al., 2000), the parietal cortex, and small portions of the prefrontal and premotor cortex along the arcuate sulcus (Rizzolatti et al., 1983) as well as the orbitofrontal cortex. Some of these regions were also activated by near space stimulations (see section 4.2.2). However, a large portion of this functional network preferred far space (Figure 4.2[7], black activations) while other portions preferred near space. In other words, these cortical regions coded for both near and far space, but showed stronger activations for far than for near stimuli. Areas preferring far space to near space included the entire visual striate and extra-striate cortex: areas V1, V2, V3, V3A, and V4; posterior (PIP), lateral (LIP), and caudal (CIP, parietal reach region PRR) intraparietal sulcus; and the medial (5v) and lateral (7a, 7ab, and 7b) parietal convexity. These activations extended towards the parieto-occipital cortex (including areas V6A and V6) and temporal cortex: medial and superior temporal areas MT and MST. Importantly, some of these cortical regions coded exclusively for far space stimulations and not for near space stimulations (Figure 4.2[7], blue colour scale activations). Interestingly, stimulations with fronto-parallel visual coherent movements also elicited strong cortical activations mainly in the occipital striate and extra-striate areas, and in the temporal cortex (Figure 4.2[5]), similar to the far space network, possibly suggesting an interaction between stimulation size and physical location within far space.

4.2.4 Functional coupling between *stricto sensu* body margin and peripersonal space

From a neurophysiological point of view, the skin and PPS are very strongly coupled, whereby areas coding for near space at all levels of the visual hierarchy also code for tactile stimulations on the skin. In the following discussion, we will consider visuo-tactile convergence, which corresponds to the fact that visual and tactile stimulations played separately elicit activations in the same neurons areas but without necessarily integrating the information when presented simultaneously. Then, we will relate visuo-tactile convergence to visuo-tactile integration, which corresponds to the active process by which the neuronal response to simultaneously presented visual and tactile stimuli is different from the sum of the neuronal responses to the individual unisensory stimulations. We then address the possible behavioural relevance of this coupling.

The convergence of visual and tactile information is observed at very early stages of the visual pathway as heteromodal somatosensory projections onto the visual cortex have been described (Wallace et al., 2004). The extent of this widespread heteromodal visuo-tactile convergence has recently been demonstrated at different levels of the visual processing hierarchy using monkey fMRI (Guipponi et al., 2015a), including in striate and extra-striate visual cortices (V1, V2, V3,V3A, Figure 4.2[3]). These results confirm and extend previous studies describing heteromodal modulation of primary sensory cortices (Macaluso et al., 2000; Amedi et al., 2001; Schroeder and Foxe, 2005; Vasconcelos et al., 2011; Iurilli et al., 2012). Visual and tactile convergence is also described in higher order associative cortical areas— for example, in the posterior parietal cortex in areas PIP, VIP, 7b, and parietal opercular area 7op (Hikosaka et al., 1988; Duhamel et al., 1998; Bremmer et al., 1999, 2000, 2002b, 2002a; Avillac et al., 2004, 2005; Schlack et al., 2005; Sereno and Huang, 2006; Avillac et al., 2007; Rozzi et al., 2008; Guipponi et al., 2013), as well as in the anterior parietal cortex (Hikosaka et al., 1988; Huang et al., 2012) or in the temporal cortex including the superior temporal sulcus STS and medial superior temporal area MST (Benevento et al., 1977; Bruce et al., 1981; Hikosaka et al., 1988; Beauchamp et al., 2004; Barraclough et al., 2005; Beauchamp, 2005; Guipponi et al., 2015a). Specifically, visuo-tactile cortical convergence (Figure 4.2[3]) is observed in the insular and peri-insular cortex within area Pi and in a region anterior to SII/PV in the upper bank of the lateral sulcus (Augustine, 1996; Guipponi et al., 2015a), in the somatosensory area 2, and in cingulate areas 24c-d (Laurienti et al., 2003; Cléry et al., 2017a). In the frontal part of the brain, visuo-tactile convergence activates prefrontal areas 46v/9, precentral F4/F5, 44 at the fundus of the inferior branch of the arcuate sulcus and prefrontal GrF/ProM (Graziano et al., 1994, 1997; Fogassi et al., 1996; Graziano and Gandhi, 2000; Petrides et al., 2005; Graziano and Cooke, 2006; Belmalih et al., 2009), and the inferior orbitofrontal cortex with areas 11 and 13 (Rolls and Baylis, 1994; Rolls, 2004).

Visuo-tactile convergence often results in multisensory integration (see Cléry et al., 2015b). Visual information thus gets combined with tactile information to signal to the brain more than just tactile stimulation. This results in either speeded up neuronal responses or enhanced detection of low intensity stimuli (or both) (Avillac et al., 2004, 2007). In the context of PPS, the functional binding between the two sensory modalities results in an anticipated recruitment of tactile processes, by visual stimuli, contributing to the prediction of the consequences of the visual modality onto the skin. This is discussed further below.

4.2.5 Functional cortical network involved in impact prediction (near to the skin)

In this section, we investigate the possible functional coupling between near and far space coding and the prediction of impact of an approaching object onto the face. This is based on the description of the overlap of activations derived from different studies by Cléry et al. and Guipponi et al. In Cléry et al. (2018), the dynamical approach of the smaller cube into near space elicited strong activations in both hemispheres (Figure 4.2[8]): in visual areas (V1, V2, V3), in the temporal cortex (MT, MST, TPOr, IPa), in the parietal cortex (VIP, PIP, 7a, 7b, 7 ab), in the prefrontal and premotor cortex (FEF, F4, F5, 46p), and in the posterior orbitofrontal cortex (12). This approach of an object towards near space can be considered as an intrusion into the PPS. Interestingly, the approach of a small cube towards near space produced very similar activation patterns as the approach of a big cube within far space, suggesting that the risk of intrusion into PPS is coded in the same way, whether it is triggered by a far stimulus or by a close-by stimulus. Importantly, this 'approach' network includes both near space- and far space-coding areas (Cléry et al., 2018).

The visual dynamic stimulations that predict an impact onto the face activate a network very close to the approach network just described (Figure 4.2[4]). Looming stimuli have been shown to trigger stereotyped defence responses (Schiff et al., 1962; Ball and Tronick, 1971; King and Cowey, 1992; Vagnoni et al., 2012), enhance reaction times (Canzoneri et al., 2012; Kandula et al., 2015; de Haan et al., 2016; De Paepe et al., 2016), and enhance sensitivity to a second heteromodal stimulus (Cléry et al., 2015a). Impact prediction processing elicited activations in prefrontal/premotor cortex (areas F4, 44, 45b, and 46), cingulate cortex (area 24c-d), insular cortex (area IPro), parietal cortex (LOP, VIP, and MIP), temporal cortex (MT, MST, FST, and TEO) and striate and extra-striate cortex (areas V1, V2, V3, V3a, and V4) (Cléry et al., 2017b). This functional impact prediction network strongly overlaps with the visuo-tactile convergence network (Guipponi et al., 2015a), providing the functional substrates of the anticipated touch by looming stimuli.

Figure 4.3 represents the functional overlap between the visuo-tactile convergence network, the visual impact prediction network, the near space visual network and the far to near space visual network. Real 3D stimulations resulted in the largest activations (Dosso and Kingstone, 2018). This functional overlap takes place mainly in parietal (VIP, MIP, AIP), temporal (FST/MST) and frontal areas (45B/12). This suggests that these regions are not only at the core of the PPS encoding and defence behaviour, but also possibly serve different functions using external cues (on the body or approaching the body) to dynamically change and adapt the PPS size to protective and proactive behaviour.

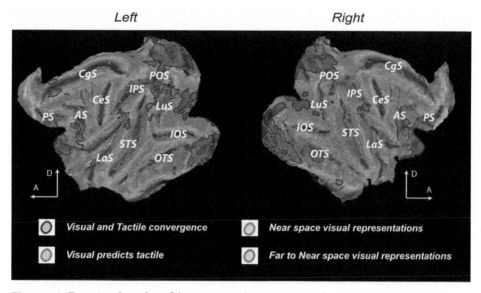

Figure 4.3 Functional overlap of the visuo-tactile convergence (blue), impact prediction (pink), near space (orange) and looming stimuli (green) networks. Contrasts are performed with a level of significance set at p<0.001 uncorrected level, (t>3.1 blue, pink, and green margins) and at p<0.05 corrected for multiple comparison (FWE, t>4.8 orange margin).

4.3 Multiple body parts, multiple spaces: functional perspectives

While PPS was initially considered unitary and homogenous, recent evidence supports the idea of multiple PPS, with specificities associated with the body parts they are anchored to (e.g. hand, head, trunk, Serino et al., 2015a, 2015b). These PPS interact with each other, objects within one PPS modulating other PPS, depending on their relative positioning to the stimulated PPS, as well as on the direction tuning specificities of the stimulus (Serino et al., 2015b). For example, an object coming into the trunk PPS enlarges both head and hand PPS, while the reverse is not true. The velocity of the looming stimuli modulates these different PPS and rely in part on multisensory processes (Noel et al., 2018). Indeed, 30% of human parietal and temporal neurons (essentially in the post-central gyrus, insula, and parahippocampus gyrus) showing audio-tactile multisensory integrations had their response modulated by the distance from the stimuli to the trunk (Bernasconi et al., 2018).

A recent review (Kastner et al., 2017) shows that, compared to humans, the posterior parietal cortex (PPC) is well conserved in monkeys for spatial processing but not for non-spatial processing, which involves more complex processes in humans. The PPC has a complex somatotopic representation of the face and body (Sereno and Huang, 2006, 2014; Huang and Sereno, 2007; Huang et al., 2012, 2017). In particular, the human IPS shows large activations dedicated to the face (central-anterior part dedicated to the upper face, ventral and dorsal parts dedicated to the lower face) and smaller regions for the shoulders when stimulated with visual, tactile, multisensorial stimuli, looming stimuli (Huang et al., 2017). An interesting perspective is whether this parietal homunculus, unfolding in visuo-tactile regions, corresponds to the functional substrates of the multiple PPS just described. The size of the PPS and its complexity could be associated with the corresponding parietal homunculus distorted representation and its associated computational complexity. A similar prediction holds in the non-human primate. Indeed, a similar tactile homunculus can be described. For example, centre of the face tactile stimulations widely activate VIP, periphery of the face stimulations activate VIP, MIP, and PIP, and shoulder stimulations activate MIP (Wardak et al., 2016). Cortical correlates of blinks, possibly reflecting the sensory consequences of eyelid closing, are described at the limit between VIP and MIP (not shown on Figure 4.4, Guipponi et al., 2015b)—that is, very close to the face representation. The inferior parietal

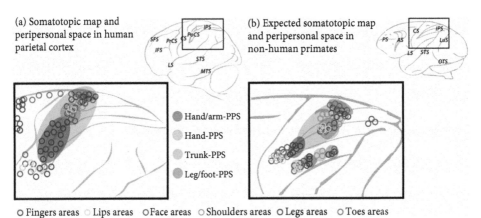

(a) Somatotopic map and peripersonal space in human parietal cortex

(b) Expected somatotopic map and peripersonal space in non-human primates

● Hand/arm-PPS

● Hand-PPS

● Trunk-PPS

● Leg/foot-PPS

○ Fingers areas ○ Lips areas ○ Face areas ○ Shoulders areas ○ Legs areas ○ Toes areas

Figure 4.4 Somatotopic organization of parietal peripersonal space (PPS) representation in humans (a) and non-human primates (b).

lobule and area 5 shows a rostro-caudal somatotopic organization from the mouth to the lower limbs (Rozzi et al., 2008; Seelke et al., 2012). Importantly, in both humans and monkeys, these cortical regions are responsive to both tactile and visual stimulations, and can thus subserve a PPS representation, along a somatotopic organization of parietal PPS representations, with a head-PPS essentially encoded by VIP, a hand-PPS encoded by areas AIP, and a trunk-PPS encoded by the areas 5 and/or 7b (Figure 4.4).

4.4 Multiple influences, multiples roles: functional perspective

All this taken together, the different studies cited above show that the encoding of different sensory cues of high relevance to PPS (visual, tactile, impact prediction, and near space cues) are subserved by a common core cortical network with specificity to each function. In the past years, evidence has accumulated in favour of multiple possible functions of PPS modulated by diverse sources. Indeed, social interactions, emotions, or action contexts shape these psychological and functional spaces (for reviews Cléry et al., 2015b; di Pellegrino and Làdavas, 2015; Cléry and Ben Hamed, 2018). Despite this accumulated data, the neural bases of these observations still need to be understood. Figure 4.5 explores possible functional

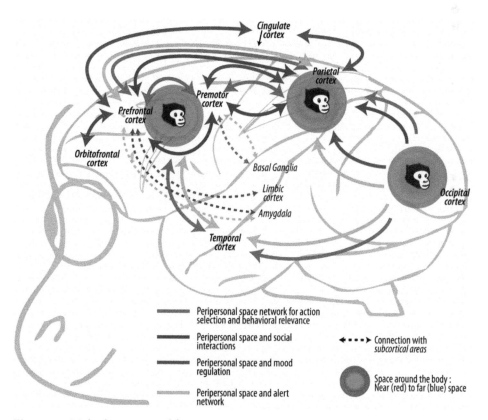

Figure 4.5 Multiple sources and functional networks can reshape PPS, coinciding with multiple functions associated with PPS.

Image adapted from Cléry et al. Neuronal bases of peripersonal and extrapersonal spaces, their plasticity and their dynamics: knowns and unknowns. Neuropsychologia. 2015 Apr;70:313–26. doi: 10.1016/j.neuropsychologia.2014.10.022.

networks contributing to PPS dynamic resizing. Based on the data presented throughout this chapter, occipital cortex encodes space at large, without any expected homogenous functional specificity with respect to the body. As one progresses towards the parietal, temporal, and prefrontal cortices, space representation is expected to become more and more specific. Numerous studies show that tool use, whether involving a physical interaction with our body (Iriki et al., 1996; Farnè and Làdavas, 2000; Maravita and Iriki, 2004; Farnè et al., 2005; Serino et al., 2007; Costantini et al., 2011; Galli et al., 2015; Biggio et al., 2017) or not (Goldenberg and Iriki, 2007; Bassolino et al., 2010; Serino et al., 2015a), increases the size of PPS by the incorporation of the tool in its representation (for more details, see Martel et al., 2016; Cléry and Ben Hamed, 2018). As a result, one would predict that such a plasticity recruits the affordance network, involving temporal regions defining tool identity, as well as parieto-frontal reach and grasp regions (Thill et al., 2013; Murata et al., 2016). The PPS, in its relevance to actions and goal-directed behaviour, is expected to recruit the occipital cortex, then parietal, premotor, and prefrontal areas with reciprocal influences, as well as the temporal cortex and the basal ganglia (Figure 4.5, red colour).

PPS is also modulated by the social context, be it our first impression of a person, perceived morality, or cooperativeness (Teneggi et al., 2013; Iachini et al., 2014, 2015, 2016; Cartaud et al., 2018; Pellencin et al., 2018). This suggests the involvement of the orbitofrontal cortex and the amygdala in a dynamic reshape of the PPS (Kennedy et al., 2009; Perry et al., 2016; Wabnegger et al., 2016). The functional network at the origin of PPS resizing by social cues is expected to recruit the occipital cortex, parietal, premotor, prefrontal, and orbitofrontal areas, as well as the amygdala, with reciprocal influences (Figure 4.5, blue colour). Whether the embodiment of other's body representations (Ishida et al., 2015; Maister et al., 2015; Teramoto, 2018) recruits this social-PPS interaction network or a distinct network centred onto body-representation (see discussion in Cléry et al., 2018) remains unclear and warrants future investigation.

PPS increases with anxiety or threatening stimuli (Poliakoff et al., 2007; Van Damme et al., 2009; Taffou and Viaud-Delmon, 2014; Ferri et al., 2015; de Haan et al., 2016). One would thus expect a contribution of the limbic cortex in this emotionally reshaping of PPS. This emotional-PPS interaction is expected to recruit the occipital cortex, the parietal, cingulate, premotor and prefrontal areas as well as the limbic cortex with reciprocal influences (Figure 4.5, green colour).

Last, the impact prediction network suggests the existence of an alert network allowing shorter reaction times and increased sensitivity (Canzoneri et al., 2012; Cléry et al., 2015a; Kandula et al., 2015) to avoid impact on the body. This network is proposed to involve the temporal cortex and the amygdala, which helps to identify the looming stimulus and its valence. This network is expected to be set up very early on in life as newborns are able to make sense of the multisensory cue combinations moving towards them (Orioli et al., 2018). The motor system is suggested to play a role in predictive mechanism by anticipating the consequences of an action, the near becoming far or the far becoming near, dynamically as a function of the context (Bisio et al., 2017). The alert-PPS interaction is expected to recruit the occipito-parieto-prefrontal network but also premotor areas, temporal areas, the basal ganglia, and the amygdala (Figure 4.5, orange colour).

4.5 Conclusion

The space around us is not a unitary and static space but rather consists of dynamic spaces shaped by the kind of stimulus (static, moving), the valence (negative or positive) and the interactions with the environment (social, inanimate) involving numerous brain areas and circuits. A core network spans the occipital, parietal, and prefrontal/premotor cortex with multiple connections with other cortical or subcortical areas. The exact nature of these functional interactions and their neural bases remains to be explored.

References

Amedi A, Malach R, Hendler T, Peled S, Zohary E (2001) Visuo-haptic object-related activation in the ventral visual pathway. *Nat Neurosci* 4:324–330.

Augustine JR (1996) Circuitry and functional aspects of the insular lobe in primates including humans. *Brain Res. Rev.* 22:229–244.

Avillac M, Denève S, Olivier E, Pouget A, Duhamel J-R (2005) Reference frames for representing visual and tactile locations in parietal cortex. *Nat Neurosci* 8:941–949.

Avillac M, Hamed SB, Duhamel J-R (2007) Multisensory integration in the ventral intraparietal area of the macaque monkey. *J. Neurosci* 27:1922–1932.

Avillac M, Olivier E, Denève S, Ben Hamed S, Duhamel J-R (2004) Multisensory integration in multiple reference frames in the posterior parietal cortex. *Cogn Process* 5:159–166.

Ball W, Tronick E (1971) Infant responses to impending collision: optical and real. *Science* 171:818–820.

Barraclough NE, Xiao D, Baker CI, Oram MW, Perrett DI (2005) Integration of visual and auditory information by superior temporal sulcus neurons responsive to the sight of actions. *J. Cogn Neurosci* 17:377–391.

Bassolino M, Serino A, Ubaldi S, Làdavas E (2010) Everyday use of the computer mouse extends peripersonal space representation. *Neuropsychologia* 48:803–811.

Beauchamp MS (2005) See me, hear me, touch me: multisensory integration in lateral occipital-temporal cortex. *Curr Opin in Neurobiol* 15:145–153.

Beauchamp MS, Lee KE, Argall BD, Martin A (2004) Integration of auditory and visual information about objects in superior temporal sulcus. *Neuron* 41:809–823.

Belmalih A, Borra E, Contini M, Gerbella M, Rozzi S, Luppino G (2009) Multimodal architectonic subdivision of the rostral part (area F5) of the macaque ventral premotor cortex. *J. Comp Neurol* 512:183–217.

Benevento LA, Fallon J, Davis BJ, Rezak M (1977) Auditory-visual interaction in single cells in the cortex of the superior temporal sulcus and the orbital frontal cortex of the macaque monkey. *Exp Neurol* 57:849–872.

Bernasconi F, Noel J-P, Park HD, Faivre N, Seeck M, Spinelli L, Schaller K, Blanke O, Serino A (2018) Audio-tactile and peripersonal space processing around the trunk in human parietal and temporal cortex: an intracranial EEG study. *Cereb Cortex* 28:3385–3397.

Biggio M, Bisio A, Avanzino L, Ruggeri P, Bove M (2017) This racket is not mine: the influence of the tool-use on peripersonal space. *Neuropsychologia* 103:54–58.

Bisio A, Garbarini F, Biggio M, Fossataro C, Ruggeri P, Bove M (2017) Dynamic shaping of the defensive peripersonal space through predictive motor mechanisms: when the "near" becomes "far." *J. Neurosci* 37:2415–2424.

Bremmer F, Duhamel J-R, Ben Hamed S, Graf W (2000) Stages of self-motion processing in primate posterior parietal cortex. *Int Rev Neurobiol* 44:173–198.

Bremmer F, Duhamel J-R, Ben Hamed S, Graf W (2002a) Heading encoding in the macaque ventral intraparietal area (VIP). *Eur. J. Neurosci* 16:1554–1568.

Bremmer F, Graf W, Ben Hamed S, Duhamel JR (1999) Eye position encoding in the macaque ventral intraparietal area (VIP). *NeuroReport* 10:873–878.

Bremmer F, Klam F, Duhamel J-R, Ben Hamed S, Graf W (2002b) Visual–vestibular interactive responses in the macaque ventral intraparietal area (VIP). *Eur. J. Neurosci* 16:1569–1586.

Bremmer F, Schlack A, Kaminiarz A, Hoffmann KP (2013) Encoding of movement in near extrapersonal space in primate area VIP. *Front. Behav Neurosci* 7:8.

Brozzoli C, Ehrsson HH, Farnè A (2014) Multisensory representation of the space near the hand: from perception to action and interindividual interactions. *Neuroscientist* 20:122–135.

Brozzoli C, Gentile G, Bergouignan L, Ehrsson HH (2013) A shared representation of the space near oneself and others in the human premotor cortex. *Curr Biol* 23:1764–1768.

Bruce C, Desimone R, Gross CG (1981) Visual properties of neurons in a polysensory area in superior temporal sulcus of the macaque. *J. Neurophysiol* 46:369–384.

Bufacchi RJ, Iannetti GD (2018) An action field theory of peripersonal space. *Trends Cogn Sci* 22:1076–1090.

Canzoneri E, Magosso E, Serino A (2012) Dynamic sounds capture the boundaries of peripersonal space representation in humans. *PLoS ONE* 7:e44306.

Caprara I, Premereur E, Romero MC, Faria P, Janssen P (2018) Shape responses in a macaque frontal area connected to posterior parietal cortex. *NeuroImage* 179:298–312.

Cartaud A, Ruggiero G, Ott L, Iachini T, Coello Y (2018) Physiological response to facial expressions in peripersonal space determines interpersonal distance in a social interaction context. *Front. Psychol* 9.

Chen X, Sanayei M, Thiele A (2014) Stimulus roving and flankers affect perceptual learning of contrast discrimination in macaca mulatta. *PLoS ONE* 9:e109604.

Cléry J, Amiez C, Guipponi O, Wardak C, Procyk E, Ben Hamed S (2017a) Reward activations and face fields in monkey cingulate motor areas. *J. Neurophysiol* 119:1037–1044.

Cléry J, Ben Hamed S (2018) Frontier of self and impact prediction. *Front. Psychol* 9.

Cléry J, Guipponi O, Odouard S, Pinède S, Wardak C, Hamed SB (2017b) The prediction of impact of a looming stimulus onto the body is subserved by multisensory integration mechanisms. *J. Neurosci* 37:10656–10670.

Cléry J, Guipponi O, Odouard S, Wardak C, Ben Hamed S (2015a) Impact prediction by looming visual stimuli enhances tactile detection. *J. Neurosci* 35:4179–4189.

Cléry J, Guipponi O, Odouard S, Wardak C, Ben Hamed S (2018) Cortical networks for encoding near and far space in the non-human primate. *NeuroImage* 176:164–178.

Cléry J, Guipponi O, Wardak C, Ben Hamed S (2015b) Neuronal bases of peripersonal and extrapersonal spaces, their plasticity and their dynamics: knowns and unknowns. *Neuropsychologia* 70:313–326.

Cohen YE, Russ BE, Gifford GW, Kiringoda R, MacLean KA (2004) Selectivity for the spatial and nonspatial attributes of auditory stimuli in the ventrolateral prefrontal cortex. *J. Neurosci* 24:11307–11316.

Cooke DF, Graziano MSA (2004) Sensorimotor integration in the precentral gyrus: polysensory neurons and defensive movements. *J. Neurophysiol* 91:1648–1660.

Costantini M, Ambrosini E, Sinigaglia C, Gallese V (2011) Tool-use observation makes far objects ready-to-hand. *Neuropsychologia* 49:2658–2663.

de Haan AM, Smit M, Stigchel SV der, Dijkerman HC (2016) Approaching threat modulates visuotactile interactions in peripersonal space. *Exp Brain Res* 234:1875–1884.

De Paepe AL, Crombez G, Legrain V (2016) What's coming near? The influence of dynamical visual stimuli on nociceptive processing. *PLoS ONE* 11:e0155864.

di Pellegrino G, Làdavas E (2015) Peripersonal space in the brain. *Neuropsychologia* 66:126–133.

Dong WK, Chudler EH, Sugiyama K, Roberts VJ, Hayashi T (1994) Somatosensory, multisensory, and task-related neurons in cortical area 7b (PF) of unanesthetized monkeys. *J. Neurophysiol* 72:542–564.

Dosso JA, Kingstone A (2018) Social modulation of object-directed but not image-directed actions. *PLoS ONE* 13:e0205830.

Duhamel J-R, Bremmer F, Ben Hamed S, Graf W (1997) Spatial invariance of visual receptive fields in parietal cortex neurons. *Nature* 389:845–848.

Duhamel J-R, Colby CL, Goldberg ME (1998) Ventral intraparietal area of the macaque: congruent visual and somatic response properties. *J. Neurophysiology* 79:126–136.

Farnè A, Iriki A, Làdavas E (2005) Shaping multisensory action-space with tools: evidence from patients with cross-modal extinction. *Neuropsychologia* 43:238–248.

Farnè A, Làdavas E (2000) Dynamic size-change of hand peripersonal space following tool use. *NeuroReport* 11:1645–1649.

Ferri F, Tajadura-Jiménez A, Väljamäe A, Vastano R, Costantini M (2015) Emotion-inducing approaching sounds shape the boundaries of multisensory peripersonal space. *Neuropsychologia* 70:468–475.

Fiave PA, Sharma S, Jastorff J, Nelissen K (2018) Investigating common coding of observed and executed actions in the monkey brain using cross-modal multi-variate fMRI classification. *NeuroImage* 178:306–317.

Fitzgerald PJ, Lane JW, Thakur PH, Hsiao SS (2004) Receptive field properties of the macaque second somatosensory cortex: evidence for multiple functional representations. *J. Neurosci* 24:11193–11204.

Fitzgerald PJ, Lane JW, Thakur PH, Hsiao SS (2006) Receptive field (RF) properties of the macaque second somatosensory cortex: rf size, shape, and somatotopic organization. *J. Neurosci* 26:6485–6495.

Fogassi L, Gallese V, Buccino G, Craighero L, Fadiga L, Rizzolatti G (2001) Cortical mechanism for the visual guidance of hand grasping movements in the monkey: a reversible inactivation study. *Brain* 124:571–586.

Fogassi L, Gallese V, Fadiga L, Luppino G, Matelli M, Rizzolatti G (1996) Coding of peripersonal space in inferior premotor cortex (area F4). *J. Neurophysiol* 76:141–157.

Gallese V, Murata A, Kaseda M, Niki N, Sakata H (1994) Deficit of hand preshaping after muscimol injection in monkey parietal cortex. *NeuroReport* 5:1525–1529.

Galli G, Noel J-P, Canzoneri E, Blanke O, Serino A (2015) The wheelchair as a full-body tool extending the peripersonal space. *Front. Psychol* 6.

Gentilucci M, Fogassi L, Luppino G, Matelli M, Camarda R, Rizzolatti G (1988) Functional organization of inferior area 6 in the macaque monkey. I. Somatotopy and the control of proximal movements. *Exp Brain Res* 71:475–490.

Gerbella M, Belmalih A, Borra E, Rozzi S, Luppino G (2010) Cortical connections of the macaque caudal ventrolateral prefrontal areas 45A and 45B. *Cereb Cortex* 20:141–168.

Gerbella M, Borra E, Tonelli S, Rozzi S, Luppino G (2013) Connectional heterogeneity of the ventral part of the macaque area 46. *Cereb Cortex* 23:967–987.

Goldenberg G, Iriki A (2007) From sticks to coffee-maker: mastery of tools and technology by human and non-human primates. *Cortex* 43:285–288.

Graziano MS, Gandhi S (2000) Location of the polysensory zone in the precentral gyrus of anesthetized monkeys. *Exp Brain Res* 135:259–266.

Graziano MS, Yap GS, Gross CG (1994) Coding of visual space by premotor neurons. *Science* 266:1054–1057.

Graziano MSA, Cooke DF (2006) Parieto-frontal interactions, personal space, and defensive behavior. *Neuropsychologia* 44:845–859.

Graziano MSA, Hu XT, Gross CG (1997) Visuospatial properties of ventral premotor cortex. *J. Neurophysiol* 77:2268–2292.

Graziano MSA, Reiss LAJ, Gross CG (1999) A neuronal representation of the location of nearby sounds. *Nature* 397:428–430.

Guipponi O, Cléry J, Odouard S, Wardak C, Ben Hamed S (2015a) Whole brain mapping of visual and tactile convergence in the macaque monkey. *NeuroImage* 117:93–102.

Guipponi O, Odouard S, Pinède S, Wardak C, Ben Hamed S (2015b) fMRI cortical correlates of spontaneous eye blinks in the nonhuman primate. *Cereb Cortex* 25:2333–2345.

Guipponi O, Wardak C, Ibarrola D, Comte J-C, Sappey-Marinier D, Pinède S, Hamed SB (2013) Multimodal convergence within the intraparietal sulcus of the macaque monkey. *J. Neurosci* 33:4128–4139.

Hikosaka K, Iwai E, Saito H, Tanaka K (1988) Polysensory properties of neurons in the anterior bank of the caudal superior temporal sulcus of the macaque monkey. *J. Neurophysiol* 60:1615–1637.

Huang R-S, Chen C, Sereno MI (2017) Mapping the complex topological organization of the human parietal face area. *NeuroImage* 163:459–470.

Huang R-S, Chen C, Tran AT, Holstein KL, Sereno MI (2012) Mapping multisensory parietal face and body areas in humans. *Proc. Natl. Acad. Sci. U.S.A.* 109:18114–18119.

Huang R-S, Sereno MI (2007) Dodecapus: An MR-compatible system for somatosensory stimulation. *NeuroImage* 34:1060–1073.

Hyvärinen J, Poranen A (1974) Function of the parietal associative area 7 as revealed from cellular discharges in alert monkeys. *Brain* 97:673–692.

Iachini T, Coello Y, Frassinetti F, Ruggiero G (2014) Body space in social interactions: a comparison of reaching and comfort distance in immersive virtual reality. *PLoS ONE* 9.

Iachini T, Coello Y, Frassinetti F, Senese VP, Galante F, Ruggiero G (2016) Peripersonal and interpersonal space in virtual and real environments: effects of gender and age. *J. Environ. Psychol* 45:154–164.

Iachini T, Pagliaro S, Ruggiero G (2015) Near or far? It depends on my impression: moral information and spatial behavior in virtual interactions. *Acta Psychol (Amst)* 161:131–136.

Iriki A, Tanaka M, Iwamura Y (1996) Coding of modified body schema during tool use by macaque postcentral neurones. *NeuroReport* 7:2325–2330.

Ishida H, Nakajima K, Inase M, Murata A (2009) Shared mapping of own and others' bodies in visuotactile bimodal area of monkey parietal cortex. *J. Cogn Neurosci* 22:83–96.

Ishida H, Suzuki K, Grandi LC (2015) Predictive coding accounts of shared representations in parieto-insular networks. *Neuropsychologia* 70:442–454.

Iurilli G, Ghezzi D, Olcese U, Lassi G, Nazzaro C, Tonini R, Tucci V, Benfenati F, Medini P (2012) Sound-driven synaptic inhibition in primary visual cortex. *Neuron* 73:814–828.

Kaas JH, Nelson RJ, Sur M, Lin CS, Merzenich MM (1979) Multiple representations of the body within the primary somatosensory cortex of primates. *Science* 204:521–523.

Kandula M, Hofman D, Dijkerman HC (2015) Visuo-tactile interactions are dependent on the predictive value of the visual stimulus. *Neuropsychologia* 70:358–366.

Kastner S, Chen Q, Jeong SK, Mruczek REB (2017) A brief comparative review of primate posterior parietal cortex: a novel hypothesis on the human toolmaker. *Neuropsychologia* 105:123–134.

Kennedy DP, Gläscher J, Tyszka JM, Adolphs R (2009) Personal space regulation by the human amygdala. *Nat Neurosci* 12:1226–1227.

King SM, Cowey A (1992) Defensive responses to looming visual stimuli in monkeys with unilateral striate cortex ablation. *Neuropsychologia* 30:1017–1024.

Krubitzer L, Clarey J, Tweedale R, Elston G, Calford M (1995) A redefinition of somatosensory areas in the lateral sulcus of macaque monkeys. *J. Neurosci* 15:3821–3839.

Laurienti PJ, Wallace MT, Maldjian JA, Susi CM, Stein BE, Burdette JH (2003) Cross-modal sensory processing in the anterior cingulate and medial prefrontal cortices. *Hum. Brain Mapp.* 19:213–223.

Macaluso E, Frith CD, Driver J (2000) Modulation of human visual cortex by crossmodal spatial attention. *Science* 289:1206–1208.

Maister L, Cardini F, Zamariola G, Serino A, Tsakiris M (2015) Your place or mine: shared sensory experiences elicit a remapping of peripersonal space. *Neuropsychologia* 70:455–461.

Maravita A, Iriki A (2004) Tools for the body (schema). *Trends Cogn Sci* 8:79–86.

Martel M, Cardinali L, Roy AC, Farnè A (2016) Tool-use: an open window into body representation and its plasticity. *Cogn Neuropsychol* 33:82–101.

Matelli M, Luppino G (2001) Parietofrontal circuits for action and space perception in the macaque monkey. *NeuroImage* 14:S27–S32.

Miller LE, Montroni L, Koun E, Salemme R, Hayward V, Farnè, A (2018) Sensing with tools extends somatosensory processing beyond the body. *Nature* 561(7722):239–242.

Murata A, Gallese V, Luppino G, Kaseda M, Sakata H (2000) Selectivity for the shape, size, and orientation of objects for grasping in neurons of monkey parietal area AIP. *J. Neurophysiol* 83:2580–2601.

Murata A, Wen W, Asama H (2016) The body and objects represented in the ventral stream of the parieto-premotor network. *J. Neurosci. Res.* 104:4–15.

Murray EA, Mishkin M (1984) Relative contributions of SII and area 5 to tactile discrimination in monkeys. *Behav. Brain Res* 11:67–83.

Nelson RJ, Sur M, Felleman DJ, Kaas JH (1980) Representations of the body surface in postcentral parietal cortex of Macaca fascicularis. *J. Comp Neurol* 192:611–643.

Noel J-P, Blanke O, Magosso E, Serino A (2018) Neural adaptation accounts for the dynamic resizing of peri-personal space: evidence from a psychophysical-computational approach. *J. Neurophysiol* 119:2307–2333.

Ó Scalaidhe SP, Wilson FAW, Goldman-Rakic PS (1997) Areal segregation of face-processing neurons in prefrontal cortex. *Science* 278:1135–1138.

Orioli G, Bremner AJ, Farroni T (2018) Multisensory perception of looming and receding objects in human newborns. *Curr Biol* 28:R1294–R1295.

Pellencin E, Paladino MP, Herbelin B, Serino A (2018) Social perception of others shapes one's own multisensory peripersonal space. *Cortex* 104:163–179.

Perry A, Lwi SJ, Verstaen A, Dewar C, Levenson RW, Knight RT (2016) The role of the orbitofrontal cortex in regulation of interpersonal space: evidence from frontal lesion and frontotemporal dementia patients. *Soc Cogn Affect Neurosci* 11:1894–1901.

Petrides M, Cadoret G, Mackey S (2005) Orofacial somatomotor responses in the macaque monkey homologue of Broca's area. *Nature* 435:1235–1238.

Poliakoff E, Miles E, Li X, Blanchette I (2007) The effect of visual threat on spatial attention to touch. *Cognition* 102:405–414.

Pons TP, Garraghty PE, Cusick CG, Kaas JH (1985) A sequential representation of the occiput, arm, forearm and hand across the rostrocaudal dimension of areas 1, 2 and 5 in macaque monkeys. *Brain Res* 335:350–353.

Preuss TM, Goldman-Rakic PS (1989) Connections of the ventral granular frontal cortex of macaques with perisylvian premotor and somatosensory areas: anatomical evidence for somatic representation in primate frontal association cortex. *J. Comp Neurol* 282:293–316.

Rizzolatti G, Luppino G (2001) The cortical motor system. *Neuron* 31:889–901.

Rizzolatti G, Matelli M (2003) Two different streams form the dorsal visual system: anatomy and functions. *Exp Brain Res* 153:146–157.

Rizzolatti G, Matelli M, Pavesi G (1983) Deficits in attention and movement following the removal of postarcuate (area 6) and prearcuate (area 8) cortex in macaque monkeys. *Brain* 106 (Pt 3):655–673.

Rizzolatti G, Scandolara C, Matelli M, Gentilucci M (1981) Afferent properties of periarcuate neurons in macaque monkeys. I. Somatosensory responses. *Behav Brain Res* 2:125–146.

Rolls ET (2004) Convergence of sensory systems in the orbitofrontal cortex in primates and brain design for emotion. *Anat. Rec., Part A Discov. Mol. Cell. Evol. Biol.* 281A:1212–1225.

Rolls ET, Baylis LL (1994) Gustatory, olfactory, and visual convergence within the primate orbitofrontal cortex. *J. Neurosci* 14:5437–5452.

Romanski LM, Averbeck BB, Diltz M (2005) Neural representation of vocalizations in the primate ventrolateral prefrontal cortex. *J. Neurophysiol* 93:734–747.

Rozzi S, Ferrari PF, Bonini L, Rizzolatti G, Fogassi L (2008) Functional organization of inferior parietal lobule convexity in the macaque monkey: electrophysiological characterization of motor, sensory and mirror responses and their correlation with cytoarchitectonic areas. *Eur. J. Neurosci* 28:1569–1588.

Schiff W, Caviness JA, Gibson JJ (1962) Persistent fear responses in rhesus monkeys to the optical stimulus of "looming." *Science* 136:982–983.

Schlack A, Sterbing-D'Angelo SJ, Hartung K, Hoffmann K-P, Bremmer F (2005) Multisensory space representations in the macaque ventral intraparietal area. *J. Neurosci* 25:4616–4625.

Schroeder CE, Foxe J (2005) Multisensory contributions to low-level, "unisensory" processing. *Curr Opin Neurobiol* 15:454–458.

Seelke AMH, Padberg JJ, Disbrow E, Purnell SM, Recanzone G, Krubitzer L (2012) Topographic Maps within Brodmann's Area 5 of macaque monkeys. *Cereb Cortex* 22:1834–1850.

Sereno MI, Huang R-S (2006) A human parietal face area contains aligned head-centered visual and tactile maps. *Nat Neurosci* 9:1337–1343.

Sereno MI, Huang R-S (2014) Multisensory maps in parietal cortex. *Curr Opin in Neurobiol* 24:39–46.

Serino A, Bassolino M, Farnè A, Làdavas E (2007) Extended multisensory space in blind cane users. *Psychol. Sci* 18:642–648.

Serino A, Canzoneri E, Marzolla M, di Pellegrino G, Magosso E (2015a) Extending peripersonal space representation without tool-use: evidence from a combined behavioral-computational approach. *Front. Behav Neurosci* 9.

Serino A, Noel J-P, Galli G, Canzoneri E, Marmaroli P, Lissek H, Blanke O (2015b) Body part-centered and full body-centered peripersonal space representations. *Sci Rep* 5.

Spaccasassi C, Romano D, Maravita A (2019) Everything is worth when it is close to my body: how spatial proximity and stimulus valence affect visuo-tactile integration. *Acta Psychol* 192:42–51.

Taffou M, Viaud-Delmon I (2014) Cynophobic fear adaptively extends peri-personal space. *Front. Psychiatry* 5. Teneggi C, Canzoneri E, di Pellegrino G, Serino A (2013) Social modulation of peripersonal space boundaries. *Curr Biol* 23:406–411.

Teramoto W (2018) A behavioral approach to shared mapping of peripersonal space between oneself and others. *Sci Rep* 8:5432.

Tessari A, Tsakiris M, Borghi AM, Serino A (2010) The sense of body: a multidisciplinary approach to body representation. *Neuropsychologia* 48:643–644.

Thill S, Caligiore D, Borghi AM, Ziemke T, Baldassarre G (2013) Theories and computational models of affordance and mirror systems: an integrative review. *Neurosci Biobehav Rev* 37:491–521.

Vagnoni E, Lourenco SF, Longo MR (2012) Threat modulates perception of looming visual stimuli. *Curr Biol* 22:R826–R827.

Van Damme S, Gallace A, Spence C, Crombez G, Moseley GL (2009) Does the sight of physical threat induce a tactile processing bias?: Modality-specific attentional facilitation induced by viewing threatening pictures. *Brain Res* 1253:100–106.

Vasconcelos N, Pantoja J, Belchior H, Caixeta FV, Faber J, Freire MAM, Cota VR, Anibal de Macedo E, Laplagne DA, Gomes HM, Ribeiro S (2011) Cross-modal responses in the primary visual cortex encode complex objects and correlate with tactile discrimination. *Proc. Natl. Acad. Sci. U.S.A.* 108:15408–15413.

Vuilleumier P, Valenza N, Mayer E, Reverdin A, Landis T (1998) Near and far visual space in unilateral neglect. *Ann Neurol* 43:406–410.

Wabnegger A, Leutgeb V, Schienle A (2016) Differential amygdala activation during simulated personal space intrusion by men and women. *Neuroscience* 330:12–16.

Wallace MT, Ramachandran R, Stein BE (2004) A revised view of sensory cortical parcellation. *PNAS* 101:2167–2172.

Wardak C, Guipponi O, Pinède S, Hamed SB (2016) Tactile representation of the head and shoulders assessed by fMRI in the nonhuman primate. *J. Neurophysiol* 115:80–91.

Weiss PH, Marshall JC, Wunderlich G, Tellmann L, Halligan PW, Freund HJ, Zilles K, Fink GR (2000) Neural consequences of acting in near versus far space: a physiological basis for clinical dissociations. *Brain* 123 Part 12:2531–2541.

5

Visuo-tactile predictive mechanisms of peripersonal space

H.C. Dijkerman and W.P. Medendorp

5.1 Introduction

In our interactions with the environment, we often make physical contact with objects and people. We use tools, shake hands, and can be comforted by the touch of a friend. However, bodily contacts also pose risks—for example, when objects are sharp, or people have bad intentions. Because bodily integrity is so critical for survival, it is essential that we can predict the consequences of potential bodily contact in our interactions. Comprehensive evidence shows a multisensory representation of the area immediately surrounding our body, termed the 'peripersonal' space (di Pellegrino & Làdavas, 2015), or field (Bufacchi & Iannetti, 2018). Here we suggest that cross-modal prediction is an important aspect of multimodal peripersonal space, a feed-forward mechanism to anticipate the tactile consequences of bodily contact with an external object. A predictive multisensory mechanism would have considerable evolutionary value, because it would allow the observer to anticipate possible bodily consequences of contact with the external stimulus and programme appropriate actions and responses—for example, to avoid bodily harm. In this chapter, we will discuss literature from neuroscience, neuropsychology, and cognitive science in support of such a predictive mechanism and indicate some directions for future research. While we discuss mainly visuo-tactile prediction for peripersonal space coding, it should be noted that input from other sensory modalities, such as auditory input, may also provide information that is important for predicting the bodily consequences of contact with an external stimulus. We therefore anticipate that similar principles apply to audio-tactile prediction. Indeed, several studies have used auditory stimuli to probe peripersonal space mechanisms and examples of these will be discussed in the relevant sections as well.

5.2 Neural basis of visuo-tactile coding

Neurophysiological studies have described neurons in posterior parietal (area 7a, VIP) and (dorsal) premotor areas with bimodal receptive fields, anchored to head or body, responding to tactile stimuli on the skin and to visual stimuli nearby (Bremmer et al., 2001; Duhamel, Colby, & Goldberg, 1998; Graziano & Cooke, 2006; Rizzolatti et al., 1981) (see Figure 5.1). Furthermore, these bimodal neurons respond in a nonlinear fashion to inputs from both modalities (super- or sub-additive), suggesting they integrate both sources of information (Avillac, Ben Hamed, & Duhamel, 2007; Bernasconi et al., 2018; Gentile, Petkova, & Ehrsson, 2011). In visual terms, the neurons are sensitive to visual motion (approaching or

H.C. Dijkerman and W.P. Medendorp, *Visuo-tactile predictive mechanisms of peripersonal space* In: *The World at Our Fingertips.*
Edited by Frédérique de Vignemont, Andrea Serino, Hong Yu Wong, and Alessandro Farnè, Oxford University Press (2021).
© Oxford University Press. DOI: 10.1093/oso/9780198851738.003.0005.

(a)

VIP   PZ

(b)

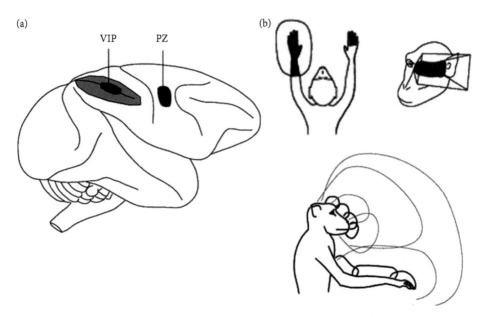

Figure 5.1 a) Premotor and posterior parietal areas involved in multimodal coding of peripersonal space (PPS). b) Examples of visual (bottom) and tactile (top) receptive fields in parietal area PZ.

Reproduced from Graziano, M. S., & Cooke, D. F. (2006). Parieto-frontal interactions, personal space, and defensive behavior. Neuropsychologia, 44(6), 845–859. http://doi.org/S0028-3932(05)00313-1 [pii]10.1016/j.neu ropsychologia.2005.09.009

moving away) (Bremmer et al., 2001; Duhamel et al., 1998), particularly to looming objects (Cléry et al., 2017; Graziano & Cooke, 2006) and optic flow during self-motion (Bremmer et al., 2002; Graziano & Cooke, 2006). From a motor perspective, they have been linked both to actions such as reaching and grasping (Brozzoli et al., 2010; Rizzolatti et al., 1981) as well as to creating a safety zone around the body, thereby coding for defensive actions (Cooke & Graziano, 2004; Graziano, 2009; Graziano & Cooke, 2006).

In humans, functional imaging has revealed activity in the parieto-frontal cortex and the lateral occipital complex (LOC) for stimuli approaching the hand in near compared with far space (Brozzoli et al., 2011; Makin, Holmes, & Zohary, 2007) (see Figure 5.2). Brozzoli, Gentile, & Ehrsson (2012) used the illusion of gaining ownership over a rubber hand to study remapping of peripersonal space. They observed that premotor cortex was associated with remapping of peripersonal space, whereas posterior parietal cortex (PPC) was linked to position sense. Furthermore, Ferri et al. (2015) reported that individual variation in peripersonal space boundaries was related to intertrial variability in premotor cortex.

For stimuli near the head, Sereno and Huang (2006) observed that the superior part of the post-central gyrus contains a tactile map of the face aligned with a near-face visual map. This observation was further extended to a visual-tactile representation of the entire body (Huang et al., 2012).

Transcranial magnetic stimulation (TMS) studies of the motor cortex have shown that approaching auditory and visual stimuli near the body modulate corticospinal excitation, providing a link with motor control (Finisguerra et al., 2015; Makin et al., 2009; Serino, Annella, & Avenanti, 2009). In support, a combined transcranial direct-current stimulation (tDCS)-TMS study suggests that inhibitory cathodal premotor, but not posterior parietal

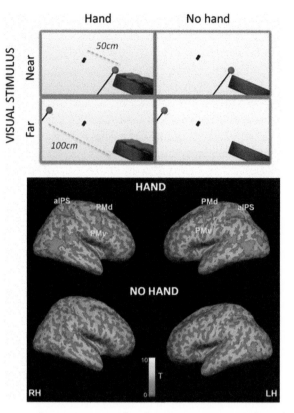

Figure 5.2 Involvement of posterior parietal and premotor areas in coding objects near the hand. An adaptation paradigm was used with presenting visual stimuli repeatedly in two conditions, with the hand retracted (bottom) or with the hand extended (top), in near and far space.

Reproduced with kind permission from Brozzoli, C., Gentile, G., Petkova, V. I., & Ehrsson, H. H. (2011). FMRI adaptation reveals a cortical mechanism for the coding of space near the hand. The Journal of Neuroscience : The Official Journal of the Society for Neuroscience, 31(24), 9023–31. Copyright © 2011, Claudio Brozzoli, Giovanni Gentile, Valeria I. Petkova, and H. Henrik Ehrsso.

tDCS, abolishes this modulatory effect of approaching stimuli on corticospinal activation (Avenanti, Annela, & Serino, 2012).

5.3 Sensory remapping

One complexity of these visuo-tactile predictions relates to the underlying frames of references. While visual information of an object arrives in eye-centred (retinotopic) coordinates, preparing which part of the body is intruded by tactile input requires a prediction in a body-centred, somatotopic reference frame. In other words, an object that generates the same retinal input could impact different parts of the body, depending on the eye, head, and body position. How are these coordinate transformations serving sensory remapping implemented by the predictive mechanisms in the brain? Single-unit recordings have shown that the processing in PPC involves a mixture of reference frames, including eye-, head-, body-centred and intermediate coordinates (Avillac et al., 2005; Chang & Snyder, 2010; McGuire & Sabes,

2009). Modelling work suggests that this variability reflects the distributed use of different reference frames and gains modulations in multilayer networks (Blohm & Crawford, 2009; Pouget, Deneve, & Duhamel, 2002; Testolin, De Filippo De Grazia, & Zorzi, 2017). More specifically, such a neural architecture provides PPC with a mechanism to implicitly create multiple modes of representation at the population level, with predictions from one format to all others. With this computational perspective, the brain could generate spatially veridical predictions from one sensory modality to another, encoding visual targets in somatotopic coordinates, and vice versa. Indeed, previous imaging studies have also provided evidence for the automatic remapping of sensory inputs, mediated by PPC (Azañón et al., 2010; Badde, Röder, & Heed, 2014; Macaluso, Frith, & Driver, 2002; Buchholz et al. 2013).

It has further been suggested that this sensory remapping could be routed by neuronal oscillations, which could transiently couple retinotopic and somatopic maps by high-frequency synchronized oscillations (Buchholz, Jensen, & Medendorp, 2011, 2013; Ruzzoli & Soto-Faraco, 2014; Thut, Miniussi, & Gross, 2012; van Atteveldt et al., 2014).

Because this coordination of firing patterns can change on small timescales, it provides a flexible mechanism for functional connectivity in large-scale networks (Fries, 2005). While high-frequency oscillations are thought to provide the spectral channels for communication, low-frequency rhythms in alpha and beta bands have been associated with changes in cortical excitability (Buchholz, Jensen, & Medendorp, 2014), which can be observed in anticipation of a stimulus (van Ede et al., 2011). Ruzzoli and Soto-Faraco (2014) demonstrated a causal link between attention-related parietal alpha oscillatory activity and the external spatial coding of touch. Furthermore, beta band activity has been found to be suppressed in somatopic coordinates in central and parietal regions during anticipation of a tactile stimulus, thereby serving as a mechanism for tactile prediction and sensory gating (Bauer et al., 2006; Buchholz et al., 2014; van Ede et al., 2011).

Other support for predictive sensory remapping in time comes from audiovisual looming experiments. For example, Maier et al. (2008) showed, in monkeys, that audiovisual looming signals elicit increased gamma-band coherence between auditory cortex and the superior temporal sulcus, two areas involved in integrating behaviourally relevant auditory-visual signals.

5.4 Functional aspects of visuo-tactile processing

There are a number of behavioural and clinical manifestations that reveal the functional implications of these visuo-tactile maps. Functional implications may involve an influence of visual (or auditory) nearby external stimuli on processing of bodily stimuli (tactile or nociceptive). For example, cross-modal attention studies report perceptual facilitation when both an endogenous and an exogenous visual cue near the hand is followed by a tactile stimulus on the hand (Driver & Spence, 1998; Eimer, van Velzen, & Driver, 2002; Spence, Pavani, & Driver, 2000). Clinical support for these observations is found in studies reporting about cross-modal extinction (Làdavas & Farnè, 2004). Cross-modal attention effects are further enhanced when the visual cues are linked to pain stimuli and other threats (Van Damme, Crombez, & Eccleston, 2004; Van Damme & Legrain, 2012). This enhanced attention to pain is larger in peripersonal compared with extrapersonal space (De Paepe, Crombez, & Legrain, 2016; De Paepe et al, 2014). Indeed, a single case study reported that damage to parietal areas resulted in pathological pain when an object approached the affected hand (Hoogenraad, Ramos, & van Gijn, 1994).

Moreover, the nature of the visual stimulus may also affect the distance at which this occurs (e.g. the peripersonal space boundary). Various studies suggest that visual social cues influence visuo-tactile representations of peripersonal space. For example, a study by Heed et al. (2010) showed that performance on a visuo-tactile cross-modal congruency task is influenced by the presence of a partner, but only when he was situated near the participant and performed the same task. Similarly, Teneggi et al. (2013) showed that the boundary of peripersonal space is influenced by the presence of another person and whether this person behaves cooperatively or competitively. Maister et al. (2014) showed that sharing a sensory experience (seeing another face being touched synchronously with the own face being touched) results in a remapping of the other person's peripersonal space onto the participant's own space. Indeed, increasing peripersonal space through tool use also resulted in an enlarged interpersonal distance (Quesque et al., 2016). Overall, this suggests a direct link between social contextual factors and peripersonal space.

5.5 Visuo-tactile prediction in peripersonal space

It may be clear from the review so far that peripersonal visuo-tactile representations play a role in a wide range of functional domains. The question is whether these functional domains involve one and the same type of mechanism or different ones. Here we suggest that these different functions exploit the predictive value afforded by a visual stimulus presented close to the observer—namely, the occurrence of an impending bodily sensory stimulus. Furthermore, we suggest that visuo-tactile prediction has several additional characteristics (see Figure 5.3). 1) Multimodal peripersonal space representation is achieved through repeated spatiotemporal coupling of visual and tactile events. That is, when nearby visual and tactile stimuli frequently occur together, they are bound into a joint representation. 2) The distance at which visual stimuli influence bodily processing depends on the predicted sensory consequences. When the consequences are anticipated to be negative, peripersonal space will increase. 3) If predicting sensory consequences of the visual stimulus close to the observer is a central mechanism, being able to recognize its features, and its affective and social identity will influence the extent of peripersonal space.

It is important to state that we do not claim that peripersonal space only involves one type of predictive mechanism. We do however propose that visuo-tactile prediction is an important principle for multisensory peripersonal space coding. This may involve different mechanisms depending on different types of input. For example, we may have spatial or temporal prediction concerning the upcoming tactile stimulus, but we may also have prediction of the features of tactile stimulus (sharp, blunt, hard, soft), or even its affective value (pleasant, unpleasant), all based on different aspects of the visual stimulus.

It is clear that prediction is an important general neurocognitive mechanism subserving a wide range of functions, from motor-to-sensory predictions (Clark, 2013; den Ouden, Kok, & de Lange, 2012; Friston, Kilner, & Harrison, 2006; Wolpert & Flanagan, 2001; Shadmehr, Smith, & Krakauer, 2010; Atsma et al., 2016) to perception (Summerfield & de Lange, 2014) and bodily self-consciousness (Apps & Tsakiris, 2013). Indeed, one of the first papers on multisensory neurons was already suggestive of this idea. Hyvärinen & Poranen (1974) suggested 'anticipatory activation' in parietal neurones: visual activation that appeared before the neuron's tactile RF was touched (p. 675). Here we propose that this mechanism to evaluate the tactile consequences of visual signals allows selection of appropriate responses to the visual stimulus (e.g. approach or avoid it) (Brozzoli et al., 2010; Graziano, 2009;

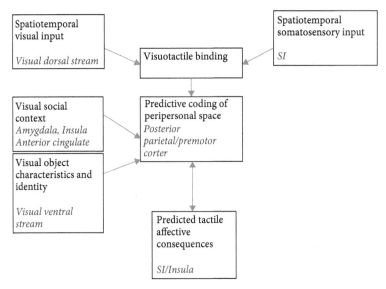

Figure 5.3 An overview of the proposed visuo-tactile predictive system for coding peripersonal space (PPS). Central is a visuo-tactile representation of PPS in parieto-premotor areas. This develops through visuo-tactile spatiotemporal binding and is modulated by visual information about object characteristics involving ventral stream processing and visual social and affective cues (left). The visuo-tactile representation allows prediction of the *bodily* consequences of contact with the visual object/person (bottom), which is used to modulate PPS.

Graziano & Cooke, 2006) at an early stage, thereby allowing the observer to respond fast and effectively.

Considerable evidence for this visuo-tactile prediction already exists (Carlsson et al., 2000; Cléry et al., 2015; Kandula, Hofman, & Dijkerman, 2014; Liu, Denton, & Nelson, 2007). In an early study, Carlsson et al. (2000) observed that anticipated tickling activated a somatosensory network including contralateral primary somatosensory cortex. In two behavioural studies, Gray & Tan (2002) and Kandula et al. (2014) reported that moving visual stimuli facilitated responses to a tactile target if they occurred at the location and time predicted by a moving visual stimulus. Indeed, Cléry et al. (2015) showed that tactile sensitivity is enhanced at the time and location of contact with an approaching visual stimulus on the face. Similarly, Voudouris & Fiehler (2017) showed enhanced tactile sensitivity on the left hand during reaches towards that hand with the right hand. Huang, Chen, & Sereno, (2018) used visually looming stimuli in combination with airpuffs on the face to study underlying neural processing. They reported that, when visual and tactile stimuli were perceived to be in sync (e.g. the timing of the tactile stimulus followed logically on the looming visual stimulus) enhanced neural processing was observed in areas such as MT+, V6A, 7b, SII, LIP+, FEF and aPCu (anterior part of the precuneus). Thus, when the tactile stimulus occurred at the predicted time of impact based on the looming visual stimulus, neural processing was enhanced in several cortical areas.

Indeed, this spatial and temporal impact prediction appears to be linked to multisensory integration mechanisms. Cléry et al. (2017) showed that the neural network involved in multisensory integration is also more active when the location and timing of tactile stimuli are predicted by approaching visual stimuli. Moreover, these brain areas show nonlinear multisensory responses (mainly super-additivity) suggesting multisensory integration.

Further evidence for multisensory integration in peripersonal space comes from a study by Kandula (2020). She used approaching visual stimuli at different distances in isolation or together with tactile stimuli on the hand. Participants had to respond to either the onset of the movement of the visual stimulus or to the tactile stimulus on the hand in unimodal or combined visuo-tactile conditions. Response times in the combined visuo-tactile conditions showed race model violations when the visual stimulus was presented within peripersonal space (at 30 and 50 cm from the hand), but not when further away from the body (70 and 80 cm). This suggests that multisensory integration levels off, or disappears, the further the stimuli stimulus is presented from the body.

Another study investigated the temporal course of neural processing of multisensory integration in peripersonal space (Bernasconi et al., 2018). Using an audiotactile paradigm in combination with intracranial EEG recordings in epilepsy patients, they showed that that 19% of the electrodes exhibited multisensory integration, while 30% of the electrodes were sensitive to the distance between sound and body (e.g. peripersonal space effect). Both multisensory and peripersonal space responses had similar spatiotemporal characteristics. An early response was found in insular cortex (50 ms), followed by later responses in the pre- and post-central gyrus (200 ms). The authors concluded that peripersonal space representation is based on a spatial modulation of multisensory integration.

While it is clear that multisensory integration plays a role in PPS representation, how it relates to visuo-tactile prediction has not yet been established. Here we suggest that perhaps super-additivity (see Cléry et al., 2017) might be a multisensory integration mechanism consistent with visuo-tactile prediction. Thus, a nearby visual stimulus, in combination with a tactile stimulus, may result in an enhanced neural response to the tactile stimulus.

5.6 Development of peripersonal space prediction

An important question is how this visuo-tactile predictive system develops. There are some indications that newborns can distinguish between visual stimuli approaching their body and those moving towards another area of space, with a preference for stimuli in near space (Orioli et al., Filippetti, Gerbino, Dragovic, & Farroni, 2018). However, knowing how nearby stimuli impact on the body presumably involves learning in the first few years of life. An example of how this could be done comes from a robotics study (Roncone et al., 2015). Based on child development literature, they suggest that learning visuo-tactile associations involves several stages. First, self-touch (or double touch) results in concurrent tactile-motor activation, allowing development of correspondence between these two signals. By including proprioceptive signals as well, invariance with respect to the position of the limb is achieved. Second, by adding vision about the limb, the visual input could be linked to the tactile consequences of the moving arm that touches another body part. By adding proprioceptive input from the head and neck, transformation from eye to body part centred reference frames is achieved. Finally, the tactile-visuo-proprioceptive association learned in the second stage is generalized and applied to external objects approaching the body. This is achieved by linking an area of artificial skin on the robot to the visual space extending from this area. For every object entering this visual space, the distance is recorded and also whether it eventually touched this particular part of the artificial skin. Ultimately, this results in a set of probabilities that are updated and carry information about the likelihood of stimuli in the environment contacting a particular area of the artificial skin. The authors then linked this representation with an avoidance controller that triggered it, allowing the robot to maintain a margin of safety around the body. By reversing the sign in the controller, it could also be used for performing reaching movements to objects near the robot. The study of Roncone et al. (2015) showed that this approach allowed an ICub robot to avoid or reach to objects. An important question that remains, however, is whether serial stages are required, or whether this all develops at the same time.

Another possibility is that these visuo-tactile predictions may depend on low-level associative learning mechanisms. In terms of learning, a visual object near the body is often followed by a tactile impact of that object, resulting in binding of the two stimuli. As a consequence, a visual stimulus near the body will activate a prediction of the likely tactile consequences, based on the output of an internal model. Similar cross-modal learning-induced expectations have been found for auditory and visual stimuli in the infant brain (Emberson, Richards, & Aslin, 2015).

5.7 Visual contextual factors

If predicting sensory consequences is a central mechanism, factors other than spatio-temporal prediction—for example, recognition of our visual environment—must be important as well. Visual processing may lead to identification of the several relevant object features such as its surface texture (soft or rough), shape, or hardness. This may be used not only to predict the likely tactile consequences when touching the object (e.g. for programming the appropriate grip force when grasping the object (Gordon et al., 1991), but also to maintain a sufficient safety margin when reaching past the object (de Haan et al., 2014; Menger et al., 2012; Tresilian, 1998). Furthermore, it may allow identification of threatening approaching objects or animals such as barking dogs or spiders. Indeed, several studies have shown that such threatening approaching stimuli in peripersonal space result in faster response times to subsequent tactile stimuli (de Haan et al., 2016; Taffou & Viaud-Delmon, 2014) Likewise, visual identification of the other person may influence the tactile processes during social interaction. One could hypothesize that predicting the tactile consequences of a touch by another person may be an important mechanism for maintaining an appropriate distance to this person. Thus, when you anticipate unpleasant bodily consequences, you may maintain a larger interpersonal distance than when you anticipate neutral tactile experiences. Various social cues have been shown to influence peripersonal space boundaries including social cooperation with a confederate (Teneggi et al., 2013) and perceived morality of the other (Pellencin et al., 2018). Interestingly, a recent study by Cartaud et al., (2018) suggests that peripersonal (action) space and interpersonal space are similarly sensitive to emotional expressions of another person suggesting a common underlying mechanism.

With respect to the neural processes underlying social and affective influences on peripersonal visuo-tactile mechanisms, visual processing of emotion-evoking stimuli has received considerable attention in recent years. It has been linked closely to bodily experience—for example, in the context of complex decision making (Damasio, 1996) or visual perceptual recognition (Barrett & Bar, 2009; Damaraju et al., 2009). These affective processes not only depend on interoceptive input, but also on stimuli that can be applied to the body surface, such as pain or pleasant stroking (Craig, 2002; Loken et al. 2009; van Stralen et al., 2014). Here we suggest that the link between visual processing of social and affective information and bodily experience is not only used for complex decision making or visual recognition, but also for modulating peripersonal space. At a neural level, this involves modulation of parieto-frontal multimodal representations of peripersonal space by the affective/social neural network (Lloyd, Morrison, & Roberts, 2006). This modulation may involve both implicit processing of visual stimuli through a fast subcortical route including the amygdala and the periadquaductal grey (for threat) (Zald, 2003; Mobbs et al., 2007) and cerebellum (for positive emotions) (Schutter, Enter, & Hoppenbrouwers, 2009) as well as higher level cortical mechanisms. Indeed, amygdala damage leads to impaired interpersonal spatial regulation (Kennedy et al., 2009). Furthermore, a recent neuroimaging study reported that peripersonal space intrusions by approaching social (faces) and non-social (animals) stimuli activated the fronto-parietal peripersonal space network in addition to the midbrain threat areas (periaquaductal grey) (Vieira, Pierzchajlo, & Mitchell, 2019). Further evidence for the role of subcortical mechanisms in peripersonal space representation comes from a series of studies on the hand blink reflex. Sambo, Liang et al. (2012) showed that stimulation of the median nerve when the hand is near the face, in peripersonal space, results in an eye blink reflex. This reflex is mediated by brain stem mechanisms and modulated by top-down influences, such as cognitive expectation (Sambo, Forster et al., 2012), anxiety (Sambo & Iannetti, 2013), interpersonal interactions, and empathy (Fossataro et al., 2016).

Evidence for the cortical route comes from studies that show involvement of visual ventral, posterior parietal and primary somatosensory areas when perceiving threatening stimuli (Adolphs et al., 2000). In particular, the superior parietal cortex is active in the presence of a visually threatening person in peripersonal space (Lloyd & Morrison, 2008).

5.8 The role of predicted bodily consequences

It is well established that peripersonal space boundaries are dynamic and influenced by a range of variables—for example, the use of tools (Farne & Ladavas, 2000) and the velocity of the approaching stimulus (Fogassi et al., 1996; Kandula, 2020; Noel et al., 2018). In this section, we explore another variable, which we consider important: the role of the bodily consequences. We suggest that the acuity and the emotional valence of predicted tactile consequences influence the extent of peripersonal space. When tactile acuity is enhanced, often distances between tactile stimuli are perceived as larger (e.g. an identical Cartesian distance between two tactile stimuli is judged to be larger on the hand than on the forearm, referred to as 'Weber's illusion' (Anema, Wolswijk, Ruis, & Dijkerman, 2008; Taylor-Clarke, Jacobsen, & Haggard, 2004), and possibly resulting in an increased peripersonal space. This hypothesis predicts variations in the extent of peripersonal space, depending on tactile spatial acuity, which varies widely between different body parts (Weinstein, 1968). It may also contribute to our understanding of peripersonal spatial abnormalities in a range of clinical populations. That is, if somatosensory consequences of contact with the nearby stimulus are reduced, or even absent, then this may have consequences for how stimuli in the area surrounding the impaired body part are processed. For example, Makin, Wilf, Schwartz, & Zohary (2010) observed that peripersonal space is somewhat neglected around the amputated body part in amputees. Furthermore, hemispatial neglect and somatosensory deficits often co-occur and have been linked functionally (Meyer et al., 2016; Vallar, 1997), and this would be consistent with the idea of a contralesional tactile deficit resulting in impaired representation of peripersonal space on that side. Thus, a lack of tactile input on the contralesional side might result in a reduced relevance, and therefore salience of visual and auditory stimuli on that side as the somatosensory consequences of contact are reduced or even absent. While we do not claim that neglect is solely caused by somatosensory impairments (indeed, patients can have neglect without somatosensory deficits and vice versa) (Meyer et al., 2016), we suggest that somatosensory impairments may exacerbate neglect because of a lack of visuo-tactile prediction on that side. Furthermore, neglect may also worsen tactile impairments as a lack of awareness of visual and auditory stimuli on the neglected side results in a reduction in predicted tactile input and therefore more severe somatosensory deficits. Evidence for this comes from studies showing that reducing neglect using caloric stimulation or prism adaptation can also result in improved somatosensory function (Bottini et al., 2005; Dijkerman, Webeling, Ter Wal, Groet, & Van Zandvoort, 2004; Facchin, Sartori, Luisetti, De Galeazzi, & Beschin, 2019).

Regarding emotional valence, when the consequences are anticipated to be negative (e.g. painful, or harmful), peripersonal space may increase (de Haan et al., 2016; De Paepe et al., 2016; Poliakoff, Miles, Li, & Blanchette, 2007). The effect of positive affective stimuli on peripersonal space can be less easily explained. It is possible that these stimuli result in a reduced peripersonal space representation because the predicted consequences are pleasant (e.g. affective touch), or that affective value of the stimulus makes it more relevant to the observer, thus resulting in an increased peripersonal space. A recent study provides evidence for the latter (Spaccasassi, Romano, & Maravita, 2019).

At a neural level, changing the tactile input, or its affective component, must influence cortical representations of peripersonal space in posterior parietal and premotor cortices (see also Figure 5.3). While the first may involve direct connections between anterior and posterior parietal cortices, the latter may additionally rely on processing of pain and pleasant stimulation in affective areas such as the insular cortex (Craig, 2002; Olausson et al., 2002), which is also known to have extensive connections with the PPC (Dijkerman & de Haan, 2007).

5.9 Comparison with other models of peripersonal space

Several other authors have suggested different mechanisms for peripersonal space and it might be useful to discuss how the idea of visuo-tactile prediction as a central mechanism might compare with these models. De Vignemont & Iannetti (2015) suggested a dual model of peripersonal space, with a functional distinction between whether the space is used for bodily protection or for goal-directed action. They illustrated this by showing that anxiety expands protective space, while it reduces working space. Furthermore, tool use may expand working space, but not necessarily protective space. While the idea of two spaces, a protective and a working space, seems at odds with a single predictive visuo-tactile mechanism, this may not necessarily be the case. Indeed, de Vignemont & Iannetti (2015) themselves suggest that there may be two ways in which *the expectation created by visual information* influences somatosensory processing. Either a visual stimulus moves towards the observer, or the observer moves towards the external stimulus. The former might be more relevant for protective space, while the latter is linked to interacting with the object—namely, working space. Thus, the general principle of using visual input for predicting somatosensory consequences can be used for both working and defensive peripersonal space. De Vignemont and Iannetti (2015) correctly suggested that processing of visual input for defensive purposes needs to occur faster, with less need for precise detail, and indeed may involve subcortical routes (Makin et al., 2012).

In their recent review, Noel, Blanke, & Serino (2018) proposed to go beyond classical multisensory integration principles and use modern computational understandings of multisensory integration to provide a mechanistic understanding of bodily self-consciousness. Crucially, they describe peripersonal space as a stochastic body space. Peripersonal space here is linked to the likelihood that an object will come into contact with the body, which could be different for different body parts. In other words, different body parts have different likelihood gradients for the multisensory processes that occur. Overall, this is entirely consistent with the idea of visual input predicting the possible consequences of contact. Perhaps what is missing here is the valence of the contact—for example, whether it is pleasant or harmful. Moreover, the link with possible actions has not been defined. The latter is central in another recent review. Bufacchi & Iannetti (2018) suggested that there is not one peripersonal space with a sharp boundary, but rather it contains 'a series of continuous fields that describe the behavioural relevance of action aimed at creating or avoiding contact between objects and the body' (p. 1076). Overall, this idea is consistent with a mechanism in which visual input is used to predict possible consequence of contact with the body in order to respond flexibly. A few aspects perhaps differ from our proposal. First, Bufacchi & Iannetti (2018) emphasize the action component, specifying how the possible actions that the observer can perform shape peripersonal space. As a result, this aspect, while mentioned in the present chapter in relation to visuo-tactile prediction, is better specified in the review of Bufacchi & Iannetti (2018). Second, they propose that peripersonal space is a gradual field rather than a space with a fixed and sharp boundary. This fits well with the idea of the stochastic space of Noel et al. (2018) and might allow further specification of contact prediction. Third, our proposal suggests that the type of predicted somatosensory stimulus is relevant for peripersonal space coding (e.g. is it innocuous, pleasant or harmful). This aspect has not been explicitly described by either Noel et al. (2018) or Bufacchi & Iannetti (2018), but is in our opinion crucial for the selection of an appropriate action.

Overall, while the proposed mechanism for peripersonal space as described in this chapter is not necessarily inconsistent with other models, there are differences in what would be considered to be central to peripersonal space. Thus, with respect to the notion of de Vignemont & Iannetti (2015), one could test whether both reaching and defensive peripersonal space involves visuo-tactile prediction. Most studies testing visuo-tactile prediction involve dynamic visual stimuli coming towards the observer, something de Vignemont and Iannetti would consider to activate defensive peripersonal space. Perhaps a reverse paradigm, involving the timing of tactile stimuli depending on the velocity of the approaching hand, could be used to test visuo-tactile prediction in reaching peripersonal space. If there are no effects, this would discount the role of visuo-tactile prediction as a general mechanism. Similarly, with respect to the stochastic body space proposal of Noel, Blanke and Serino (2018), one could test this against the current proposal by manipulating the likelihood of bodily contact and comparing this with the role of the valence of the somatosensory stimuli. If peripersonal space is just dependent on the likelihood, then changing the valence (making it pleasant or painful) should not have any effect on peripersonal space boundaries. This can also be used to distinguish our proposal of visuo-tactile prediction from the idea of Bufacchi & Iannetti (2018) on contact-related behavioural relevance. Moreover, to further test the proposal of Bufacchi & Iannetti (2018) against our ideas, one could assess whether peripersonal space fields are more influenced by the range of action possibilities available to the observer or the valence of the tactile stimulus predicted by the visual stimulus.

To summarize, we suggest that the predictive value afforded by a visual stimulus presented close to the observer—namely, the occurrence of an impending bodily sensory stimulus—is a central mechanism for constructing peripersonal space, and we have discussed this in light of the current literature on peripersonal space and compared it with several other models. We suggest that a visuo-tactile predictive mechanism has several additional characteristics.

First, a representation of peripersonal space is achieved by binding visual and tactile stimuli together as they occur frequently in close spatial and temporal proximity. Second, the distance at which visual stimuli influence bodily processing depends on the predicted sensory consequences. When the consequences are anticipated to be negative, peripersonal space will increase. Finally, if predicting sensory consequences of the visual stimulus close to the observer is a central mechanism, being able to recognize its features, and its affective and social identity, will influence the extent of peripersonal space.

5.10 Conclusions and future directions

In this review, we suggest that visuo-tactile prediction is a central mechanism to peripersonal space representation. This one mechanism may explain the role of peripersonal spatial coding in a range of functional domains, including visuomotor control, attention, and social cognition. While there is ample evidence, as reviewed here and elsewhere (Cardinali, Brozzoli, & Farnè, 2009; Làdavas & Farne, 2004; Làdavas & Serino, 2008; Macaluso & Maravita, 2010; Spence et al., 2004) for the role of visuo-tactile representations in peripersonal space, several aspects of the visuo-tactile prediction hypothesis require further investigation. The idea that the development of visuo-tactile coding depends on developing an internal model through associative learning should be tested, including spatial and temporal constraints as well as contingency rules. Associative learning principles may also lead to investigations of whether newly learned visuo-tactile associations modulate peripersonal space. Furthermore, while it is clear that visual object identity information and visual social cues influence visuo-tactile coding in peripersonal space, it remains to be tested whether these cues actually contribute to prediction of bodily consequences. It could be possible that they mainly influence motor responses to nearby targets without the need for prediction of tactile or painful consequences.

Another aspect of the proposed hypothesis that is tentative concerns the neural basis of the visuo-tactile prediction. The studies reviewed here show that the visuo-tactile coding itself is based on a fronto-parietal network involving the PPC and the premotor area. This fronto-parietal system may be modulated by visual identity information from the ventral stream, while the affective/social neural network (including insula, amygdala, anterior cingulate, and orbitofrontal cortex) may provide modulation in a social context (see also Figure 5.3). Again, the evidence for this is circumstantial and a direct test of this proposed neural organization remains to be performed. For example, transient disturbance using TMS of ventral stream processing should affect parieto-frontal visuo-tactile processing when the identity of the visual stimulus in peripersonal space is relevant (see Verhagen et al., 2012) for a similar approach for visuomotor grasping).

Finally, the proposed visuo-tactile predictive coding of peripersonal space may provide some further pointers for understanding certain aspects of some clinical disorders. This may not only be the case for hemispatial neglect, in which somatosensory deficits and visual inattention of one side of space often co-occur (Vallar, 2007). Also, in autism spectrum disorders or schizophrenia, abnormalities in cross-modal binding (Noel, Stevenson, & Wallace, 2018; Stevenson et al., 2014) may be related to difficulties in modulating social interpersonal distance. Combined with insight into the modulation of visuo-tactile coding in peripersonal space, understanding how visuo-tactile predictive coding is disturbed in these patient groups may also provide some clues for a possible remediation of these deficits.

Acknowledgements

HCD was supported by an Netherlands Organization for Scientific Research (NWO) Vici grant (453-10-003). WPM was supported by the European Research Council (EU-ERC-283567) and by an NWO vici grant (453-11-001).

References

Adolphs, R., Damasio, H., Tranel, D., Cooper, G., & Damasio, A. R. (2000). A role for somatosensory cortices in the visual recognition of emotion as revealed by three-dimensional lesion mapping. *J. Neurosci, 20*(7), 2683–2690.

Anema, H. A., Wolswijk, V. W. J., Ruis, C., & Dijkerman, H. C. (2008). Grasping Weber's illusion: the effect of receptor density differences on grasping and matching. *Cogn Neuropsychol, 25*(7–8). doi:org/10.1080/02643290802041323

Apps, M. A. J., & Tsakiris, M. (2013). The free-energy self: a predictive coding account of self-recognition. *Neurosc Biobehav Rev, 41*, 85–97. doi:org/10.1016/j.neubiorev.2013.01.029

Atsma, J., Maij, F., Koppen, M., Irwin, D. E., & Medendorp, W. P. (2016). Causal inference for spatial constancy across saccades. *PLoS Comput Biol, 12*(3), e1004766. doi:10.1371/journal.pcbi.1004766

Avenanti, A., Annela, L., & Serino, A. (2012). Suppression of premotor cortex disrupts motor coding of peripersonal space. *NeuroImage, 63*(1), 281–288. doi:10.1016/j.neuroimage.2012.06.063

Avillac, M., Ben Hamed, S., & Duhamel, J. R. (2007). Multisensory integration in the ventral intraparietal area of the macaque monkey. *J. Neurosci, 27*(8), 1922–1932. doi:org/10.1523/JNEUROSCI.2646-06.2007

Avillac, M., Denève, S., Olivier, E., Pouget, A., & Duhamel, J.-R. (2005). Reference frames for representing visual and tactile locations in parietal cortex. *Nat Neurosci, 8*(7), 941–99. doi:org/10.1038/nn1480

Azañón, E., Longo, M. R., Soto-Faraco, S., & Haggard, P. (2010). The posterior parietal cortex remaps touch into external space. *Curr Biol, 20*(14), 1304–1309. doi:org/10.1016/j.cub.2010.05.063

Badde, S., Röder, B., & Heed, T. (2014). Flexibly weighted integration of tactile reference frames. *Neuropsychologia, 70*, 367–374. doi:org/10.1016/j.neuropsychologia.2014.10.001

Barrett, L. F., & Bar, M. (2009). See it with feeling: affective predictions during object perception. *Philos Trans R Soc Lond B Biol Sci, 364*(1521), 1325–1334. doi:10.1098/rstb.2008.0312

Bauer, M., Oostenveld, R., Peeters, M., & Fries, P. (2006). Tactile spatial attention enhances gamma-band activity in somatosensory cortex and reduces low-frequency activity in parieto-occipital areas. *J. Neurosci, 26*(2), 490–501. doi:org/10.1523/JNEUROSCI.5228-04.2006

Bernasconi, F., Noel, J.-P., Park, H. D., Faivre, N., Seeck, M., Spinelli, L., ... Serino, A. (2018). Audio-tactile and peripersonal space processing around the trunk in human parietal and temporal cortex: an intracranial EEG study. *Cereb Cortex*, (September), 3385–3397. doi:org/10.1093/cercor/bhy156

Blohm, G., & Crawford, J. D. (2009). Fields of gain in the brain. *Neuron, 64*(5), 598–600. doi:org/10.1016/J.NEURON.2009.11.022

Bottini, G., Paulesu, E., Gandola, M., Loffredo, S., Scarpa, P., Sterzi, R., ... Vallar, G. (2005). Left caloric vestibular stimulation ameliorates right hemianesthesia. *Neurology, 65*, 1278–1283. doi:10.1212/01.wnl.0000182398.14088.e8

Bremmer, F., Duhamel, J. R., Ben Hamed, S., & Graf, W. (2002). Heading encoding in the macaque ventral intraparietal area (VIP). *Eur. J. Neurosci*, *16*(8), 1554–1568. doi:10.1046/j.1460-9568.2002.02207.x

Bremmer, F., Schlack, A., Duhamel, J. R., Graf, W., & Fink, G. R. (2001). Space coding in primate posterior parietal cortex. *NeuroImage*, *14*(1, Pt 2), S46–51. doi:10.1006/nimg.2001.0817

Brozzoli, C., Cardinali, L., Pavani, F., & Farnè, A., (2010). Action-specific remapping of peripersonal space. *Neuropsychologia*, *48*(3), 796–802. doi:10.1016/j.neuropsychologia.2009.10.009

Brozzoli, C., Gentile, G., & Ehrsson, H. H. (2012). That's near my hand! Parietal and premotor coding of hand-centered space contributes to localization and self-attribution of the hand. *J. Neurosci*, *32*(42), 14573–14582. doi:org/10.1523/JNEUROSCI.2660-12.2012

Brozzoli, C., Gentile, G., Petkova, V. I., & Ehrsson, H. H. (2011). FMRI adaptation reveals a cortical mechanism for the coding of space near the hand. *J. Neurosci*, *31*(24), 9023–9031. doi:org/10.1523/JNEUROSCI.1172-11.2011

Buchholz, V. N., Jensen, O., & Medendorp, W. P. (2011). Multiple reference frames in cortical oscillatory activity during tactile remapping for saccades. *J. Neurosci*, *31*(46), 16864–16871. doi:org/10.1523/JNEUROSCI.3404-11.2011

Buchholz, V. N., Jensen, O., & Medendorp, W. P. (2013). Parietal oscillations code nonvisual reach targets relative to gaze and body. *J. Neurosci*, *33*(8), 3492–3499. doi:org/10.1523/JNEUROSCI.3208-12.2013

Buchholz, V. N., Jensen, O., & Medendorp, W. P. (2014). Different roles of alpha and beta band oscillations in anticipatory sensorimotor gating. *Front. Hum. Neurosci*, *8*, 446. doi:org/10.3389/fnhum.2014.00446

Bufacchi, R. J., & Iannetti, G. D. (2018). An action field theory of peripersonal space. *Trends Cogn Sci*, 1–15. doi:org/10.1016/j.tics.2018.09.004

Cardinali, L., Brozzoli, C., & Farnè, A. (2009). Peripersonal space and body schema: two labels for the same concept? *Brain Topogr*, *21*(3–4), 252–60. doi:org/10.1007/s10548-009-0092-7

Carlsson, K., Petrovic, P., Skare, S., Petersson, K. M., & Ingvar, M. (2000). Tickling expectations: neural processing in anticipation of a sensory stimulus. *J. Cogn Neurosci*, *12*(4), 691–703.

Cartaud, A., Ruggiero, G., Ott, L., Iachini, T., & Coello, Y. (2018). Physiological response to facial expressions in peripersonal space determines interpersonal distance in a social interaction context. *Front. Psychol*, *9*, 657. doi:10.3389/fpsyg.2018.00657

Chang, S. W. C., & Snyder, L. H. (2010). Idiosyncratic and systematic aspects of spatial representations in the macaque parietal cortex. *Proc. Natl. Acad. Sci. U.S.A.*, *107*(17), 7951–7956. doi:org/10.1073/pnas.0913209107

Clark, A. (2013). Whatever next? Predictive brains, situated agents, and the future of cognitive science. *Behav. Brain Sci*, *36*(3), 181–204. doi:org/10.1017/S0140525X12000477

Cléry, J., Guipponi, O., Odouard, S., Pinède, S., Wardak, C., & Ben Hamed, S. (2017). The prediction of impact of a looming stimulus onto the body is subserved by multisensory integration mechanisms. *J. Neurosci*, *37*(44), 0610–0617. doi:org/10.1523/JNEUROSCI.0610-17.2017

Cléry, J., Guipponi, O., Odouard, S., Wardak, C., & Ben Hamed, S. (2015). Impact prediction by looming visual stimuli enhances tactile detection. *J. Neurosci*, *35*(10), 4179–4189. doi:org/10.1523/JNEUROSCI.3031-14.2015

Cooke, D. F., & Graziano, M. S. (2004). Sensorimotor integration in the precentral gyrus: polysensory neurons and defensive movements. *J. Neurophysiol*, *91*(4), 1648–1660. doi:10.1152/jn.00955.2003

Craig, A. D. (2002). How do you feel? Interoception: the sense of the physiological condition of the body. *Nat Rev Neurosci, 3*, 655–666.

Damaraju, E., Huang, Y. M., Barrett, L. F., & Pessoa, L. (2009). Affective learning enhances activity and functional connectivity in early visual cortex. *Neuropsychologia, 47*(12), 2480–2487. doi:10.1016/j.neuropsychologia.2009.04.023

Damasio, A. R. (1996). The somatic marker hypothesis and the possible functions of the prefrontal cortex. *Philos Trans R Soc Lond B Biol Sci, 351*(1346), 1413–1420. doi:org/10.1098/rstb.1996.0125

de Haan, A. M., Smit, M., Van der Stigchel, S., & Dijkerman, H. C. (2016). Approaching threat modulates visuotactile interactions in peripersonal space. *Exp Brain Res.* doi:org/10.1007/s00221-016-4571-2

de Haan, A. M., Van der Stigchel, S., Nijnens, C. M., Dijkerman, H. C., (2014). The influence of object identity on obstacle avoidance reaching behaviour. *Acta Psychol, 150*, 94–99. doi:org/10.1016/j.actpsy.2014.04.007

De Paepe, A. L., Crombez, G., & Legrain, V. (2016). What's coming near? The influence of dynamical visual stimuli on nociceptive processing. *PloS ONE, 11*(5), e0155864. doi:org/10.1371/journal.pone.0155864

De Paepe, A. L., Crombez, G., Spence, C., & Legrain, V. (2014). Mapping nociceptive stimuli in a peripersonal frame of reference: evidence from a temporal order judgment task. *Neuropsychologia, 56*(1), 219–228. doi:org/10.1016/j.neuropsychologia.2014.01.016

den Ouden, H. E. M., Kok, P., & de Lange, F. P. (2012). How prediction errors shape perception, attention, and motivation. *Front. Psychol, 3* (December), 548. doi:org/10.3389/fpsyg.2012.00548

de Vignemont, F., & Iannetti, G. D. (2015). How many peripersonal spaces? *Neuropsychologia, 70*, 327–334. doi:org/10.1016/j.neuropsychologia.2014.11.018

di Pellegrino, G., & Làdavas, E. (2015). Peripersonal space in the brain. *Neuropsychologia, 66*, 126–133. doi:org/10.1016/j.neuropsychologia.2014.11.011

Dijkerman, H. C., Webeling, M., Ter Wal, J. M., Groet, E., & Van Zandvoort, M. J. E. (2004). A long-lasting improvement of somatosensory function after prism adaptation, a case study. *Neuropsychologia, 42*(12). doi:10.1016/j.neuropsychologia.2004.04.004

Dijkerman, H. C., & de Haan, E. H. (2007). Somatosensory processes subserving perception and action. *Behav. Brain Sci, 30*(2), 139–189. doi:10.1017/S0140525X07001392.

Driver, J., & Spence, C. (1998). Attention and the crossmodal construction of space. *Trends Cogn Sci, 2*(7), 254–262. doi:org/10.1016/S1364-6613(98)01188-7

Duhamel, J. R., Colby, C. L., & Goldberg, M. E. (1998). Ventral intraparietal area of the macaque: congruent visual and somatic response properties. *J. Neurophysiol, 79*(1), 126–136.

Eimer, M., van Velzen, J., & Driver, J. (2002). Cross-modal interactions between audition, touch, and vision in endogenous spatial attention: ERP evidence on preparatory states and sensory modulations. *J. Cogn Neurosci, 14*(2), 254–271.

Emberson, L. L., Richards, J. E., & Aslin, R. N. (2015). Top-down modulation in the infant brain: learning-induced expectations rapidly affect the sensory cortex at 6 months. *Proc. Natl. Acad. Sci. U.S.A., 112*(31), 201510343. doi:org/10.1073/pnas.1510343112

Facchin, A., Sartori, E., Luisetti, C., De Galeazzi, A., & Beschin, N. (2019). Effect of prism adaptation on neglect hemianesthesia. *Cortex, 113*, 298–311. doi:10.1016/j.cortex.2018.12.021

Farnè, A., Làdavas, E., (2000). Dynamic size-change of hand peripersonal space following tool use. *NeuroReport, 11*(8), 1645–1649.

Ferri, F., Costantini, M., Huang, Z., Perrucci, M. G., Ferretti, A., Romani, G. L., & Northoff, G. (2015). Intertrial variability in the premotor cortex accounts for individual differences in peripersonal space. *J. Neurosci, 35*(50), 16328–16339. doi:org/10.1523/JNEUROSCI.1696-15.2015

Finisguerra, A., Canzoneri, E., Serino, A., Pozzo, T., & Bassolino, M. (2015). Moving sounds within the peripersonal space modulate the motor system. *Neuropsychologia, 70*, 421–428. doi:org/10.1016/j.neuropsychologia.2014.09.043

Fogassi, L., Gallese, V., Fadiga, L., Luppino, G., Matelli, M., & Rizzolatti, G. (1996). Coding of peripersonal space in inferior premotor cortex (area F4). *J. Neurophysiol, 76*(1), 141–157.

Fossataro, C., Sambo, C., Garbarini, F., & Iannetti, P. G. (2016). Interpersonal interactions and empathy modulate perception of threat and defensive responses. *Sci Rep, 6.* doi:org/10.1038/srep19353

Fries, P. (2005). A mechanism for cognitive dynamics: neuronal communication through neuronal coherence. *Trends Cogn Sci, 9*(10), 474–480. doi:org/10.1016/j.tics.2005.08.011

Friston, K., Kilner, J., & Harrison, L. (2006). A free energy principle for the brain. *J. Physiol, Paris, 100*(1–3), 70–87. doi:10.1016/j.jphysparis.2006.10.001

Gentile, G., Petkova, V. I., & Ehrsson, H. H. (2011). Integration of visual and tactile signals from the hand in the human brain: an FMRI study. *J. Neurophysiol, 105*(2), 910–922. doi:10.1152/jn.00840.2010

Gordon, A. M., Forssberg, H., Johansson, R. S., & Westling, G. (1991). Visual size cues in the programming of manipulative forces during precision grip. *Exp Brain Res, 83*(3), 477–482.

Graziano, M. S. (2009). *The intelligent movement machine.* New York: Oxford University Press.

Graziano, M. S., & Cooke, D. F. (2006). Parieto-frontal interactions, personal space, and defensive behavior. *Neuropsychologia, 44*(6), 845–859. doi:10.1016/j.neuropsychologia.2005.09.009

Gray, R., & Tan, H. Z. (2002). Dynamic and predictive links between touch and vision. *Exp Brain Res, 145*(1), 50–55. doi:org/10.1007/s00221-002-1085-x

Heed, T., Habets, B., Sebanz, N., & Knoblich, G. (2010). Others' actions reduce crossmodal integration in peripersonal space. *Curr Biol, 20*(15), 1345. doi:10.1016/j.cub.2010.05.068–1349

Hoogenraad, T. U., Ramos, L. M. P., & van Gijn, J. (1994). Visually induced central pain and arm withdrawal after right parietal lobe infarction. *J. Neurology. Neurosurg. Psychiatry, 57*, 850–852.

Huang, R.-S., Chen, C.-F., Tran, A. T., Holstein, K. L., & Sereno, M. I. (2012). Mapping multisensory parietal face and body areas in humans. *Proc. Natl. Acad. Sci. U.S.A., 109*(44), 18114–18119. doi:org/10.1073/pnas.1207946109

Huang, R.-S., Chen, C., & Sereno, M. I. (2018). Spatiotemporal integration of looming visual and tactile stimuli near the face. *Hum. Brain Mapp., 39*(5), 2156–2176. doi:org/10.1002/hbm.23995

Hyvärinen, J., & Poranen, A. (1974). Function of parietal associative area 7 as revealed from cellular discharges in alert monkeys. *Brain, 97*, 673–692.

Kandula, M. (2020). *Properties of the peripersonal space in behaving humans* (Doctoral thesis. Utrecht University, Utrecht the Netherlands). Retrieved from http://dspace.library.uu.nl/bitstream/handle/1874/396186/5ea81dd8711e5.pdf?sequence=1&isAllowed=y

Kandula, M., Hofman, D., & Dijkerman, H. C. (2014). Visuo-tactile interactions are dependent on the predictive value of the visual stimulus. *Neuropsychologia,* 1–9. doi:org/10.1016/j.neuropsychologia.2014.12.008

Kennedy, D. P., Glascher, J., Tyszka, J. M., & Adolphs, R. (2009). Personal space regulation by the human amygdala. *Nat Neurosci, 12*(10), 1226–1227. doi: 0.1038/nn.2381

Làdavas, E., & Farnè, A. (2004). Visuo-tactile representation of near-the-body space. *J Physiol, Paris, 98*(1–3), 161–170. doi:10.1016/j.jphysparis.2004.03.007

Làdavas, E., & Serino, A. (2008). Action-dependent plasticity in peripersonal space representations. *Cogn Neuropsychol, 25*(7–8), 1099–1113.

Liu, Y., Denton, J. M., & Nelson, R. J. (2007). Neuronal activity in monkey primary somatosensory cortex is related to expectation of somatosensory and visual go-cues. *Exp Brain Res, 177*(4), 540–550. doi:org/10.1007/s00221-006-0702-5

Lloyd, D. M., & Morrison, C. I. (2008). 'Eavesdropping' on social interactions biases threat perception in visuospatial pathways. *Neuropsychologia, 46*(1), 95–101. Doi:org/10.1016/j.neuropsychologia.2007.08.007

Lloyd, D., Morrison, I., & Roberts, N. (2006). Role for human posterior parietal cortex in visual processing of aversive objects in peripersonal space. *J. Neurophysiol, 95*(1), 205–214. doi:10.1152/jn.00614.2005

Loken, L. S., Wessberg, J., Morrison, I., McGlone, F., & Olausson, H. (2009). Coding of pleasant touch by unmyelinated afferents in humans. *Nat Neurosci, 12*(5), 547–548. doi:10.1038/nn.2312

Macaluso, E., Frith, C. D., & Driver, J. (2002). Crossmodal spatial influences of touch on extrastriate visual areas take current gaze direction into account. *Neuron, 34*(4), 647–658.

Macaluso, E., & Maravita, A. (2010). The representation of space near the body through touch and vision. *Neuropsychologia, 48*(3), 782–795. doi:org/10.1016/j.neuropsychologia.2009.10.010

Maier, J. X., Chandrasekaran, C., & Ghazanfar, A. A. (2008). Integration of bimodal looming signals through neuronal coherence in the temporal lobe. *Curr Biol, 18*(13), 963–968. doi:10.1016/j.cub.2008.05.043

Maister, L., Cardini, F., Zamariola, G., Serino, A., & Tsakiris, M. (2014). Your place or mine: shared sensory experiences elicit a remapping of peripersonal space. *Neuropsychologia, 70*, 455–461. doi:org/10.1016/j.neuropsychologia.2014.10.027

Makin, T. R., Holmes, N. P., Brozzoli, C., & Farnè, A. (2012). Keeping the world at hand: rapid visuomotor processing for hand-object interactions. *Exp Brain Res. Experimentelle Hirnforschung. Expérimentation Cérébrale, 219*(4), 421–428. doi:org/10.1007/s00221-012-3089-5

Makin, T. R., Holmes, N. P., Brozzoli, C., Rossetti, Y., & Farne, A. (2009). Coding of visual space during motor preparation: approaching objects rapidly modulate corticospinal excitability in hand-centered coordinates. *J. Neurosci, 29*(38), 11841–11851. doi:org/10.1523/JNEUROSCI.2955-09.2009

Makin, T. R., Holmes, N. P., & Zohary, E. (2007). Is that near my hand? Multisensory representation of peripersonal space in human intraparietal sulcus. *J. Neurosci, 27*(4), 731–740. doi:org/10.1523/JNEUROSCI.3653-06.2007

Makin, T. R., Wilf, M., Schwartz, I., & Zohary, E. (2010). Amputees 'neglect' the space near their missing hand. *Psychol. Sci, 21*(1), 55–57.

McGuire, L. M. M., & Sabes, P. N. (2009). Sensory transformations and the use of multiple reference frames for reach planning. *Nat Neurosci, 12*(8), 1056–1061. doi:org/10.1038/nn.2357

Menger, R., Van der Stigchel, S., & Dijkerman, H. C. (2012). How obstructing is an obstacle? The influence of starting posture on obstacle avoidance. *Acta Psychol, 141*(1), 1–8. doi:org/10.1016/j.actpsy.2012.06.006

Meyer, S., De Bruyn, N., Lafosse, C., Van Dijk, M., Michielsen, M., Thijs, L., … Verheyden, G. (2016). somatosensory impairments in the upper limb poststroke: distribution and association with motor function and visuospatial neglect. *Neurorehabil Neural Repair, 30*(8), 731–742. doi:10.1177/1545968315624779

Meyer, S., Kessner, S. S., Cheng, B., Bönstrup, M., Schulz, R., Hummel, F. C., … Verheyden, G. (2016). Voxel-based lesion-symptom mapping of stroke lesions underlying somatosensory deficits. *NeuroImage: Clin, 10*, 257–266. doi:10.1016/j.nicl.2015.12.005

Mobbs, D., Petrovic, P., Marchant, J. L., Hassabis, D., Weiskopf, N., Seymour, B., ... Frith, C. D. (2007). When fear is near: threat imminence elicits prefrontal-periaqueductal gray shifts in humans. *Science, 317*(5841), 1079–1083. doi:10.1126/science.1144298

Noel, J.-P., Blanke, O., Magosso, E., & Serino, A. (2018). Neural adaptation accounts for the dynamic resizing of peripersonal space: evidence from a psychophysical-computational approach. *J. Neurophysiol, 119*(6), 2307–2333. doi:org/10.1152/jn.00652.2017

Noel, J.-P., Stevenson, R. A., & Wallace, M. T. (2018). Atypical audiovisual temporal function in autism and schizophrenia: similar phenotype, different cause. *Eur. J. Neurosci, 47*(10), 1230–1241. doi:org/10.1111/ejn.13911

Noel, J.-P., Blanke, O., & Serino, A. (2018). From multisensory integration in peripersonal space to bodily self-consciousness: from statistical regularities to statistical inference. *Ann N. Y. Acad. Sci., 1426*(1), 146–165. doi:org/10.1111/nyas.13867

Olausson, H., Lamarre, Y., Backlund, H., Morin, C., Wallin, B. G., Starck, G., ... Bushnell, M. C. (2002). Unmyelinated tactile afferents signal touch and project to insular cortex. *Nat Neurosci, 5*(9), 900–904.

Orioli, G., Filippetti, M. L., Gerbino, W., Dragovic, D., & Farroni, T. (2018). Trajectory discrimination and peripersonal space perception in newborns. *Infancy, 23*(2), 252–267. doi:org/10.1111/infa.12207

Pellencin, E., Paola, M., Herbelin, B., & Serino, A. (2018). Social perception of others shapes one's own multisensory peripersonal space. *Cortex,* 1–17. doi:10.1016/j.cortex.2017.08.033

Poliakoff, E., Miles, E., Li, X., & Blanchette, I. (2007). The effect of visual threat on spatial attention to touch. *Cognition, 102*(3), 405–414. doi:org/10.1016/j.cognition.2006.01.006

Pouget, A., Deneve, S., & Duhamel, J.-R. (2002). A computational perspective on the neural basis of multisensory spatial representations. *Nat Rev Neurosci, 3*(9), 741–747. doi:org/10.1038/nrn914

Quesque, F., Ruggiero, G., Mouta, S., Santos, J., Iachini, T., & Coello, Y. (2016). Keeping you at arm's length: modifying peripersonal space influences interpersonal distance. *Psychol. Res.* doi:org/10.1007/s00426-016-0782-1

Rizzolatti, G., Scandolara, C., Matelli, M., & Gentilucci, M. (1981). Afferent properties of periarcuate neurons in macaque monkeys. II. Visual responses. *Behav. Brain Res, 2*(2), 147–163.

Roncone, A., Hoffmann, M., Pattacini, U., & Metta, G. (2015). Learning peripersonal space representation through artificial skin for avoidance and reaching with whole body surface. In IEEE/RSJ International Conference on Intelligent Robots and Systems (IROS) 3366–3373, IEEE. doi:10.1109/IROS.2015.7353846

Ruzzoli, M., & Soto-Faraco, S. (2014). Alpha stimulation of the human parietal cortex attunes tactile perception to external space. *Curr Biol, 24*(3), 329–332. doi:org/10.1016/j.cub.2013.12.029

Sambo, C. F., Forster, B., Williams, S. C., & Iannetti, G. D. (2012). To blink or not to blink: fine cognitive tuning of the defensive peripersonal space. *J. Neurosci, 32*(37), 12921–12927. doi:org/10.1523/JNEUROSCI.0607-12.2012

Sambo, C. F., & Iannetti, G. D. (2013). Better safe than sorry? The safety margin surrounding the body is increased by anxiety. *J. Neurosci, 33*(35), 14225–14230. doi:org/10.1523/JNEUROSCI.0706-13.2013

Sambo, C. F., Liang, M., Cruccu, G., & Iannetti, G. D. (2012). Defensive peripersonal space: the blink reflex evoked by hand stimulation is increased when the hand is near the face. *J. Neurophysiol, 107*(3), 880–889. doi:org/10.1152/jn.00731.2011

Schutter, D. J., Enter, D., & Hoppenbrouwers, S. S. (2009). High-frequency repetitive transcranial magnetic stimulation to the cerebellum and implicit processing of happy facial expressions. *J. Psychiatry Neurosci, 34*(1), 60–65.

Sereno, M. I., & Huang, R.-S. (2006). A human parietal face area contains aligned head-centered visual and tactile maps. *Nat Neurosci, 9*(10), 1337–1343. doi:org/10.1038/nn1777

Serino, A., Annella, L., & Avenanti, A. (2009). Motor properties of peripersonal space in humans. *PLoS ONE, 4*(8), e6582. doi:org/10.1371/journal.pone.0006582

Shadmehr, R., Smith, M. A., & Krakauer, J. W. (2010). Error correction, sensory prediction, and adaptation in motor control. *Annu Rev Neurosci, 33*(1), 89–108. doi:org/10.1146/annurev-neuro-060909-153135

Spaccasassi, C., Romano, D., & Maravita, A. (2019). Everything is worth when it is close to my body: how spatial proximity and stimulus valence affect visuo-tactile integration. *Acta Psychol (Amst), 192*, 42–51. doi:10.1016/j.actpsy.2018.10.013

Spence, C., Pavani, F., & Driver, J. (2000). Crossmodal links between vision and touch in covert endogenous spatial attention. *J. Exp Psychol Hum Percept Perform, 26*(4), 1298–1319.

Spence, C., Pavani, F., Maravita, A., & Holmes, N. (2004). Multisensory contributions to the 3-D representation of visuotactile peripersonal space in humans: evidence from the crossmodal congruency task. *J. Physiol, Paris, 98*(1–3), 171–189. doi:10.1016/j.jphysparis.2004.03.008

Stevenson, R. A., Siemann, J. K., Woynaroski, T. G., Schneider, B. C., Eberly, H. E., Camarata, S. M., & Wallace, M. T. (2014). Evidence for diminished multisensory integration in autism spectrum disorders. *J. Autism Dev Disord*, 3161–3167. doi:org/10.1007/s10803-014-2179-6

Summerfield, C., & de Lange, F. P. (2014). Expectation in perceptual decision making: neural and computational mechanisms. *Nat Rev Neurosci*, (October). doi:org/10.1038/nrn3838

Taffou, M., & Viaud-Delmon, I. (2014). Cynophobic fear adaptively extends peri-personal space. *Front. Psychiatry, 5*(September), 3–9. doi:org/10.3389/fpsyt.2014.00122

Taylor-Clarke, M., Jacobsen, P., & Haggard, P. (2004). Keeping the world a constant size: object constancy in human touch. *Nat Neurosci, 7*(3), 219–220.

Teneggi, C., Canzoneri, E., Pellegrino, G., Serino, A., Giuridiche, S., Cicu, A., … Paris, I. (2013). Report social modulation of peripersonal space boundaries. *Curr Biol, 23*(5), 406–411. doi:org/10.1016/j.cub.2013.01.043

Testolin, A., De Filippo De Grazia, M., & Zorzi, M. (2017). The role of architectural and learning constraints in neural network models: a case study on visual space coding. *Front. Comput. Neurosci, 11*, 13. doi:org/10.3389/fncom.2017.00013

Thut, G., Miniussi, C., & Gross, J. (2012). The functional importance of rhythmic activity in the brain. *Curr Biol, 22*(16), R658–663. doi:org/10.1016/j.cub.2012.06.061

Tresilian, J. R. (1998). Attention in action or obstruction of movement? A kinematic analysis of avoidance behavior in prehension. *Exp Brain Res, 120*(3), 352–368.

Vallar, G. (1997). Spatial frames of reference and somatosensory processing: a neuropsychological perspective. *Philos Trans R Soc Lond B Biol Sci, 352*(1360), 1401–1409. doi:org/10.1098/rstb.1997.0126

Vallar, G. (2007). A hemispheric asymmetry in somatosensory processing. *Behav Brain Sci, 30*(2), 223–224.

van Atteveldt, N., Murray, M. M., Thut, G., & Schroeder, C. E. (2014). Multisensory integration: flexible use of general operations. *Neuron, 81*(6), 1240–1253. doi:org/10.1016/J.NEURON.2014.02.044

Van Damme, S., Crombez, G., & Eccleston, C. (2004). The anticipation of pain modulates spatial attention: evidence for pain-specificity in high-pain catastrophizers. *Pain, 111*(3), 392–399. doi:org/10.1016/j.pain.2004.07.022

Van Damme, S., & Legrain, V. (2012). How efficient is the orienting of spatial attention to pain? An experimental investigation. *Pain, 153*(6), 1226–31. doi:org/10.1016/j.pain.2012.02.027

van Ede, F., de Lange, F., Jensen, O., & Maris, E. (2011). Orienting attention to an upcoming tactile event involves a spatially and temporally specific modulation of sensorimotor alpha- and beta-band oscillations. *J. Neurosci*, *31*(6), 2016–2024. doi:org/10.1523/JNEUROSCI.5630-10.2011

van Stralen, H. E., van Zandvoort, M. J. E., Hoppenbrouwers, S. S.,Vissers, L. M. G., Kappelle, L. J., & Dijkerman, H. C. (2014). Affective touch modulates the rubber hand illusion. *Cognition*, *131*(1), 147–158. doi:org/10.1016/j.cognition.2013.11.020

Verhagen, L., Dijkerman, H. C., Medendorp, W. P., & Toni, I. (2012). Cortical dynamics of sensorimotor integration during grasp planning. *J. Neurosci*, *32*(13), 4508–19. doi:org/10.1523/JNEUROSCI.5451-11.2012

Vieira, J. B., Pierzchajlo, S. R., & Mitchell, D. G. V. (2019). Neural correlates of social and non-social personal space intrusions: role of defensive and peripersonal space systems in interpersonal distance regulation. *Soc Neurosci*, 1–16. doi:10.1080/17470919.2019.1626763

Voudouris, D., & Fiehler, K. (2017). Spatial specificity of tactile enhancement during reaching. *Atten Percept Psychophys*, *79*(8), 2424–2434. doi:org/10.3758/s13414-017-1385-7

Weinstein, S. (1968). Intensive and extensive aspects of tactile sensitivity as a function of body part, sex, and laterality. In D. R. Kenskalo (ed.), *The skin senses* (pp. 195–222). Springfield, Illinois: Charles C Thomas.

Wolpert, D. M., & Flanagan, J. R. (2001). Motor prediction. *Curr Biol*, *11*(18), R729–R732.

Zald, D. H. (2003). The human amygdala and the emotional evaluation of sensory stimuli. *Brain Res. Rev.*, *41*(1), 88–123. doi:org/S0165017302002485

6

Functional actions of hands and tools influence attention in peripersonal space

Catherine L. Reed and George D. Park

6.1 Introduction

Human perceptual and attentional systems operate to help us perform functional and adaptive actions in the world around us, such as grabbing our cup of coffee or using a tennis racquet to hit a ball outside our reach. Peripersonal space generally refers to the space near and surrounding the body (Previc, 1998; Rizzolatti, Scandolara, Matelli, & Gentilucci, 1981) and is the region in which our visual system and the body can best interact to perform actions. It includes the space near the hands, or 'peri-hand space', as well as reachable space, and potentially even space just outside body reach. Behaviourally, processing in peripersonal space plays a crucial role in coordinating defensive and non-defensive actions (Cléry, Guipponi, Wardak, & Hamed, 2015; de Vignemont & Iannetti, 2015; Graziano & Cooke, 2006; Graziano & Gross, 1998; Lourenco, Longo, & Pathman, 2011; Makin, Holmes, Brozzoli, Rossetti, & Farnè, 2009). Extrapersonal space refers to the far space away from the body and well beyond reach (Previc, 1998). Monkey and human studies indicate that peripersonal and extrapersonal space are dissociated from each other by separate neural systems (Committeri et al., 2007; Cowey, Small, & Ellis, 1994; Halligan & Marshall, 1991; Rizzolatti, Fadiga, Fogassi, & Gallese, 1997; Rizzolatti, Matelli, & Pavesi, 1983).

Compared with extrapersonal space, a defining feature of peripersonal space is that it is the space in which actions are typically performed. To successfully perform action, neural systems must integrate information about spatial position, vision, touch, and proprioception. Attention to such information in peripersonal space determines whether objects near or approaching the body may be candidates for grasping (e.g. food) or deflecting (e.g. threats). Motor and proprioceptive information about arm position is enhanced by stereovision and the 3D spatial position of the visual object relative to the body. In particular, the visual system can use stereopsis to permit fine coordination between the hand and eye, and evolutionarily important actions such as eating, tool making, and transforming one object into something else. Importantly, attentional mechanisms help the system coordinate what multimodal systems are most relevant for upcoming action.

In this chapter, we address how the functional capabilities of our hands and tools, and their actions, may influence the distribution of attention within peripersonal space. Although a number of studies have considered the role of attention and the possibility for action in space near the body and tool, the different types of paradigms used to study peripersonal space and the different definitions of peripersonal space regions make it difficult to reach conclusions (see Wu, this volume, for discussion). To address these issues, we review research primarily from our laboratory to examine how the functional capabilities and actions of the hand and

Catherine L. Reed and George D. Park, *Functional actions of hands and tools influence attention in peripersonal space* In: *The World at Our Fingertips*. Edited by Frédérique de Vignemont, Andrea Serino, Hong Yu Wong, and Alessandro Farnè, Oxford University Press (2021).
© Oxford University Press. DOI: 10.1093/oso/9780198851738.003.0006.

tool influence the distribution of attention in peripersonal space in the service of action. Our behavioural and neurophysiological paradigms use variations of a common target detection paradigm as an attentional probe so that we can examine the distribution of attention around functional and non-functional regions of the hand and tool. Further, to address whether functional capabilities and action affect different regions of peripersonal space similarly, our studies examine the role of effector—function and task relevance/action in different regions of peripersonal space—near the hand, reaching, and interactions between peri- and extrapersonal space. From these combined studies, we propose that the functional capabilities of the hand and tools direct attention to action-relevant regions of peripersonal space and, to some extent, extrapersonal space.

6.2 Peripersonal space near the hand

Researchers such as Previc (1998) have defined different regions of space extending away from body based on behavioural and neural data. Peripersonal space near the hand (i.e. peri-hand space) encompasses the region surrounding all parts of the hand, including the region just next to palm in graspable space. We proposed that visuospatial attention to a visual stimulus is facilitated or prioritized by current hand proximity and, more specifically, to the functional capabilities of our hands. Hand position and the functional ability to act may facilitate the visual processing of space by providing references for upcoming actions.

6.2.1 Potential for action near the hand

Neurophysiological research has provided one neural mechanism that differentiates the processing of space near the hand from other regions of peripersonal space and relates it to action. The difference in attentional processing between extrapersonal space and peri-hand space may be attributed, at least in part, to the *combined* contributions from visual and bimodal neurons that respond to both visual and tactile stimuli (Graziano & Gross, 1993) as well as proprioceptive information (Graziano, 1999). These visuo-tactile bimodal neurons are found in brain regions associated with action and action planning, such as the ventral premotor cortex, putamen, ventral intraparietal sulcus, and cortical areas 6 and 7b (Graziano, 2001). These cortical and subcortical regions form a multimodal neural network to coordinate visual and tactile motor systems when interaction with the world is required (Fadiga, Fogassi, Gallese, & Rizzolatti, 2000).

One of the first human behavioural studies to demonstrate that the ability to perform upcoming actions could influence processing in peri-hand space was from Reed and colleagues (Reed, Grubb, & Steele, 2006). Researchers manipulated the relative positions of hands and visual stimuli in a predictive covert-orienting task. Participants held one hand next to one of two lateralized target locations. Hand location changed attention to the peripersonal space near the hand: participants were faster to detect targets appearing next to the palm, in *graspable space*, compared to when the hand was positioned far from the stimulus, outside of graspable space. They further showed that this biasing of attention to graspable space was specific to the hand and could not be attributed to visual reference frames provided by non-hand, visual anchors located next to the visual stimulus (Figure 6.1a). In addition, attention advantages were attenuated when the hand was moved away from the visual stimulus, when hand positions were visually blocked, and when a fake hand eliminated proprioceptive

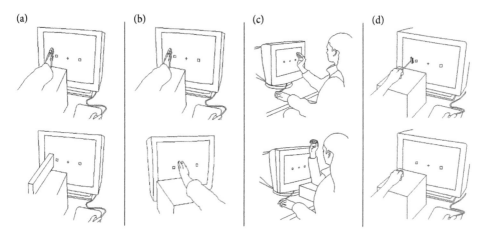

Figure 6.1 Hand proximity conditions (Reed et al., 2006, 2010): a) Palm vs Visual Anchor; b) Palm vs Back of hand; c) Palm vs Forearm; and d) Tool vs Hand in extrapersonal space.

Adapted by permission from Springer Nature: Reed, C.L., Betz, R., Garza, J.P. et al. Grab it! Biased attention in functional hand and tool space. Attention, Perception, & Psychophysics 72, 236–245 (2010) doi:10.3758/APP.72.1.236.

inputs for hand location but maintained visual hand location. This pattern of attentional fa-cilitation of targets appearing in peri-hand space was consistent with the properties of visuo-tactile bimodal neurons: hand-centred (i.e. has decreasing strength as the target moves away from the hand) and multimodal (i.e. sensitive to both visual and haptic/proprioceptive in-formation). The conditions under which attentional facilitation for targets in grasping space near the hand suggested a relation between the single-cell monkey studies and these behav-ioural effects. However, these experiments did not distinguish whether this attentional facili-tation was specific only to graspable space per se or space around the hand.

Subsequent research has shown hand proximity to influence a variety of attentional and cognitive processes (Brockmole, Davoli, Abrams, & Witt, 2013). For example, visual stimuli presented to peri-hand space are found to elicit attentional shifting (Lloyd, Azanon, & Poliakoff, 2010), attentional prioritization of space (Davoli & Brockmole, 2012), slower at-tentional disengagement (Abrams et al., 2008), and more accurate detection and discrimin-ation of visual stimuli (Dufour & Touzalin, 2008). This attentional performance advantage appears to be influenced by hand location, even when the hand is completely irrelevant to the task (Abrams et al., 2008; Cosman & Vecera, 2010; Reed et al., 2006; Tseng, Bridgeman, & Juan, 2012).

6.2.2 Peri-hand space versus functional parts of hands

Although the mere proximity of the hand appears to facilitate processing of visual infor-mation in peri-hand space, the hand's functional capabilities and potential for action in-fluence the distribution of attention. Functionality is determined by the action intentions or goals of the user and the appropriate physical affordances or functional parts available to achieve those intentions or goals. As such, a hand's functionality is based on the user's intention to act (e.g. grab a cup) and the functional parts available (e.g. hand's grasping fin-gers). Therefore, hand proximity effects on attention may be associated to the space where the hands can functionally act.

To distinguish whether visual processing was facilitated in grasping space compared to any space near the hand, Reed, Betz, Garza, & Roberts (2010) compared visual processing in the space near the hand's palm (i.e. in grasping space) to space equidistantly behind the hand (Figure 6.1b). Using a covert orienting paradigm, targets appearing near the palm produced faster responses than targets appearing to the back of the hand. In addition, palm-side, grasping space facilitation was found regardless of whether orienting cues were predictive or non-predictive of target location (Garza, Strom, Wright, Roberts, & Reed, 2013). Further, the hand proximity effect was eliminated when the visual stimulus was presented adjacent to the forearm (Figure 6.1c). Other behavioural studies have collaborated these findings by showing diminished processing of information on the outside or non-functioning side of the hands (Davoli & Brockmole, 2012). Thus, the topology of the facilitated space around the hand appears to be defined, in part, by the hand's potential to grasp visually presented objects.

6.2.3 Specificity of functional interactions in peri-hand space

As hand functionality is predicated on the action intentions of the user, other studies have shown that different hand postures for specific functional actions can result in different attentional biases and processing (Bush & Vecera, 2014; Thomas, 2013). In Thomas (2013), researchers demonstrated that specific hand postures and affordances differentially affected visual global motion and form perception. A power grasp posture (i.e. using all fingers) improved relative performance on a visual motion detection task, but a precision grasp posture (i.e. using primarily thumb and index fingers) improved relative performance on a form perception task. Therefore, the visual system may alter processing based on the hand's potential to perform specific functional actions. While power grasps for fast and forceful interactions with objects initiate temporal sensitivity, precision grasps for specific movements with parts of objects initiate spatial sensitivity. Further, Bush and Vecera (2014) demonstrated that the proximity of one versus two hands in power grasp postures can affect the relative benefits of form over temporal processing. The proximity of a single hand configured in a power grasp posture produces a narrower focus of attention than two hands, which can relatively enhance form processing. These studies are consistent with the idea that hand proximity and the ways we use our hands affect visual processing through differential contributions from the M and P pathways, which in turn may lead to the activation of dorsal versus ventral stream processing (Davoli, Brockmole, & Goujon, 2012; Gozli, West, & Pratt, 2012; Tseng, Bridgeman, & Juan, 2012).

6.2.4 Attention affects two distinct stages of process: multisensory integration and task relevance

Although single-cell recording studies in monkeys indicate early, sensory-related integration of hand position and visual processing, more recent research using electrophysiology/event-related potentials (EEG/ERP) in humans has indicated that attention also influences more cognitive levels of processing. Several EEG/ERP studies have provided evidence that hand proximity manipulations affect attention to integrate information from vision and the body as well as attention to action-relevant stimuli. Simon-Dack et al. (2009) first showed a human correspondence at a neural level to the monkey single-cell bimodal findings. They presented lights on or above the fingertips in a sustained attention task and found enhanced N1 and P3 ERP amplitudes when the light was *on* the fingers. Subsequent EEG/ERP studies documented that hand proximity manipulations in peri-hand space affect attention at two distinct stages of processing—at early multimodal integration stages and later cognitive stages involving motivation and task goals (Reed, Clay, Kramer, Leland, & Hartley, 2017; Reed, Garza, & Vyas, 2018; Reed, Leland, Brekke, & Hartley, 2013; Vyas, Garza, & Reed, 2019). This coincides with studies finding that functional actions associated with an object and the action intentions of the observer influence early and low-level visual and attentional processes in the brain (Bortoletto, Mattingley, & Cunnington, 2011; Goslin, Dixon, Fischer, Cangelosi, & Ellis, 2012).

Manipulations of attention and hand proximity are often associated with amplitude modulations of the N1 and P3 ERP components. The N1, a negative amplitude deflection occurring between 140 and 200 ms measured from occipitoparietal electrodes, is associated with visuo-tactile or visuo-proprioceptive integration, visual attention, and stimulus discrimination processes during early sensory processing in the visual and parietal cortices (Eimer, 1994; Kennett, Eimer, Spence, & Driver, 2001; Simon-Dack et al., 2009). Alternatively, the P3, a positive inflection on the ERP waveform occurring between 300 and 500 ms measured from parietal electrodes, is associated with event categorization and higher-order cognitive influences such as task relevance, attentional distribution, and motivation (Kok, 2001). This information is thought to be communicated across the frontoparietal action network (FPAN) (Ptak, 2012).

Importantly, these distinct attentional processes are also related to action. Reed and colleagues used paradigms that distinguished between hand proximity effects for stimuli requiring action (i.e. targets) and those that did not (i.e. non-targets) in a 50/50 go/no go target detection task during which participants placed a hand either near or far from the monitor (Figure 6.2). Hand proximity affected both N1 components recorded from parietal-occipital electrodes and P3 components recorded from parietal electrodes. Contralateral N1 amplitudes were greater for targets compared to non-targets for the near hand (i.e. Palm condition in this study) condition only. In addition, increased contralateral P3 amplitudes were found for targets presented near the hand compared to targets far from the hand; no such effect was found for non-targets. Thus, hand proximity manipulations affected attention and multimodal integration at both early sensory discrimination stages and more cognitive stages for visual stimuli requiring subsequent action.

EEG/ERP methods also allow us to distinguish what aspects of the N1 and P3 hand proximity effects are influenced by the potential for action. Reed et al. (2006) showed behaviourally an attenuation in target detection facilitation when visual and proprioceptive inputs were reduced. When vision of hand is blocked and a physical barrier is placed between the hand and the visual stimulus in a similar target detection ERP study, hand condition differences for N1 amplitudes disappear, but the P3 differences and behavioural differences are maintained (Reed et al., 2018). Further, the examination of older individuals (e.g. ages 50–80 years) with decreasing sensory capabilities provides a way to investigate how variable sensory input may affect processing in near-hand space. Aging appears to decrease hand condition differences for sensory-related N1 amplitudes. Although hand proximity differences were maintained for the task-related P3 amplitudes, the scalp topographies of older adults showed extensive frontal lobe recruitment in conjunction with the parietal activation. These data suggest that, compared to younger adults, the brains of older individuals rely less on multisensory integration processes and more strongly on fronto-parietal attention networks. These two studies suggest that the later attentional component that reflect goals and

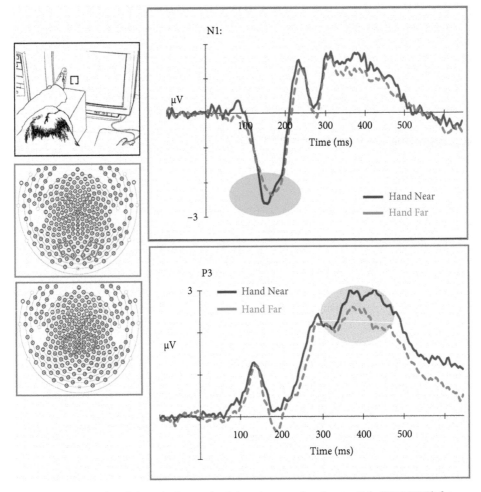

Figure 6.2 In Reed et al. (2013), electrophysiology/event-related potentials (EEG/ERP) data were measured when hands were placed near vs far from the visual stimulus. a) Hand Near condition; b) Electrode cap; c) Electrode clusters for N1 waveforms; d) Electrode clusters for P3 waveforms; e) N1 grand average waveform: blue = Hand Hear, red = Hand Far; P3 grant average waveform: blue = Hand Near, red = Hand Far.

Reproduced from Reed C. et al. Attention's grasp: early and late hand proximity effects on visual evoked potentials. Front Psychol. 2013; 4: 420. Published online 2013 Jul 12. doi: 10.3389/fpsyg.2013.00420 Copyright © 2013 Reed, Leland, Brekke and Hartley. Licensed under a creative commons attribution.

task relevance helps drive the behavioural hand proximity differences even when sensory inputs are attenuated. In addition, the distinction between early multimodal integration and later task-relevant attention may help explain some discrepancies in the behavioural litera-ture because behavioral measures reflect the outcomes of both processes.

Finally, to understand the role of hand function, separately from hand proximity, we compared two 'hand near' conditions: Palm, in which the visual stimulus was presented in grasping space, and Back of Hand, in which the visual stimulus was presented equidistantly behind the hand. This comparison of processing in the front and back of the hand has rele-vance for single-cell studies that show facilitation around the whole hand (e.g. Graziano &

Gross, 1998). Consistent with the monkey study patterns, we found that the presence of the visual stimulus equidistant from the front or back of the hand did not produce differential N1 amplitudes. This suggests that early multisensory integration processing near the hand was not specific to grasping and could be relevant to all functional actions (e.g. Makin et al., 2009). Of interest, however, was in the P3 target versus non-target differences for Palm versus Back of Hand conditions. Although the target waveforms were similar for both Palm and Back of Hand conditions (i.e. visual targets in grasping space were not enhanced per se), the non-target P3 amplitude was relatively more suppressed in the Palm condition, effectively accentuating target vs non-target neural response. Thus, task-relevant processing emerges later. In this paradigm, which did not require defensive action, the brain selectively emphasized the processing of to-be-acted upon stimuli near the palm.

6.2.5 Further considerations: peripersonal space near the hand

Considering behavioural and neurophysiological data using common target detection paradigms, it appears that attention aids in the body's functional interactions with visual objects in peripersonal space near the hand. Nonetheless, a variety of attentional and cognitive processes appear influenced by hand proximity alone (Brockmole et al., 2013). Further, the EEG/ERP data for our target detection paradigms reveal that such attentional effects may arise from an early, sensory-related multisensory integration component and a later, more cognitive attentional component. It is possible that the ability of these different types of attention to flexibly combine relevant processes may help explain how proprioception may influence processing in some studies for which proprioception may be relevant to the task and action but not in other studies for which proprioception is less relevant. In addition, the differential attention mechanisms may explain why a barrier may not affect performance but still influence multisensory integration processes (Farnè, Demattè, & Làdavas, 2003; Kitagawa & Spence, 2005). Evolutionarily, this dissociation of multisensory integration processes and task relevance processes allows our neural systems to be predisposed to emphasize objects in near peripersonal space that are candidates for action, but also to be able to override this predisposition to perform other actions.

The EEG/ERP studies also help to explain some differential findings regarding attentional enhancement in different regions of peri-hand space. The early N1 component appears to reflect sensory-related integration of vision and touch all around peri-hand space, consistent with the operation of visual and bimodal neurons recorded in single-cell monkey studies. The later P3 component reflects more task-relevant attention that allows for differential findings depending on the task's emphasis on specific functional capabilities of the hand. Thus, our paradigms implicitly emphasize the importance of grasping, but other studies emphasize the need for defensive actions (e.g. Graziano & Gross, 1998; Makin et al., 2009; Rossetti, Romano, Bolognini, & Maravita, 2015). Given that behaviour reflects the culmination of early and later processing, these dual attentional mechanisms can account for these different findings.

In summary, behavioural, neuropsychological, and physiological data all indicate that attention aids in the body's functional interactions with visual objects in peripersonal space. However, when using tools in our hands, the body's functional affordances and interactions with objects can drastically change and modify the corporal boundaries of space. These changes in attention due to tool use can thus provide insight into the properties of peri- and extrapersonal space. In the next section, we consider how the function of hands and tools may guide attention into peripersonal reaching space and extrapersonal space.

6.3 Extending peripersonal space: attention, functionality and tool use

In everyday life, we do not just use our hands to perform actions near our bodies. We also frequently use tools close to our bodies. Even if a tool may afford reach into extrapersonal space, we tend to use tools in reachable peripersonal space when possible to improve any number of performance factors, such as visual and motor accuracy, physical leverage, and postural stability. Consider the difference between hammering a nail in close to the body versus away. As attention in peri-hand space is facilitated relative to extrapersonal space, attention during tool use in peripersonal space may also be facilitated. Further, the relevant functional parts of a given tool can be different depending on the goals and intentions of the user. The same hammer used for pounding nails can also be used to retrieve a nail that is out of reach in extrapersonal space. Since the goals of most tool-use studies have been to observe how attention to extrapersonal tool space can exhibit those properties of peri-hand space, there is a lack of research for understanding the role of specific functional tool actions and tool use in peripersonal space. However, studies from Park and Reed (2015; 2020) and Park, Strom, and Reed (2013) have begun to explore this topic.

6.3.1 Tools extend attention to functional space

The functionality of a tool, like the hands, is determined by the user's intention or goal and physical affordances or functional parts for achieving it. For example, a garden rake's functionality is based on the user's intentions to retrieve or pull objects at a distance and the rake's physical affordances or functional parts to do so, such as the rake's prongs, shaft length, shaft stiffness, etc. When perceiving a tool's functionality, visual attention may prioritize the processing of those characteristics (e.g. object dimensions and functional parts) in the tool space that are relevant for the intended actions.

In Reed et al. (2010), the functional part of a tool was found to be critical when using tools to extend hand proximity effects beyond the hand's reach. Using the same covert, visual orienting paradigm, Reed and colleagues examined the allocation of spatial attention to a tool's functional part when participants held a small rake positioned next to potential visual target locations located outside of reach (Figure 6.1d). After a period of functional experience with the rake (i.e. manipulating sand in a Zen garden), participants showed facilitated processing for targets presented on the functional side of the rake (i.e. the prong side) but not on the rake's backside, or for the hand positioned at the same distance in extrapersonal space but not holding a rake. The results suggested that functional use of a tool can extend the hand proximity effects from peri-hand space to extended peripersonal space (or extrapersonal space) and that a functioning effector, whether hand or tool, is important for facilitated processing, not just proximity.

6.3.2 Facilitation of attention to tool parts in peri- versus extrapersonal space

Although previous studies showed that attention is drawn to the functional part of the tool, and not necessarily to its absolute length (Farnè, Iriki, & Làdavas, 2005), additional questions remained whether this attentional effect was equivalent in both peri- and extrapersonal space. In Park et al. (2013), the functional part of a tool was modified to be at the tool's middle part as opposed to what is normally the tool's end. This manipulation of a tool's functional part was then assessed in both peri- and extrapersonal space for changes in visuospatial attention to the tool space (Figure 6.3). Visual attention was measured to the space near a handheld tool's functionally identical middle and end parts. The tool was a 45-cm long wooden stick with shallow cups attached to the middle and end of the shaft. Participants performed a 50/50 go/no go target discrimination response time (RT) task while the tool was held in peri- space or extrapersonal space, before and after tool practice occurred in the corresponding space. Functional tool practice involved a hockey-like game using only the tool's middle, functional part, to move a puck through an obstacle course and shoot for a goal as many times as possible within 3 minutes. Distance from the hockey table controlled tool practice in peri- or extrapersonal space. Because tool practice only involved the tool's middle part, attention should have improved for that tool space. However, an attentional bias was observed for the tool's end in both peripersonal space and extrapersonal space regardless of tool practice. Importantly though, a redistribution of attention (i.e. reduction in RT difference between the tool's middle and end) was found only after tool practice in peripersonal space. The results suggested that changes in attentional distribution during tool use may be more amenable to the effects of tool practice and functionality changes when operating in peripersonal space.

Park and Reed (2015) confirmed, in part, that differential attentional distribution is a function of tool actions in peri- or extrapersonal space (Figure 6.4). As in Park et al. (2013), participants performed a target discrimination RT task while the tool was held in peripersonal space or extrapersonal space, before and after tool practice occurred in the corresponding space. The tool was a 30-cm long wooden stick with a shallow cup at the end. Functional tool practice involved a 3-minute hockey-like game using the tool's functional end. When the tool was practised and tested in extrapersonal space (Figure 6.4c), participants indicated no attentional differences between the tool's end or the tool-holding hand prior to the tool practice. But after tool practice, attention to targets near the tool's end improved (i.e. faster RTs)

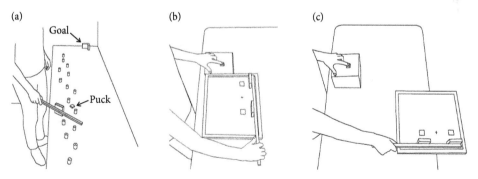

Figure 6.3 From Park et al. (2013). The functional part of a tool was modified to occur at the tool's middle part and not at the tool's end. Functional tool practice (a) involved using only the tool's middle part in a 3-minute hockey-like game. Visual attention was tested to the tool's middle and end parts with the tool held in (b) peripersonal and (c) extrapersonal space, before and after functional practice in the corresponding space.

Reproduced from Park, G. D., Strom, M., & Reed, C. L. (2013). To the end! Distribution of attention along a tool in peri- and extrapersonal space. Experimental Brain Research, 227(4), 423-432. doi: 10.1007/s00221-013-3439-y

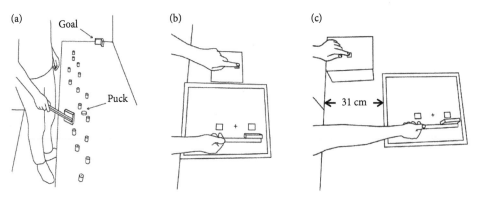

Figure 6.4 From Park and Reed (2015). Changes to the distribution of attention to tool space was measured as a function of tool actions in peripersonal or extrapersonal space. Visual attention was measured to the space near a tool's functional end and tool-occupied hand. Functional tool practice (a) involved using the tool's end in a 3-minute hockey-like game. Visual attention was tested with the tool held in (b) peripersonal and (c) extrapersonal space, before and after functional practice in the corresponding space.

Reproduced from Park G., Reed C. Haptic over visual information in the distribution of visual attention after tool-use in near and far space. Exp Brain Res. 2015 Oct;233(10):2977-88. doi: 10.1007/s00221-015-4368-8. Epub 2015 Jul 1.

with no change in performance at the tool-holding hand. The results were consistent with previous tool-use studies in far peripersonal space or extrapersonal space where changes in visuo-tactile integration were observed only after active tool practice (Farnè et al., 2005; Maravita et al., 2002).

When the tool was practised and tested in peripersonal space (Figure 6.4b), an initial attentional bias toward the tool's end was found prior to any tool practice similar to the after-practice effects found for extrapersonal space tool use. After tool practice in peripersonal space, no change occurred for attention towards the tool's end, but attention towards the tool-occupied hand improved. This improvement resulted in no observable attention difference between the tool-holding hand and the tool's end, suggesting an even distribution of visual attention along the tool. Therefore, peripersonal space tool use may potentially initiate earlier the attentional effects of extrapersonal space tool use and facilitate an even distribution of attention to encompass the entirety of the tool.

6.3.3 Distribution of attention as a function of tool actions in space

Above, we focused on how the functional part of a tool might affect the distribution of attention near it, when positioned in peri- versus extrapersonal space. However, an additional question is whether the tool user's functional action intentions or goals influence attentional distribution differently depending on specific directional tool actions executed in peri- or extrapersonal space. Park & Reed (2020) manipulated whether functional tool actions involved either pushing objects remaining in extrapersonal space or pulling objects towards the body in peripersonal space. Visuospatial attention was measured to three sections along a novel, curved tool to indicate attentional distributions: the tool's functional end and middle shaft, and the tool-occupied hand (Figure 6.5a). The curved tool design allowed for a central fixation point that was equal distance to the three target locations and controlled for any inherent functional parts for a specific tool action. Further, tool space testing occurred only while the tool was extended into extrapersonal space to show that attentional changes were due to tool use within a body space and not necessarily RT testing within that space. Finally, the specific tool actions for tool practice were controlled in terms of the number and direction of tool action movements within a specific body space. In Experiment 1, participants used the tool's end to push 60 individual poker chips to the far end of a table such that all tool actions remained in extrapersonal space (Figure 6.5b) or to pull poker chips towards the body into peripersonal space (Figure 6.5c). In Experiment 2, tool actions were manipulated such that participants either pulled objects only within extrapersonal space or pushed objects within peripersonal space. This allowed post-hoc comparisons to determine whether tool-related attentional changes were due to the kind of directional motor action (push vs pull) or the operating space (far vs near) in which they occurred.

Results indicated that that kind of functional tool actions performed relative to the body can affect how attention is distributed to the tool space. For participants who used the tool to move objects in extrapersonal space, regardless of pushing or pulling tool actions, attention improved (i.e. faster RTs) for targets presented only at the tool's end, with no RT changes at the tool's middle or tool-occupied hand. For participants who used the tool to pull objects from far to near (i.e. peripersonal space), attention improved for targets presented at the tool's end and for the tool's middle, suggesting a distributed tool space attention. Although no significant RT changes were observed for pushing actions within peripersonal space, overall results suggested that peripersonal tool use may better facilitate distributed tool-related attentional changes than extrapersonal space tool use. As the only tool study (as known) that has utilized multiple tool action types, the results potentially explain why some tool studies

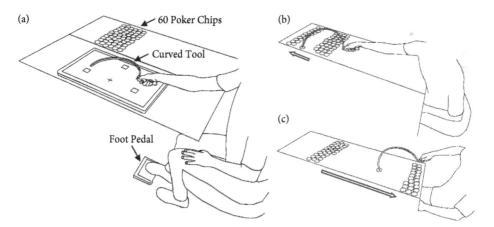

Figure 6.5 In Park & Reed (2020), visual attention was measured to the tool space due to specific functional tool actions in space relative to the body. Visual attention using a target discrimination task with foot pedal response times was measured to the space near a curved tool's functional end, middle shaft, and tool-occupied hand (a), before and after tool practice. Tool practice involved using the tool's end to move 60 individual poker chips to either (b) the far end of the table in far space, or (c) towards the body into near space.

From Park G., Reed C. Functional tool actions in near and far space affect the distribution of visual attention along the tool. J Exp Psychol Hum Percept Perform, 2020 Apr; 46(4): 388-404.

find attention to distribute only towards the tool's functional end (Holmes, 2012; Holmes, Calvert, & Spence, 2004) or towards the entire tool space (Bonifazi et al., 2007).

6.3.4 Further considerations for tool use in peripersonal and extrapersonal space

In the above studies, as well as other studies (e.g. Farnè et al., 2005; Maravita et al., 2002), functional experience with a tool is necessary to affect the distribution of attention. Given that the major behavioural effects occurred in peripersonal space and not extrapersonal space, these findings suggest that functional tool interaction affects the processes mediating peripersonal space, allowing attention to be distributed more generally along the tool. Is it possible that the flexible, multimodal nature of peripersonal space makes this possible, or might this result be reliant on the space in which functional tool use occurs? Our results suggest that functional tool use in peripersonal space appears to provide facilitated or differential attentional processing compared to tool use in extrapersonal space. For tool use in peripersonal space, changes in attention to functional parts were found to be more amendable or nuanced. Also, distributions of attention between the tool-occupied hand and the tool's end appeared to initiate earlier in peripersonal space. Finally, different functional tool actions involving body space resulted in different attentional distributions to the tool. Functional actions involving the tool's end to push or probe objects resulted in attentional improvements only towards the tool's functional end. Conversely, functional actions involving the tool's end to pull or retrieve objects towards the body resulted in attentional improvements to not only the tool's end but to the tool's middle as well, suggesting a more evenly distributed attention to tool space. At minimum, future tool-use studies must consider the kinds of tool actions performed relative to the tool user's body space when assessing tool-related attentional changes.

6.4 Conclusions

In conclusion, our attentional systems operate to help us perform functional and adaptive actions in the world around us. The studies reviewed here used common target detection paradigms to assess the distribution of attention near the hand and tool in different regions of peri- and extrapersonal space. Together they show that visual attention is facilitated or differentiated by both the function of the effector—hand or tool—and their actions in all regions of peripersonal space. Although neural mechanisms such as bimodal neurons may enhance the processing of visual information presented in near-hand regions of peripersonal space, functional experience and the relevance of the space for upcoming actions more strongly directs attention within regions of peripersonal space. And while some aspects of functionality can be extended into extrapersonal space, the multimodal nature of peripersonal space allows it to be more modifiable in the service of action.

Although the proximity of the hand facilitates the visual processing of space, a critical component to this effect is the hand's functional capabilities. The functionality of a hand or tool is defined by the action goals of the user and the available functional affordances or parts available to achieve the goals. Functionality thus corresponds to the findings of recent ERP studies suggesting that hand proximity effects involve both late cognitive (i.e. motivation and task goals) and early sensory discrimination stages. Finally, recent tool-use studies demonstrate how the distribution of attention to tool space can change as a result of tool functionality and operations in peri- or extrapersonal space.

A special feature of peripersonal space is that it appears to flexibly adapt to changing action capabilities, whether by changes to the body—for example, paralysis (Scandola, Aglioti, Bonente, Avesani, & Moro, 2016), or experience using handheld tools (Berti & Frassinetti, 2000; Brockmole et al., 2013; Holmes, 2012; Maravita & Iriki, 2004). Our tool studies suggest that peripersonal space is more flexible or modifiable than extrapersonal space. However, it remains unclear how flexible peripersonal space is and whether all properties of peripersonal space can expand into extrapersonal space. Relative to the hands, research with tools provide an experimentally advantageous means of understanding peripersonal space properties and the effects of functionality. Unlike the hands, the functional properties of a tool are easily manipulated and controlled. Further, tools with their ability to extend spatial reach allows for a dissociation between functionality and body space that is not necessarily available for hand studies. Therefore, tool-use studies may help reveal not only why humans are such effective tool users, but also the relationship between the brain, the body, and the actions it performs.

References

Abrams, R. A., Davoli, C. C., Du, F., Knapp, W. H., & Paull, D. (2008). Altered vision near the hands. *Cognition, 107*(3), 1035–1047. doi: 10.1016/j.cognition.2007.09.006

Berti, A., & Frassinetti, F. (2000). When far becomes near: remapping of space by tool use. *Journal of Cognitive Neuroscience, 12*(3), 415–420. doi: 10.1162/089892900562237

Bonifazi, S., Farnè, A., Rinaldesi, L., & Làdavas, E. (2007). Dynamic size-change of peri-hand space through tool-use: spatial extension or shift of the multi-sensory area. *Journal of Neuropsychology, 1*, 101–114. doi: 10.1348/174866407x180846

Bortoletto, M., Mattingley, J. B., & Cunnington, R. (2011). Action intentions modulate visual processing during action perception. *Neuropsychologia, 49*(7), 2097–2104. doi: 10.1016/j.neuropsychologia.2011.04.004

Brockmole, J. R., Davoli, C. C., Abrams, R. A., & Witt, J. K. (2013). The world within reach: effects of hand posture and tool use on visual cognition. *Current Directions in Psychological Science, 22*(1), 38–44. doi: 10.1177/0963721412465065

Bush, W. S., & Vecera, S. P. (2014). Differential effect of one versus two hands on visual processing. *Cognition, 133*(1), 232–237. doi: 10.1016/j.cognition.2014.06.014

Cléry, J., Guipponi, O., Wardak, C., & Ben Hamed, S. (2015). Neuronal bases of peripersonal and extrapersonal spaces, their plasticity and their dynamics: knowns and unknowns. *Neuropsychologia, 70*, 313–326. doi: 10.1016/j.neuropsychologia.2014.10.022

Committeri, G., Pitzalis, S., Galati, G., Patria, F., Pelle, G., Sabatini, U., ... Pizzamiglio, L. (2007). Neural bases of personal and extrapersonal neglect in humans. *Brain, 130*(Pt 2), 431–441. doi: 10.1093/brain/awl265

Cosman, J. D., & Vecera, S. P. (2010). Attention affects visual perceptual processing near the hand. *Psychological Science, 21*(9), 1254–1258. doi: 10.1177/0956797610380697

Cowey, A., Small, M., & Ellis, S. (1994). Left visuo-spatial neglect can be worse in far than in near space. *Neuropsychologia, 32*(9), 1059–1066.

Davoli, C. C., & Brockmole, J. R. (2012). The hands shield attention from visual interference. *Attention Perception & Psychophysics, 74*(7), 1386–1390. doi: 10.3758/s13414-012-0351-7

Davoli, C. C., Brockmole, J. R., & Goujon, A. (2012). A bias to detail: how hand position modulates visual learning and visual memory. *Memory & Cognition, 40*(3), 352–359. doi: 10.3758/s13421-011-0147-3

de Vignemont, F., & Iannetti, G. D. (2015). How many peripersonal spaces? *Neuropsychologia, 70*, 327–334. doi: 10.1016/j.neuropsychologia.2014.11.018

Dufour, A., & Touzalin, P. (2008). Improved visual sensitivity in the perihand space. *Experimental Brain Research, 190*(1), 91–98. doi: 10.1007/s00221-008-1453-2

Eimer, M. (1994). An ERP study on visual spatial priming with peripheral onsets. *Psychophysiology, 31*(2), 154–163.

Fadiga, L., Fogassi, L., Gallese, V., & Rizzolatti, G. (2000). Visuomotor neurons: ambiguity of the discharge or 'motor' perception? *International Journal of Psychophysiology, 35*(2–3), 165–177.

Farnè, A., Iriki, A., & Làdavas, E. (2005). Shaping multisensory action-space with tools: evidence from patients with cross-modal extinction. *Neuropsychologia, 43*(2), 238–248. doi: 10.1016/j.neuropsychologia.2004.11.010

Farnè, A., Demattè, M. L., & Làdavas, E. (2003). Beyond the window: multisensory representation of peripersonal space across a transparent barrier. *International Journal of Psychophysiology, 50*(1–2), 51–61. doi: 10.1016/S0167-8760(03)00124-7

Garza, J. P., Strom, M. J., Wright, C. E., Roberts, R. J., & Reed, C. L. (2013). Top-down influences mediate hand bias in spatial attention. *Attention, Perception, & Psychophysics, 75*(5), 819–823. doi: 10.3758/s13414-013-0480-7

Goslin, J., Dixon, T., Fischer, M. H., Cangelosi, A., & Ellis, R. (2012). Electrophysiological examination of embodiment in vision and action. *Psychological Science, 23*(2), 152–157. doi: 10.1177/0956797611429578

Gozli, D. G., West, G. L., & Pratt, J. (2012). Hand position alters vision by biasing processing through different visual pathways. *Cognition, 124*(2), 244–250. doi: 10.1016/j.cognition.2012.04.008

Graziano, M. S. (1999). Where is my arm? The relative role of vision and proprioception in the neuronal representation of limb position. *Proceedings of the National Academy of Sciences, 96*(18), 10418–10421.

Graziano, M. S. (2001). A system of multimodal areas in the primate brain. *Neuron, 29*(1), 4–6.

Graziano, M. S., & Cooke, D. F. (2006). Parieto-frontal interactions, personal space, and defensive behavior. *Neuropsychologia, 44*(13), 2621–2635.

Graziano, M. S., & Gross, C. G. (1993). A bimodal map of space: somatosensory receptive fields in the macaque putamen with corresponding visual receptive fields. *Experimental Brain Research, 97*(1), 96–109.

Graziano, M. S., & Gross, C. G. (1998). Spatial maps for the control of movement. *Current Opinion in Neurobiology, 8*(2), 195–201.

Halligan, P. W., & Marshall, J. C. (1991). Spatial compression in visual neglect: a case study. *Cortex, 27*(4), 623–629.

Holmes, N. P. (2012). Does tool use extend peripersonal space? A review and re-analysis. *Experimental Brain Research, 218*(2), 273–282. doi: 10.1007/s00221-012-3042-7

Holmes, N. P., Calvert, G. A., & Spence, C. (2004). Extending or projecting peripersonal space with tools? Multisensory interactions highlight only the distal and proximal ends of tools. *Neuroscience Letters, 372*(1–2), 62–67. doi: 10.1016/j.neulet.2004.09.024

Kennett, S., Eimer, M., Spence, C., & Driver, J. (2001). Tactile-visual links in exogenous spatial attention under different postures: convergent evidence from psychophysics and ERPs. *Journal of Cognitive Neuroscience, 13*(4), 462–478.

Kitagawa, N., & Spence, C. (2005). Investigating the effect of a transparent barrier on the crossmodal congruency effect. *Experimental Brain Research, 161*(1), 62–71. doi: 10.1007/s00221-004-2046-3

Kok, A. (2001). On the utility of P3 amplitude as a measure of processing capacity. *Psychophysiology, 38*(3), 557–577.

Lloyd, D. M., Azanon, E., & Poliakoff, E. (2010). Right hand presence modulates shifts of exogenous visuospatial attention in near perihand space. *Brain and Cognition, 73*(2), 102–109. doi: 10.1016/j.bandc.2010.03.006

Lourenco, S. F., Longo, M. R., & Pathman, T. (2011). Near space and its relation to claustrophobic fear. *Cognition, 119*(3), 448–453. doi: 10.1016/j.cognition.2011.02.009

Makin, T. R., Holmes, N. P., Brozzoli C., Rossetti, Y., & Farnè, A. (2009). Coding of visual space during motor preparation: approaching objects rapidly modulate corticospinal excitability in hand-centered coordinates. *Journal of Neuroscience, 29*(38), 11841–11851. doi: 10.1523/JNEUROSCI.2955-09.2009.

Maravita, A., & Iriki, A. (2004). Tools for the body (schema). [Review]. *Trends in Cognitive Sciences, 8*(2), 79–86. doi: 10.1016/j.tics.2003.12.008

Maravita, A., Spence, C., Kennett, S., & Driver, J. (2002). Tool-use changes multimodal spatial interactions between vision and touch in normal humans. *Cognition, 83*(2), B25–B34. doi: 10.1016/s0010-0277(02)00003-3

Park, G. D., & Reed, C. L. (2015). Haptic over visual information in the distribution of visual attention after tool-use in near and far space. *Experimental Brain Research, 233*(10), 2977–2988. doi: 10.1007/s00221-015-4368-8

Park, G. D., & Reed, C. L. (2020). Functional tool actions in near and far space affect the distribution of visual attention along the tool. *Journal of Experimental Psychology: Human Perception and Performance, 46*(4), 388–404. doi: 10.1037/xhp0000722

Park, G. D., Strom, M., & Reed, C. L. (2013). To the end! Distribution of attention along a tool in peri- and extrapersonal space. *Experimental Brain Research, 227*(4), 423–432. doi: 10.1007/s00221-013-3439-y

Previc, F. H. (1998). The neuropsychology of 3-D space. [Review]. *Psychological Bulletin, 124*(2), 123–164.

Ptak, R. (2012). The frontoparietal attention network of the human brain: action, saliency, and a priority map of the environment. *Neuroscientist, 18*(5), 502–515. doi: 10.1177/1073858411409051

Reed, C. L., Betz, R., Garza, J., & Roberts, R. (2010). Grab it! Biased attention for functional hand and tool space. *Attention, Perception, & Psychophysics, 72*, 236–245. doi: 10.3758/APP.72.1.236

Reed, C. L., Clay, S. N., Kramer, A. O., Leland, D. S., & Hartley, A. A. (2017). Attentional effects of hand proximity occur later in older adults: evidence from event-related potentials. *Psychology and Aging, 32*(8), 710–721. doi: 10.1037/pag0000207

Reed, C. L., Garza, J. P., & Vyas, D. B. (2018). Feeling but not seeing the hand: occluded hand position reduces the hand proximity effect in ERPs. *Conscious Cognition, 64*, 154–163. doi: 10.1016/j.concog.2018.04.012

Reed, C. L., Grubb, J. D., & Steele, C. (2006). Hands up: attentional prioritization of space near the hand. *Journal of Experimental Psychology: Human Perception and Performance, 32*(1), 166–177. doi: 10.1037/0096-1523.32.1.166

Reed, C. L., Leland, D. S., Brekke, B., & Hartley, A. A. (2013). Attention's grasp: early and late hand proximity effects on visual evoked potentials. *Frontiers in Psychology*, 4, 420. doi: 10.3389/fpsyg.2013.00420

Rizzolatti, G., Fadiga, L., Fogassi, L., & Gallese, V. (1997). The space around us. [Editorial material]. *Science*, *277*(5323), 190–191. doi: 10.1126/science.277.5323.190

Rizzolatti, G., Matelli, M., & Pavesi, G. (1983). Deficits in attention and movement following the removal of postarcuate (area-6) and prearcuate (area-8) cortex in macaque monkeys. *Brain*, *106*(Sep), 655–673. doi: 10.1093/brain/106.3.655

Rizzolatti, G., Scandolara, C., Matelli, M., & Gentilucci, M. (1981). Afferent properties of periarcuate neurons in macaque monkeys. I. Somatosensory responses. *Behavioural Brain Research*, *2*(2), 125–146.

Rossetti, A., Romano, D., Bolognini, N., & Maravita, A. (2015). Dynamic expansion of alert responses to incoming painful stimuli following tool use. *Neuropsychologia*, *70*, 486–494. doi: org/10.1016/j.neuropsychologia.2015.01.019

Scandola, M., Aglioti, S. M., Bonente, C., Avesani, R., & Moro, V. (2016). Spinal cord lesions shrink peripersonal space around the feet, passive mobilization of paraplegic limbs restores it. *Scientific Reports*, *6*, 24126. doi: 10.1038/srep24126

Simon-Dack, S. L., Cummings, S. E., Reetz, D. J., Alvarez-Vazquez, E., Gu, H., & Teder-Salejarvi, W. A. (2009). 'Touched' by light: event-related potentials (ERPs) to visuo-haptic stimuli in peri-personal space. *Brain Topography*, *21*(3–4), 261–268. doi: 10.1007/s10548-009-0096-3

Thomas, L. E. (2013). Grasp posture modulates attentional prioritization of space near the hands. *Frontiers in Psychology*, *4*. doi: 10.3389/fpsyg.2013.00312

Tseng, P., Bridgeman, B., & Juan, C. H. (2012). Take the matter into your own hands: a brief review of the effect of nearby-hands on visual processing. *Vision Research*, *72*, 74–77. doi: 10.1016/j.visres.2012.09.005

Vyas, D., Garza, J.P, & Reed, C.L. (2019). Hand function, not proximity, biases visuotactile integration later in object processing: An ERP study. *Consciousness Cognition*, *69*, 26–35. doi:https://doi.org/10.1016/j.concog.2019.01.007

7

Dissecting the experience of space as peripersonal

Wayne Wu

'Does seeing a big rock close to my foot differ from seeing the moon in the sky?'
—Frédérique de Vignemont (2018a)

7.1 Experiencing peripersonal space as peripersonal

Is there a controversy about *peripersonal space*? In the first instance, talk of peripersonal space reflects a choice about how to partition objective space around the subject for theoretical purposes. Here, 'peripersonal space' refers to part of objective space that has a distinctive behavioural significance for the organism. Accordingly, there is an interesting question about whether we (or our brains) represent or respond to peripersonal space so demarcated, for various sensory systems like vision, audition, and touch have neurons with spatial receptive fields that tile parts of that space or its boundaries. Some of these neural responses play a causal role in our perceptually experiencing that space.

That said, the proper criteria for theoretically demarcating peripersonal space can be debated. How should we fix the reference of the term 'peripersonal space'? Our choice will reflect diverging theoretical commitments that render different empirical evidence salient. For example, we can opt for a behavioural or functional criterion, something like space within the limits of one's effectors, say space within arm's reach. Alternatively, we can delimit peripersonal space in terms of electrophysiological properties—say, in respect of the response of neurons to the relevant part of space around the subject (for an overview of these issues, see Noel et al., and Klein for computational issues, both this volume). These criteria need not pick out the same region of space. There are important questions about boundary conditions, but I sidestep these because all agree that there is something distinctive about 'near' space around the subject's body in respect of neural, psychological, and/or behavioural response. In speaking of peripersonal space in what follows, readers can slot in their preferred space as referent.

This chapter focuses on *phenomenal response* in respect of our conscious awareness of space. Concerning the theoretically defined peripersonal space, whatever it ends up being, the question is this: Are we aware of peripersonal space *as peripersonal*? That is, is there a distinctive way that peripersonal space is perceptually experienced that differs from the perceptual experience of other parts of space? The functional and electrophysiological criteria just noted are neutral on whether this phenomenal question gets an affirmative answer if only because, on most accounts, the neural and functional do not fix all aspects of the

Wayne Wu, *Dissecting the experience of space as peripersonal* In: *The World at Our Fingertips*. Edited by Frédérique de Vignemont, Andrea Serino, Hong Yu Wong, and Alessandro Farnè, Oxford University Press (2021). © Oxford University Press.
DOI: 10.1093/oso/9780198851738.003.0007.

phenomenal. So, how might we accurately capture the distinctive phenomenal appearance of peripersonal space, if there is any?

This chapter explores the possibility of a distinctive spatial phenomenology when experiencing peripersonal space. In what follows, I will speak of *peripersonal experiences* to mark the distinctive aspectual shape of perceptual experience of space *as* peripersonal. In contrast, I will speak of representations *of* peripersonal space to mark representations of that space as objectively defined where such experiences need not ground peripersonal experiences (e.g. neural representations of that space that do not contribute to consciousness).[1] Whether we experience space as peripersonal is, I take it, part of the point of de Vignemont's question, quoted in the epigraph: do objects look different in relation to how close they are to us (our bodies)? An interesting answer to this would be that their location is experienced as peripersonal.

This chapter explores two ways of thinking about peripersonal experience. A *substantial* view takes the content of experience of peripersonal space to effectively represent that space as peripersonal. This idea is illusive, and I will spell out a concrete proposal in terms of a distinctive type of multi-model experience of objects relative to the body. In contrast, a *deflationary* view will understand peripersonal experience to be constituted by specific sensory-motor links where this implicates a distinctive role for attention that can affect how we experience space over time. In both cases, there is a distinctive type of peripersonal experience, but only on the substantial view does the perceptual system speak 'in terms of the peripersonal'.

Section 7.2 explains the difference between substantial versus deflationary aspectual shape in terms of representational characteristics of phenomenal consciousness. Section 7.3 is explicit about methodology. While introspection is used to adjudicate phenomenal claims, empirically minded theorists should distrust it, or at least be wary. This follows from the fact that one should distrust or be wary of data generated by instruments that have not been calibrated and shown reliable. We have not calibrated introspection. Accordingly, I take introspective pronouncements to provide at best materials to generate hypotheses about the shape of experience. Crucially, I only entertain hypotheses if they have plausible neural correlates. Thus, in section 7.4, I work with introspective reports that highlight the spatial experience of objects in peripersonal space around the body as an experience of such objects as near *one's* body. This spatial experience is elusive but can be made salient when it is distorted during tool use. During tool use, one's introspective sense of one's body registers bodily *expansion* relative to the space around it. This forges a link to neuroscientific work on the expansion of multimodal spatial receptive fields anchored to the body during tool use. Hence, the potential biological foundation for peripersonal experience is appropriate neural receptive field expansion. I then assess the evidence for receptive field expansion in humans, focusing on empirical work on visual extinction in neuropsychological subjects. Against standard interpretations, I discuss an alternative *attention* explanation. Section 7.5 clarifies what attention is, and section 7.6 argues that the neuropsychological evidence for receptive field expansion is equivocal because it is consistent with the attention alternative. If so, the neurobiological foundation supporting the substantial phenomenology of peripersonal experience is weakened. At this stage, we should not endorse a substantive conception of peripersonal

---

[1] What does 'representation' mean? Philosophers and scientists use the word differently, but this provides a possible common ground: a representation of peripersonal is something that carries information, in the statistical sense, about that space where this information is used by systems that generate behavior. This conforms to certain teleological analyses of representation in philosophy and science (deCharms and Zador 2000; Dretske 2000).

experience over a deflationary conception. Section 7.7 expands on what a deflationary form of peripersonal experience might be.

I emphasize that the focus of this chapter is on conscious experience and whether there is a distinctive form of peripersonal experience. Denying that there is a substantive way of experiencing space as peripersonal is consistent with allowing that there is a type of peripersonal representation that plays a distinctive role in behaviour including engagement with attention. Such behavioural influence can have phenomenal upshots but they need not amount to anything more than how we deploy attention to the space around us and what behaviour attention motivates. We are on firmer ground in respect of behaviour and representations of peripersonal space, so it is important to separate phenomenal versus functional claims about peripersonal representations.

7.2 Substantial and deflationary aspectual shape

Let me distinguish substantial versus deflationary ways of speaking of *perceiving-as*. One can be said to see a dog *as dangerous* if, in seeing the dog, one is strongly disposed to—or did—take flight. Talk of the *dangerous aspect* refers to a functional link between vision and action, a link between certain visual experiences of the dog's features—say, bared teeth—and one's motor capacities for fleeing, a type of sensory-motor coupling. Remarking on the fleeing individual, we might say that he saw the dog as dangerous. 'Shucks, Fido was just showing his teeth ... he's harmless,' the owner might complain even as the fleer's seeing the teeth triggered his running away. This coupling between what the subject sees, the teeth, and his response fixes the deflationary version of seeing-as. On this view, *to see an X as an F is to respond to one's seeing X in an F-appropriate way—i.e. appropriate to X's being an F.* To see an X as dangerous is to respond to one's seeing X by acting in ways appropriate to X's being dangerous.

The deflationary characterization provides a negative answer to the question: does the subject *visually* represent or experience the dog as dangerous? In crude metaphorical terms, does the vocabulary of the visual system trade on the term 'dangerous'? One way of characterizing the deflationary view (but not the only way) is to hold that the visual system trades on a simpler vocabulary regarding colours, shapes, and other basic visible features and their combination. The visual system might also represent higher-level features like bared teeth, but to represent bared teeth is not to represent danger. Rather, the learned correlation between bared teeth and danger leads to the fleeing response, informed by past experience. On this view, danger is not part of the visual vocabulary. Rather, talk of seeing the dog as dangerous is grounded in a learned behaviour in light of specific visual experiences.[2]

The substantial approach's contrast with the deflationary approach begins with a disagreement about visual representational resources. Both sides can agree that visual experiences draw on a representational vocabulary V. Yet, let there be meaningful talk of seeing X as F where F is not in V (as in seeing the dog as dangerous). Whereas the deflationary approach restricts itself to V and cashes out seeing-as-F in terms of the links between visual

---

[2] It might be that, for some readers, *as-dangerous* is clearly visual. If this is you, then find some other value for X in talk of seeing as-X that you think is not visible. For example, a physicist might see a trace in a vapor chamber as (that of) a neutrino, but I suspect that most of us would not want to say that 'neutrino' is in the visual vocabulary. A neutrino is, of course, literally not visible. Rather, the particle physicist has developed a type of conceptual expertise so that, in seeing a diagnostic trace, she immediately judges the trace as that of a neutrino. There is a substantive question of what is in the visual vocabulary (visual representations; for philosophical debates on this point, one might start with Siegel (2005) and the literature that her discussion sparked).

experience that represents in relation to V and behavioural capacities appropriate to F things (e.g. dangerous things), a substantial approach to visual experience goes beyond these resources. There are many substantial approaches, depending on how the visual experience is enriched. One substantial approach to seeing the dog as dangerous can enrich V to include representations of danger that go beyond representations of visible features of the dog. If one thought that visual experience involves conceptual representations of the sort that figure in thought, then one substantial approach imports the concept DANGEROUS into the visual vocabulary. Deflationary approaches will view this as unnecessary expansion of V. Another approach, closer to the one that we shall consider in section 7.5, is to imbue visual experience with other psychological states—say, emotional states such as fear. One could say, although this is only an intuitive gloss, that the dog looks dangerous because one's emotion 'penetrates' one's visual experience.[3] Two experiential capacities, vision and emotion, in some sense merge and in that way go beyond visual representational resources to explain seeing-as-dangerous.

How does this bear on questions about peripersonal experience? Consider de Vignemont's question. There is a trivial answer to it that misunderstands the force of her query—namely, to note that there are basic differences in content between seeing an object X at a distance and seeing X near one's body For instance, only the latter represents a human body. Capturing this does not require going beyond the basic visual vocabulary of objects and locations. For example, specifying spatial content does not require more than the capacity for allocentric spatial representation (see section 7.4). Consider then a different contrast brought out by thinking of seeing an X near *one's* hand and seeing an X near *another's* hand, especially where the other's hand is visually matched and oriented in a similar way to one's own. An appropriately oriented photograph would not reveal a visual difference between the two, so the two *matched cases* look the same from the agent's perspective. That is, the visual vocabulary on its own does not distinguish the two. Presumably, if there is peripersonal experience, it applies only to the case where one's own foot is present, so peripersonal experience brings in *something else* beyond the sort of representational familiarities we invoke when speaking of seeing X next to Y. So we might restate de Vignemont's question: how might experience of an object differ when it is next to a body part that is one's own versus one that is not?

7.3 A brief point about method

To plumb the contours of awareness, a common method is to have subjects introspect and tell us how they experience things. I am sceptical of introspection. This is not because I think that introspection, as a capacity, is generally unreliable (Schwitzgebel 2008; Irvine 2012). Indeed, I think there are many basic cases where it is reliable. It is where *complicated* phenomenology is introspected that I am sceptical about introspective reliability. Note that this scepticism is the healthy scepticism that a scientist should take when handed some data generated by an instrument from another lab. The scepticism should hold until one ascertains that the data has been properly generated in a well-designed experiment where the instruments are known to be reliable and have been appropriately calibrated. If this cannot be ascertained, a methodologically solid scientist sets the data aside until calibration concerns are settled (trust but verify). I treat introspection as effectively an instrument that we deploy, in a first person context, to detect and discriminate consciousness. Accordingly, healthy scepticism towards introspection sets it aside until introspection in a given context can be calibrated and shown

---

[3] For recent discussion of the penetration of experience by other psychological modalities (mostly cognition), see the essays in Zeimbekis and Raftopoulos, (2015).

to be reliable. In most contexts, where the phenomenology is complicated, no such calibration has been achieved. Hence, I suggest we set aside much introspective data (see Spener 2015 for one way to calibrate introspection; much more work is needed).[4]

Now consider introspective reports relevant to thinking about peripersonal experience. It is often said that, when using artifacts in action, the artifact is incorporated into one's *experience* of the body, often expanding one's sense of the boundaries of the body into the area of action. Here is a neuroscientist describing one instance of the phenomenon:

> You're driving home from your Saturday errands and you pull into the garage, turning at just the right moment to avoid the trash cans on one side and the bicycles on the other ... How are you so confident about where the edges of your car are? A fascinating aspect of driving is that when you sit at the steering wheel, your sense of 'body' expands to becomes the outside of the car ... your perception of 'self' becomes bounded by the bumpers and steel of the car, increasing in volume more than fortyfold. (Barth 2018, p. 60)
>
> Reproduced from Barth, Alison L. 2018. "Tool Use Can Instantly Rewire the Brain." In Think Tank: Forty Neuroscientists Explore the Biological Roots of Human Experience, edited by David J. Linden, 60–65. New Haven: Yale University Press.

Philosophers often talk of feeling objects at the end of a pencil or cane implying that the subject's sense of the body expands (NB: denying the experiential claim is consistent with allowing that tool use changes behaviour—e.g. the blind's skillful use of canes to navigate space). If we interpret such claims as regarding conscious experience, then the observations lend themselves to a substantial reading of aspectual shape in spatial experience. At least in contexts of artifact use, one experiences one's body *as* larger, *as* more extended, and so forth. Accordingly, objects originally at a distance from the body come to seem closer to one's body. Peripersonal experience alters and expands. Since we can introspect experience to have the properties described, or so it is claimed, the aspectual shape seems to be in the experience's content rather than just fixed by sensorimotor links. My current point is that such introspective data, without calibration—that is, without assessing whether we are reliable in introspecting this aspect of experience in this context—should not be used to adjudicate between a deflationary versus substantial reading about the type of experience at issue. We have no idea whether introspection in these contexts is reliable.[5]

That said, uncalibrated introspection can motivate *hypothesis generation*, although one would like hypotheses to be based on reliable data. In drawing on uncalibrated introspection, I invoke a soft cartesian assumption that, in introspecting honestly and carefully, subjects cannot be wildly off about experience. Still, the most one should do is to formulate conditional claims: if introspection is correct, then ... So, I shall use intuitive ideas about our experience of peripersonal space as an inspiration for hypothesizing about what peripersonal experience might be.

Still, such uncalibrated data should not give us deep confidence that we are getting things right (this is just healthy scepticism about any such data). Accordingly, I add a second constraint. One aim of the science of consciousness is to identify the neural basis of conscious

---

[4] Empirical approaches to consciousness take a strikingly anti-empirical approach to introspection. Rather than calibrate introspection, such approaches simply assume that introspection is reliable in the context where it is deployed. Introspection is the only instrument, it seems, where science is willing to use data despite the absence of careful calibration and assessment of reliability.

[5] There is a question as to how much weight Barth and the philosophers I have in mind would place on these observations. In this essay, I will take the reports at face value, according to the cartesian assumption about the reliability of introspection.

experience, and this begins with identifying appropriate neural correlates of experience. A natural bridge principle for identifying neural correlates links content at the level of experience to content carried by neural representations. Here is a simple implementation: if one experiences some content p, then one expects that the neural basis of that experience will also represent p. As experiential content changes, so will the content of the experience's neural basis. Philosophers speak of *supervenience*: phenomenal content supervenes on the content of the experience's neural basis, so that any change in phenomenal content entails a change in neural content. So, while I shall draw on phenomenological speculation, I shall anchor these claims in plausible neural content correlates. Specifically, I anchor talk of the 'expansion' experience in the expansion of neural spatial receptive fields. This bridging of content will allow us to identify empirical work of relevance to thinking about peripersonal experience.

7.4 A substantial phenomenal hypothesis about peripersonal experience

Recall that we are separating representations *of* peripersonal space as theoretically defined from the peripersonal experience of space *as* peripersonal. Affirming the former as playing a role in behaviour does not entail that there is a corresponding peripersonal experience (consider the contrast between visual representations in the dorsal stream that guide motor behaviour, taken to be unconscious, and visual representations that provide the substrate for visual experience in the ventral stream (Milner and Goodale 2006).[6] What might peripersonal experience be?

In speaking about the perceptual experience of space, we can speak of a phenomenal spatial field such as a visual or auditory field. The structure of the field can be described in terms of a structured reference frame, and two standard frames are *egocentric* and *allocentric* frames (see Alsmith, this volume, on egocentric representation; also (Klatzky 1998)). Consider X and Y. If one's spatial experience of Y is egocentric, then Y is mapped to the ego X as origin of the reference frame (e.g. in a cartesian coordinate system). For egocentric frames, X as origin must be, or at least normally is, the subject's body. For example, in vision X can be the eye or the head. In audition and vision, X is not itself in the spatial field as a perceived object. If the visual field is anchored to the eye as origin, the eye itself is not an object within the field in the sense that it is not itself seen. Similarly, if the ear or head serves as an egocentric origin in auditory experience, it is not heard. If one's spatial experience is allocentric and X and Y are in the perceptual field, one of these—say X—can serve as the origin of the reference frame such that Y is mapped relative to X. Unlike egocentric frames, X need not be the subject's body but, if X is part of the subject's body, then X is an object within the perceptual field. That is, the origin of allocentric fields are themselves perceived.[7]

For a case that does not seem to be captured by the dichotomy of allocentric and egocentric spatial reference frames, consider looking at one's hands and seeing a stone at arm's reach relative to one's hand. While the visible hand is one's own, since both the stone and the

---

[6] Thus, I can deny a substantial view of peripersonal experience but affirm the role of tool use in changing representations of relevance to influencing behavior where these representations do not lead to such experiences, at least in the substantial sense (cf. Cardinali et al. 2009). Thanks to Frédérique de Vignemont for prompting this reply.

[7] Touch puts pressure on this way of understanding spatial experience since tactile fields are egocentric and yet the 'origin'—namely, the body—is in a sense 'felt'. Tactile fields are more complicated to unpack, but we can set this aspect of touch aside in our discussion (on sight versus touch, see (Martin 1992).

hand are visible, the stone is experienced allocentrically, not egocentrically, mapped relative to one's hand. There are two reasons for this. First, unlike the eye or head, which can serve as an origin of an egocentric visual reference frame and is not then in the visual field, one's hand is visible and hence in the visual field. In this way, one's hand differs from standard visual egocentric origins. Second, there are the two visually matched cases of the sort that we noted earlier. One's hand is, in principle, visually like another hand. Given our hand's orientation in the visual field, we could place another person's hand in the same location to replicate how things look, matching visible surface properties. In such matched cases, we would have two visually indistinguishable situations, one with one's own hand and one with a qualitatively identical hand belonging to another. If the situation where the hand belongs to another is allocentric, then it seems that, when we see a stone next to our own hand in a visually matched situation, that experience is also allocentric. This would seem to rule out a *distinctively* peripersonal experience of a stone as near one's hand.

Yet one feels that the fact that it is *one's own* hand must make a difference to how we see things *as*. Put another way: isn't there something distinctively egocentric about seeing a stone as within reach relative to one's own body? The stone, one might say, is seen *as within reach of one's hand*. The invocation of aspectual shape suggests a possible form of visual peripersonal experience but the problem is that, given the matching case, it is not clear what the difference in experience would be that would justify talk of an experience as peripersonal.

What is clear is that the distinction between egocentric and allocentric visual spatial reference frames does not allow us to phenomenally divide the matching cases. That it is one's own hand in one of the cases is not visually apparent. Yet another fact is relevant. While one's own hand might be visually indistinguishable from another's, there is a non-visual difference: one *feels* one's own hand. That is, the way to distinguish the two experiences is to recognize that perceptual experience is generally *multimodal* (O'Callaghan, n.d.). We can then draw on somatosensory contributions to perceptual experience in this way: the visible body part when one sees an object near one's own hand is also a hand one feels in the way that one's body is given in somatosensory awareness. So, even if our visual experience of our hand can be visually matched to a similar experience of another's hand, invoking one's sense of one's own body allows us to reimpose an egocentric aspect to the multimodal experience of one's body. For we see the stone relative to a hand that we feel. In that way, the stone is seen as near one's own hand in a way distinct from the matching case of seeing another's visually matched hand.

Nevertheless, the proposed phenomenology remains tantalizingly out of reach in that it is not salient. Perhaps peripersonal experience in general is not phenomenologically salient but something that must be discovered with effort or in special circumstances. De Vignemont puts a related point as follows: 'we seem to be presented with a continuous visual field devoid of phenomenological boundary between what is close and what is far' (2018a, p. 1), between peripersonal and the extrapersonal space. Still, if the form of multimodal peripersonal experience we have arrived at exists, perhaps we can render it salient by *distorting it*. This returns us to the noted introspection of perceptual experience during tool use for, when we use tools with our body, the experiential perceptual field seems to alter because our sense of our bodies in visible space seems to alter.

To follow my methodology, let us leave introspective hypotheses and shift to the neurobiology. The striking results of Iriki et al. that have done much to inspire work on peripersonal space is relevant (Iriki, Tanaka, and Iwamura 1996). In experiments on awake behaving monkeys, Iriki et al. provided evidence that neurons with bimodal receptive fields anchored to the hand seem to expand after tool use. Specifically, the neurons in question have visual and tactile receptive fields, with the tactile receptive field fixed to the hand and the visual receptive field responsive to the area surrounding the hand. These are bimodal neural representations of peripersonal space. Iriki et al. reported that, after the animal's use of a rake to retrieve objects, the visual receptive field of the recorded neuron seemed to expand in the direction along the length of the tool. The bimodal neurons were now responsive to visual stimuli in the space surrounding the tool at locations beyond their pre-tool-use receptive fields.

My hypothesis is that the reported changes in perceptual experience during and after tool use have a neural correlate in the expansion of spatial receptive fields in similar neurons in humans modulated by tool use, neurons that turn out to be representations of peripersonal space. This is, of course, speculation, since it has not been demonstrated that such changes, if they exist in humans, are the neural basis, let alone neural correlates, of any experiential phenomenology during tool use. Still, we are operating at the level of hypothesis generation, and the current proposal is that the elusive phenomenology of seeing an object as near one's body can be modified by tool use that makes it salient by distorting it. The neurophysiology provides a potential neural basis of the modification attested to in introspective reports.

We have thus arrived at a *substantial* form of peripersonal experience in that we locate the content of seeing an object as near *one's* hand by treating peripersonal experience as multimodal and, in that way, separate it from the visually matched situation where the visible hand belongs to another.[8] That is, we merge two experiential capacities to explain what visual resources on their own cannot. On this view, we see an object as near a hand that is one's own because we can feel that hand. The content of peripersonal experience thereby registers that it is one's own hand because it is felt. This 'self-identification' is secured in how things look by integrating visual experience with how the body distinctively feels.[9] One can think of this as expanding from a basic visual vocabulary V, and we arrive at a visual experience of a spatial relation of the stone relative to the hand as peripersonal.

How is peripersonal experience of the sort postulated distinguished from the case where one holds one's own hand up relative to the moon—say, when one shapes one's hands around the moon as seen from one's vantage point? Remember that peripersonal experience is *not* salient, and we found it by identifying a case where that experience can be altered, almost like detecting an unobservable particle by perturbing it to measure its trace. The proposal is that the neural basis of peripersonal experience is in neurons with bimodal receptive fields that can be altered by tool use. These neurons will be activated by stimuli that are present in peripersonal space. Accordingly, the moon will not activate such stimuli, and so will not induce the neural basis of peripersonal experience. Hence, the moon will not be experienced as within peripersonal space. Yet, since peripersonal experience is in its typical form not salient, we cannot always introspectively discern the difference between seeing the moon

---

[8] I noted that this way of construing peripersonal spatial experience need not be the only way. For example, one might have a peripersonal visual experience of objects near the face even if one does not see one's face in having that experience. In the end, there might not be a uniform phenomenology of peripersonal experience, and this might reflect the point I made in section 7.1, that we have several theoretical options for how to define the notion of peripersonal space. I would not be surprised if, in the future, we identify several distinctive phenomenologies that are currently collated within the concept of peripersonal experience. The point of this essay is to show one way that we can explicate such experience introspectively and biologically.

[9] One can make the proposal even more substantial by invoking the idea of felt bodily ownership as a contribution to the multimodal experience either because tactile experience fixes ownership (Martin 1992) or that ownership is itself a phenomenological feature of experience (de Vignemont (2018b). For scepticism about a substantial reading of ownership, see Wu, forthcoming).

relative to one's hand held up to it and seeing a stone that is within reach near our hand. Their biological bases differ and their phenomenological structure can differ, although not always.

7.5 Attention

The previous proposal about phenomenology aims for biological plausibility by drawing on neurophysiological data that emphasizes the plasticity of neural representations of peripersonal space—namely, that they expand with tool use. To critically assess that evidence, I present an alternative interpretation of some of the empirical results as they relate to our phenomenological hypothesis. In doing so, I am not questioning (a) the existence of bimodal neurons that might constitute or contribute to distinctive representations of peripersonal space, or (b) that spatial receptive fields in neurons are malleable (Duhamel, Colby, and Goldberg 1992). The empirical evidence for these two points is robust. Rather, I am questioning the extension of the Iriki et al. (1996) result to the human domain, an extension that is endorsed by many theorists in this area and on which my phenomenal hypothesis draws as biological support. Still, one should keep in mind that the needed explanatory link correlates changes in neural representations in human sensory systems with changes in human phenomenal experience, but the electrophysiological data is obtained from nonhuman primates. That said, it is widely assumed that similar neurons would be found in humans were we able to probe the human brain by direct electrophysiological recording.

I shall call the standard interpretation of the effect of tool use on representations of peripersonal space, as observed by Iriki et al. (1996), the 'receptive field expansion hypothesis' or 'the expansion hypothesis' for short. The challenge to this account is that there is an alternative *attention hypothesis* that also explains relevant data that has not yet been ruled out (Holmes 2012). It will be important to articulate the alternative attention account, and this requires understanding what attention is.

There are two complaints about attention that are oft repeated and succinctly captured in this observation by Poldrack (quoted in (Goldhill, n.d.):

> I don't think we know what 'attention' is. It's a concept that's so broad and over-used as to be meaningless. There's lots of things people in psychology and neuroscience study that they call 'attention' that are clearly different things.

There are two complaints, one accurate and one inaccurate. The accurate complaint is undeniable. Theorists use 'attention' to cover many different processes. This conceptual profligacy should be eliminated. What is inaccurate is that we do not know what attention is. Contrast part of James's (James 1890) famous claim:

> Everyone knows what attention is. *It is the taking possession by the mind, in clear and vivid form, of one out of what seem several simultaneously possible objects or trains of thought.* Focalization, concentration of consciousness are of its essence. *It implies withdrawal from some things in order to deal effectively with others* (my emphasis).

Setting aside consciousness, the Jamesian conception of attention can be distilled down to the following: attention is the mind's taking possession of one out of potentially many targets *in order to deal effectively with it*—i.e. in order to act on it. Here's the key point: *every scientist knows this implicitly.*

James's gloss is a selection for action account of attention that explains the function of selectivity in attention. Attention is constituted by the selection of one target among many to guide behaviour (Allport 1987; Neumann 1987; Wu 2014). This makes attention a central part of agency.[10] James's definition identifies a sufficient condition for attention—namely that, when a subject mentally takes possession of some target T to act on it, then the subject thereby attends to T in the relevant modality. Thus, if a subject looks at a mug in order to point to it, pick it up, imagine it exploding, commit its identity to memory ('it's a mug') or otherwise do something in respect of it, the subject has thereby visually attended to the mug and the response is guided by this selective focus.

This condition of selection for behavioural guidance as attention cannot be objectionable to scientists since they accept it implicitly. All the standard paradigms used to probe attention assume James's sufficient condition as part of the basic methodology of how to construct an appropriate *attentional* experimental task. The reasoning is straightforward. There is no way to study attention without ensuring that experimental subjects attend in the required way. To ensure this, experimenters control attention through task instructions. Thus, if you want subjects to attend to a specific visual target, you make that target task relevant to performance. Accordingly, adequate performance occurs only if the subject visually selects the target to guide response. When you observe correct performance, you can conclude that they have selectively attended to the target. Response measures provide a way for the experimenter to track attention set by task instructions. The implicit assumption is that visual selective attention to the target in this case is just visual selection of the target to guide performance.

Scientists sometimes dismiss these points as mere semantics. This is a puzzling response. Concepts and their *correct* and *consistent* use matter if we are to pose perspicuous questions, construct clear explanations, and formulate adequate theories. If 'attention' is being used sloppily, this breeds confusion and makes integration of levels of analysis unnecessarily difficult. Sometimes, one has to police concepts because they are *technical* and have correct uses. There is nothing theoretically suspect or trivial in this.

Let me summarize the two main points. Scientists have a clear account of what attention is, one they inherited from James and made central to experimental methodology: subject selection of a target to guide behaviour. Accordingly, every scientist knows what attention is. Despite this, scientists sometimes use 'attention' to cover too many things. It should only cover one thing: the mental level selectivity of a target to guide task performance.

7.6 Visual extinction and tool use

We are now in a position to state two contrasting explanations of a central result that supports the proposed biological basis of multimodal peripersonal experience. I am going to focus on work emphasizing sensory extinction due to lesions in parietal cortex. Some neuropsychological subjects demonstrate tactile neglect in that, while they can detect tactile stimuli with high accuracy on the hand contralateral to their brain lesion—say, the left hand relative to a lesion in the right hemisphere—such detection performance is greatly reduced when they are touched on both hands. In this case, a right hand ipsilesional tactile stimulus *extinguishes* awareness of the left hand contralesional tactile stimulus. Interestingly, extinction is bimodal

---

[10] For an a priori argument that every action involves attention, see Wu, n.d.

or cross-modal in that ipsilesional visual stimuli near the hand can lead to tactile extinction of a concomitant contralesional stimulus. A plausible hypothesis is that bimodal visuotactile neurons play a role in both visual and tactile extinction. I accept this hypothesis for visual stimuli near the hand.

The crucial result is that, while a visual stimulus located *at a distance* from the hand does not strongly extinguish a tactile stimulus, *after tool use in relation to that same stimulus*, the visual stimulus will now more strongly extinguish the tactile stimulus. Drawing inspiration from the work of Ikiri et al. (1996) one hypothesis explaining this effect, and perhaps the most commonly endorsed one, is that bimodal representations of peripersonal space around the hand in humans expand with tool use. In an important early study of this effect by Farnè and Làdavas (2000), the authors concluded that:

> The results of the present study were clear in showing that the spatial extension of peri-hand space representation in humans is not fixed and, on the contrary, it can be rapidly modified by the use of a tool. (p. 1648)[11]

Given receptive field expansion, visual stimuli originally in extrapersonal space can now activate bimodal neurons that represent peripersonal space. Through activation of those neurons, now with expanded receptive fields given tool use, the extrapersonal visual stimuli generate greater cross-modal extinction. In effect, they are now within an expanded peripersonal space. The force of this proposal is that it provides a potential biological basis of the tool use-based modulation of peripersonal spatial experience that we postulated on introspective grounds.

The challenge is that attention might explain the observed extinction effects without invoking an expansion of bimodal receptive fields. If so, the attention alternative has two upshots that we can separate. The first is that it provides a competing explanation of central results in the peripersonal space literature and, second, it undercuts or at least weakens the biological support for the phenomenal hypothesis regarding peripersonal experience. In that sense, it undercuts or at least weakens the motivation for a substantial reading of peripersonal experience.

Focusing on attention is especially relevant since extinction, like neglect, is thought to be due to defects in attention—namely an inability to perceptually attend to contralesional targets. Let us connect neglect and attention as explicated earlier. Attention is selection of a target to guide behaviour. This I argued is implicitly accepted by scientists, given their experimental methodology. Human behaviour typically occurs in a context where there are too many possible targets and too many potential responses. One can pick up a mug but also point to it, trace its shape, verbally describe it, imagine it broken, commit its color to memory, plan its destruction, and so forth. One can make the same set of responses to the wine glass next to the mug or to myriad objects in one's visual field. No behaviour—that is, no execution of a potential target-response mapping—is possible unless this *competition* between possible targets and responses is resolved. Therefore, we are confronted with many possible actions when we perceive the world as we move through it, and these compete for priority in engaging behavioural production and control mechanisms (Cisek and

---

[11] Again, it does not follow from such claims that there is a correlated change in phenomenology. So, by endorsing receptive field expansion, the authors need not be committed to the phenomenal hypothesis we are considering.

Kalaska 2010). Coherent behaviour requires resolving this competition among behavioural possibilities.

There is a corollary: in each of the visually guided behaviours canvassed, one can identify visual selection of a target that provides the information needed to accurately guide performance. Given the Jamesian conception, such visual selection to guide performance is visual attention. More generally, perceptual selection to guide performance yields perceptual attention. Thus, the competition for action is effectively a competition for perceptual attention. In acting in a certain way, one thereby attends in the way needed to guide that action. Attention and action are yoked together. The resolution of competition that yields action entails appropriate perceptual selection for the guidance of behaviour and, hence, perceptual attention. This coheres with current biased competition accounts of attention (Desimone and Duncan 1995).[12]

The link between attention and action makes salient the notion of a *priority map* (Fecteau and Munoz 2006; Zelinsky and Bisley 2015). This notion is related to the notion of a *salience map* (Itti and Koch 2000) but subsumes the latter in that it integrates the salience of items in a sensory field as well as their behavioural relevance to intentions, goals, valence, and other factors. Priority maps incorporate salience and behavioural relevance. The explanatory function of the priority map is to identify the targets of attention: one attends to an item of highest priority.[13] The notions of competition and priority provide the basic ingredients to explain extinction.

To see this, let us begin with a familiar example concerning inattentional blindness— namely, the case of the gorilla and the basketball. Simons and Chabris (1999) asked subjects to count the number of basketball passes between players in white shirts while ignoring a second ball passed between players in black shirts. This context involves a two-way competition between counting actions, one focused on the whites' ball and the other on the blacks' ball. To act in this context, one follows task instructions and counts the passes of one and not the other ball. This evokes the Jamesian condition on attention, fixing visual attention to just one ball through instructions. The striking result is that, while doing this task, many subjects fail to see a gorilla that walks through the scene (see also Mack and Rock 1998). Items of higher priority—namely, task-relevant objects—can effectively exclude putatively salient items from engaging behaviour since those items are task irrelevant. The gorilla is not sufficiently prioritized so as to engage orienting mechanisms. It is then not noticed.

Similarly, we can understand visual extinction as the result of aberrant competition and resulting distortion of priority (Ptak and Fellrath 2013). In a detection task where neuropsychological subjects susceptible to extinction have to report all stimuli presented in a given trial, such subjects who are normal or near normal in detecting a single tactile stimulus on the contralesional hand (say, left hand relative to right hemisphere damage) show lower detection capacity when a second stimulus is simultaneously presented on the ipsilesional

---

[12] There is, of course, much more to discuss regarding attention, including the relation between the selection to guide performance or action conception and various models of attention. In (Wu 2014), I argue that the Jamesian conception provides what Marr would call 'a computational theory of attention'—namely, an answer to what attention is in terms of what it does. This then provides a basis for unifying much of the attentional literature at different levels of analysis. A critical dichotomy that is relevant here but would take us too far afield is the issue of control and automaticity in attention and, in my view, the role of intention in setting attention. For control and automaticity, see Wu (2013); on the role of intention in terms of divisive normalization accounts of attention and top-down modulation, see Wu (2017).

[13] This is not a tautologous claim but a causal one. Certainly, there is a concept of priority, which just means that which is attended, hence the tautology of saying that an attended item is of highest priority. In contrast, priority as a magnitude in a priority map explains the shift of attention to a target.

side. Given the task to report all stimuli in the sensory field, both stimuli should have equal priority as task-relevant targets. Yet the balance of competition between multiple stimuli, in terms of the underlying mechanisms, is shifted in such patients. Invoking the priority map as an explanatory device, we can say that, in contrast to normal circumstances where both targets would get equal priority and appropriately engage detection response ('there are two lights'), the lesion leads to a distortion in the priority map so that the ipsilesional target tends to get higher priority and, thus, tends to engage selection for report. The result is extinction of the contralesional stimulus in that it does not engage detection response.[14]

It should be noted that, in the patients studied, there is cross-modal extinction even before tool use where a visual stimulus in extrapersonal space on the ipsilesional side can out-compete the contralesional tactile stimulus. This is critical to keep in mind since it shows that extrapersonal stimuli drive visual neurons whose receptive fields cover extrapersonal space sufficient for extinction effects, even if more weakly. Accordingly, the bimodal neurons that I have invoked to support the expansion hypothesis are not necessary for cross-modal extinction. Rather, visual neurons with extrapersonal receptive fields can induce extinction even before tool use. One possibility is that tool use then increases the capacity of the ipsilesional visual stimulus, via the extrapersonal visual neurons at issue, to induce cross-modal extinction of the contralesional tactile stimulus. In terms of a priority map, the visual stimulus in extrapersonal space acquires higher priority after tool use.

We are trying to explain the tool-mediated increase in extinction by a visual stimulus that is located in extrapersonal space. Both the attention and the expansion hypotheses can emphasize an alteration in attentional processing to explain extinction after tool use—say, through distortion of the priority map that assigns a higher priority to extrapersonal stimuli. Where the hypotheses differ is in how they explain this distortion effect. On the receptive field expansion account, visual stimuli in peripersonal space are potent extinguishers of con-comitant stimuli. If a visual stimulus in extrapersonal space becomes a potent extinguisher of a competing stimulus after tool use, then this is because the receptive field of bimodal neurons expands towards the extrapersonal stimulus bringing it within peripersonal space. On this view, bimodal neurons play the critical role by altering their response properties.

Alternatively, the postulated distortion in the priority map can be the result of changes in the properties of the *unimodal* visual neurons whose receptive fields tile extrapersonal space within which the visual stimuli at issue occur.[15] Remember that visual stimuli in extrapersonal space do extinguish concomitant tactile stimuli, so this demonstrates that uni-modal visual neurons representing extrapersonal space can shift attentional processing in a way that leads to extinction. The attention account holds that tool use increases the contribu-tion of these neurons to constructing the priority map, leading to the observed extinction of contralateral tactile stimuli.

This division of mechanisms between unimodal versus bimodal neural bases is some-what coarse. After all, the egocentric visual reference frame, being centred on some part of the body (e.g. eye or head), can only be constructed by drawing on information regarding the orientation of body—say, from proprioception (e.g. head-centered coding in parietal areas such as VIP [Cohen and Andersen 2002]). As visual egocentric reference frames are

---

[14] It is important to note that we need not say something as extravagant as that the subject is literally numb to the contralesional stimulus. All that we need say is that, given that the stimulus is lower in priority, it does not affect behaviour—say, orienting and report mechanisms. This is neutral on whether the stimulus is in some sense still in consciousness.

[15] Dividing the mechanisms in this way is consistent with allowing that some ways of measuring peripersonal effects can identify unimodal influences. See Blini et al., this volume.

in that way multimodal, relying on proprioceptive information, one should be wary of talk of 'purely' unimodal neural representations that relate to perceptual experience. That said, given the neurophysiology, the expansion account focuses on those neural populations that exhibit the visuotactile signatures of the sort found by Iriki et al. (1996). Such signatures are not found throughout the visual system (see Noel et al., this volume, who argue that bimodal visual-somatosensory peripersonal neural representations are found in two areas of parietal cortex, area VIP and 7b).[16] So, the unimodal hypothesis focuses on visual neurons that are not multimodal in the way identified by Iriki and colleagues.

The challenge in assessing these conflicting explanations is that without direct recording of neural activity and fine-tuned understanding of the circuit differences between the two cases, we must leverage behavioural data to adjudicate the conflict. It might be that the behavioural data is too blunt to yield clear contrastive predictions but, if that is the case, then we are not in a position to endorse the receptive field expansion model over the attention model. Both remain live options. If we cannot rule out the attention model, we are not free to endorse the expansion model over it. We then undercut our appeal to the expansion model as biological support for the substantive conception of peripersonal experience identified on introspective grounds. This also should give investigators pause in applying an expansion account reading on the extinction data, as if that was the only plausible hypothesis.

There have been attempts to test the models against each other, and these mark important areas of research. One critical result that is taken to support the expansion as against the attention hypothesis comes from work showing that extinction by visual stimuli occurs *along the length* of the tool after tool use (Bonifazi et al. 2007) and this is taken to provide evidence against alternative attention explanations (Holmes 2012). In fact, this is not so clear. First, the effects of attention are highly task sensitive, which is not surprising if attention plays task-relative selective roles as the Jamesian conception suggests (on extinction and task dependence, see Bonato 2012). Thus, one must be careful comparing studies that deploy different tasks and how these interact with the method for measuring attentional effects. Depending on how attention is deployed, the effects might vary. This may explain the data from using a tool to pull an object *towards* the subject, which is interpreted as favouring the expansion hypothesis, and data from using a tool to point or push an object, *away* from the subject, which is interpreted as favouring the attention hypothesis. These are different tasks (see further work by Reed and co-workers using different tools, summarized in Reed and Park, this volume; they also emphasize the task-dependence of results).[17]

Second, in tool tasks, the tool is task relevant in that the subject must select the tool to manipulate a target to generate correct performance. It follows that the subject will be attending to the tool even if this selection changes its character with practice, with attention to the tool becoming automatic (on automaticity and control, see Wu 2013).[18] Attention to

---

[16] One concern, which I shall set aside although it is a substantial one, is that the areas in which bimodal neurons have been found are not areas that clearly contribute to sensory experience. For example, VIP and 7b lie in the dorsal visual stream, and this stream is considered by many to be unconscious in the sense that the visual states supported by the dorsal visual stream are unconscious visual states.

[17] Holding and using tools also alters attentional processing in visual detection. Thus, even prior to tool use, facilitation of reaction time in detection tasks occurs at the functional end of the tool (Park and Reed 2015). Reed and Park (reported in this volume) also report that the nature of the task influences the attentional facilitation. For example, pushing and pulling tasks in extrapersonal space exhibited attentional facilitation at the end of a tool while pulling tasks in peripersonal space demonstrated attentional facilitation along the length of the tool. Importantly, this suggests that different types of task can lead to different modulations of attention effects, as measured by reaction time in detection tasks.

[18] In well-practised tasks with a tool, some are inclined to say that one does not attend to the tool. It is important not to confuse automatic attention to the tool, here automatically selecting it for behaviour in the Jamesian sense,

the tool will involve both object and spatial attention, and, given that we have a functional artifact, something we might call 'functional' attention as well—say, the properties of the tool that one might be tempted to speak of as tied to the *affordances* it presents (see Wu 2008) for an argument about functional distributions of top-down attention in tool use). These complexities suggest that manipulation of attention in tool use is likely to be a complex matter.

To see this, consider the experiment by Bonifazi et al. (2007, op. cit.) who explicitly tested the expansion model against the attention model. They worked with right parietal lesioned patients on crossmodal extinction before and after tool use. In this case, subjects were tasked with using a rake to retrieve objects in extrapersonal space. Before tool use, the subjects were able to detect left contralesional tactile stimuli with normal accuracy (98% detection accuracy). Crossmodal extinction does occur in these patients before tool use in that visual stimuli outside of peripersonal space 35 cm away from the hand induce a drop in accuracy in tactile detection of contralesional stimuli to about 65%. There was a significant main effect of the position of the visual stimulus, with extinction more severe when the visual stimulus was at the hand (accuracy 42%). Since the visual stimuli 35 cm away from the hand are by hypothesis outside peripersonal space, this shows that the activation of extrapersonal visual neurons can lead to crossmodal extinction of the tactile contralesional stimulus.

In support of the expansion hypothesis, Bonifazi et al. (2007) note that after using a rake to retrieve an object (i.e. a pulling task), visual extinction along the length of the rake increased in severity so that visual stimuli at middle and far locations (35 and 60 cm from the hand) showed similar extinction effects as visual stimuli at the hand. The authors conclude that the result is not compatible with an attention-based account (they specifically target the so-called *addition hypothesis* whereby attention is directed only to the tip of the tool (as argued by Holmes et al., 2004).

Yet, with the conceptual clarifications noted earlier, there are reasons to think that an attention-based account is compatible with an increase in extinction by stimuli along the tool after tool use. First, given the task, the tool is task relevant, thus a target of selection for performance, and hence attended to even if only automatically. Second, the functional nature of the tool can lead to different distributions of attention (see Reed and Park, this volume). Third, there are historical effects in performance that influence attention across trials (Awh, Belopolsky, and Theeuwes 2012). The block of tool use trials (50 total trials) in the Bonifazi et al. (2007) study set an empirical condition that requires attention to be directed at the tool—namely, selection for guiding behaviour. The repetitive deployment of attention could lead to priming effects in subsequent trials, in this case potentially the salience of the tool that remains in view. In the Bonifazi study, lights placed just above the tool were used as putative extinction stimuli. Finally, object-based attention effects, while complicated, suggest that, at a minimum, there is a predisposition to spread spatial attention along the length of an attended object although there is also evidence that attention can be deployed along an object in a strategic way (see, for a review on whether this spread is automatic or strategic, Shomstein 2012).

These are just a few of the complexities that one must consider in interpreting results if one is assessing the involvement of attention as a putative mechanism. The suggestion then is

---

with not attending to the tool, here not selecting it for behaviour. Clearly, in any tool use, information regarding the tool must be selected. What changes, over practice, is the automaticity of such selection to guide behaviour—i.e. of attention.

that, if attention is deployed, this can itself lead to the increased salience and relevance of the tool, at least right after tool use. The point then is that both theories can make the same prediction regarding extinction, at least at a coarse grain of analysis. Both predict that we should see increased extinction effects along the length of the tool. In the one case, it is because of an expansion of the bimodal receptive fields anchored to the hand; in the other, it is because of the increased relevance of an area of space that coincides with the location of task-relevant targets and the tool. What differs is not the behavioural prediction but the proposed underlying mechanisms.

There have been additional attempts to test the two hypotheses. For example, Rossetti et al. (2015) examined skin conductance responses (SCR) to threatening stimuli directed towards the subject's body before and after tool use. They noted that SCR increases as the noxious stimuli are positioned closer to the body. Further, objects that yielded a weak SCR response because they were placed further from the body before tool use induced a stronger SCR response after tool use. The expansion hypothesis explains the result by positing an expansion of the receptive fields of bimodal neurons that are anchored to the body towards the noxious stimuli. As a control, Rossetti et al. (2015) tested whether attention on its own could lead to similar effects. They had subjects hold the tool without using it and generated observational reports regarding objects in the area around the tool. Drawing on our Jamesian sufficient condition, this task directs visual attention to targets around the tool. After this attentional task, the authors did not observe an increase in SCR when noxious stimuli were placed around the tool at locations to which visual attention was previously directed. This suggests that visual attention on its own is not sufficient to yield the shifts in SCR attributed to tool use.

This is an important control but invites a weaker conclusion—namely, that certain deployments of attention are required to induce the effect on SCR (Rossetti et al. 2015 effectively note this possibility, p. 491). After all, attention is also involved in training with the tool since one must attend to objects at the end of the tool in those tasks. Thus, we do not have contrasting conditions where attention is absent versus present. All that we are entitled to claim is that attention that is tied to tool use is correlated with changes in SCR but not attention that is decoupled from tool use.[19] So the result is informative, but it leaves open the central question: what is the mechanism mediating the attentional effect?

Where does this leave us? I myself lean towards the expansion hypothesis but it seems that the attention model I have outlined remains a live option. There is a fine-grained mechanistic difference that separates different explanations of what happens in visual extinction studies deploying tools. To rigorously test these options, I would suggest the following: (a) that researchers converge on the same task, perhaps one analogous to Iriki et al.'s (1996) original task as replicated in the extinction studies; (b) that researchers more directly probe for attention effects; (c) that researchers find a way to differentiate between what I have called the unimodal and bimodal neural bases; (d) that researchers, especially proponents of the attention account, derive clear contrastive predictions. More work is needed, keeping the two contrasting hypotheses clearly in view.

I would emphasize focusing on a uniform task context across experiments, given different attentional effects in different tasks. It would be good if the field agreed on one concrete task

---

[19] I'm grateful to Alessandro Farné for drawing my attention to this study. A similar response would apply to work from his group (Makin et al. 2009) where the relevant form of attention was exogenous attention directed towards an approaching ball. These studies rule out certain deployments of attention as capable of explaining the effect and contribute to sharpening the contrast between the two theories noted.

to unpack and deploy in experimental work testing the two mechanisms. The original raking task in the Iriki et al. (1996) study as well as similar deployments in work with extinction patients might provide the appropriate paradigm. With agreement on a task, theoretical predictions would be easier to derive. As it is, a lot of work has been done across different tasks, making it hard to evaluate the two accounts (e.g. see work on extinction and the crossmodal congruency effect, reassessed in Holmes 2012). Further work dissecting the original monkey results would also be ideal, especially since that work provides the inspiration for the expansion hypothesis.

7.7 Peripersonal experience

Let us return to the deflationary approach to peripersonal experience. In the case of experiencing something as dangerous, the deflationary proposal fixed talk of aspectual shape in terms of responses to certain visible features consonant with those features indicating danger. Given the account of attention noted earlier, responses guided by visual experiences entail a type of visual attention, given the Jamesian condition of selection for behaviour as a form of attention. Thus, if a subject's seeing a dog as dangerous is secured by the subject's fleeing (or propensity to flee) from a dog when he sees the dog's bared teeth, this implicates attention to the dog's teeth as what initiates the fleeing response. Seeing the dog as dangerous in this context is tied to a distinctive deployment of visual attention that functions to guide a specific response, one that is appropriate if the dog's teeth indicates danger. This distinguishes the fleer from the owner who instead sees the dog as friendly and pets the dog (perhaps this dog bares teeth to indicate affection). The distinctive experience is all in the experience's role in guiding a response, given a basic visual representational vocabulary.

In the case of peripersonal experience, we can highlight a similar phenomenon. The empirical work suggests that certain brain areas with sensory receptive fields exhibit a distinctive response to the area around the body. This work identifies how our brains might prioritize space around the body and how the representations of that peripersonal space feed into specific types of behaviour. If so, then again, such behaviour implicates a distinctive deployment of attention whose biological basis involves perceptual neurons that represent peripersonal space. These deployments of attention and their link to specific responses provide a basis for talking about perceiving (and attending to) space as peripersonal. The phenomenology is deflationary in that the experience that emerges from such interactions with peripersonal space is secured by how attention is deployed and what behaviours it initiates, given an underlying neural basis of representations of peripersonal space. In respect of phenomenology, this link between attention to peripersonal space and specific behaviours might be peripersonal enough to identify why the space around the body makes a difference for the subject.

Acknowledgement

I am grateful to Frèdèrique de Vignemont, Alessandro Farnè, and a third anonymous reviewer for their very helpful comments on this chapter.

References

Allport, A. 1987. "Selection for Action: Some Behavioral and Neurophysiological Considerations of Attention and Action." In *Perspectives on Perception and Action*, 395–419. Hillsdale, N.J.: Lawrence Erlbaum Associates, Publishers.

Awh, Edward, Artem V. Belopolsky, and Jan Theeuwes. 2012. "Top-down versus Bottom-up Attentional Control: A Failed Theoretical Dichotomy." *Trends in Cognitive Sciences* 16 (8): 437–443. https://doi.org/10.1016/j.tics.2012.06.010

Barth, Alison L. 2018. "Tool Use Can Instantly Rewire the Brain." In *Think Tank: Forty Neuroscientists Explore the Biological Roots of Human Experience*, edited by David J. Linden, 60–65. New Haven: Yale University Press.

Bonato, Mario. 2012. "Neglect and Extinction Depend Greatly on Task Demands: A Review." *Frontiers in Human Neuroscience* 6: 195. https://doi.org/10.3389/fnhum.2012.00195

Bonifazi, S., A. Farnè, L. Rinaldesi, and E. Làdavas. 2007. "Dynamic Size-Change of Peri-Hand Space through Tool-Use: Spatial Extension or Shift of the Multi-Sensory Area." *Journal of Neuropsychology* 1 (1): 101–114. https://doi.org/10.1348/174866407X180846

Cardinali, Lucilla, Francesca Frassinetti, Claudio Brozzoli, Christian Urquizar, Alice C. Roy, and Alessandro Farnè. 2009. "Tool-Use Induces Morphological Updating of the Body Schema." *Current Biology* 19 (12): R478–479. https://doi.org/10.1016/j.cub.2009.05.009

Cisek, Paul, and John F. Kalaska. 2010. "Neural Mechanisms for Interacting with a World Full of Action Choices." *Annual Review of Neuroscience* 33 (1): 269–98. https://doi.org/10.1146/annurev.neuro.051508.135409

Cohen, Y. E, and R. A Andersen. 2002. "A Common Reference Frame for Movement Plans in the Posterior Parietal Cortex." *Nature Reviews Neuroscience* 3 (7): 553–562.

deCharms, R. C., and A. Zador. 2000. "Neural Representation and the Cortical Code." *Annual Review of Neuroscience* 23: 613–647. https://doi.org/10.1146/annurev.neuro.23.1.613

de Vignemont, Frédérique. 2018a. "Peripersonal Perception in Action." *Synthese*, October. https://doi.org/10.1007/s11229-018-01962-4

de Vignemont, Frédérique. 2018b. *Mind the Body: An Exploration of Bodily Self-Awareness*. Oxford: Oxford University Press.

Desimone, Robert, and John Duncan. 1995. "Neural Mechanisms of Selective Visual Attention." *Annual Review of Neuroscience* 18: 193–222. https://doi.org/10.1146/annurev.ne.18.030195.001205

Dretske, F. 2000. "If You Don't Know How It Works, Then You Can't Build It." In *Perception, Knowledge and Belief: Selected Essays*, 208–226. Oxford: Oxford University Press.

Duhamel, J. R., C. L. Colby, and M. E. Goldberg. 1992. "The Updating of the Representation of Visual Space in Parietal Cortex by Intended Eye Movements." *Science* 255 (5040): 90.

Farnè, A., and E. Làdavas. 2000. "Dynamic Size-Change of Hand Peripersonal Space Following Tool Use." *NeuroReport* 11 (8): 1645–1649.

Fecteau, Jillian H., and Douglas P. Munoz. 2006. "Salience, Relevance, and Firing: A Priority Map for Target Selection." *Trends in Cognitive Sciences* 10 (8): 382–390. https://doi.org/10.1016/j.tics.2006.06.011

Goldhill, Olivia. n.d. "Psychology Will Fail If It Keeps Using Ancient Words like 'Attention' and 'Memory.'" *Quartz*. Accessed 4 July 2020. https://qz.com/1246898/psychology-will-fail-if-it-keeps-using-ancient-words-like-attention-and-memory

Holmes, Nicholas P. 2012. "Does Tool Use Extend Peripersonal Space? A Review and Re-Analysis." *Experimental Brain Research* 218 (2): 273–282. https://doi.org/10.1007/s00221-012-3042-7

Holmes, Nicholas P., Gemma A. Calvert, and Charles Spence. 2004. "Extending or Projecting Peripersonal Space with Tools? Multisensory Interactions Highlight Only the Distal and Proximal Ends of Tools." *Neuroscience Letters* 372 (1): 62–67. https://doi.org/10.1016/j.neulet.2004.09.024

Iriki, A., M. Tanaka, and Y. Iwamura. 1996. "Coding of Modified Body Schema during Tool Use by Macaque Postcentral Neurones." *NeuroReport* 7 (14): 2325–2330.

Irvine, Elizabeth. 2012. "Old Problems with New Measures in the Science of Consciousness." *British Journal for the Philosophy of Science* 63 (3): 627–48.

Itti, Laurent, and Christof Koch. 2000. "A Saliency-Based Search Mechanism for Overt and Covert Shifts of Visual Attention." *Vision Research* 40 (10–12): 1489–1506.

James, William. 1890. *The Principles of Psychology, Volume 1*. Boston, MA: Henry Holt and Co.

Klatzky, R.L. 1998. "Allocentric and Egocentric Spatial Representations: Definitions, Distinctions, and Interconnections." In *Spatial Cognition, An Interdisciplinary Approach to Representing and Processing Spatial Knowledge*, edited by Christian Freska, Christopher Habel, and Karl Wender, 1–18. Berlin: Springer.

Mack, Arien, and Irvin Rock. 1998. *Inattentional Blindness*. Cambridge, MA: MIT Press.

Makin, Tamar R., Nicholas P. Holmes, Claudio Brozzoli, Yves Rossetti, and Alessandro Farnè. 2009. "Coding of Visual Space during Motor Preparation: Approaching Objects Rapidly Modulate Corticospinal Excitability in Hand-Centered Coordinates." *Journal of Neuroscience* 29 (38): 11841–11851. https://doi.org/10.1523/JNEUROSCI.2955-09.2009

Martin, M. G. F. 1992. "Sight and Touch." In *The Contents of Experience*, 196–215. Cambridge: Cambridge University Press.

Milner, A. David, and Melvyn A. Goodale. 2006. *The Visual Brain in Action*. 2nd ed. Oxford: Oxford University Press.

Neumann, O. 1987. "Beyond Capacity: A Functional View of Attention." In *Perspectives on Perception and Action*, 361–94. Hillsdale: Lawrence Erlbaum Associates, Publishers.

O'Callaghan, Casey. n.d. *A Multisensory Philosophy of Perception*. Oxford: Oxford University Press.

Park, George D., and Catherine L. Reed. 2015. "Haptic over Visual Information in the Distribution of Visual Attention after Tool-Use in near and Far Space." *Experimental Brain Research* 233 (10): 2977–2988. https://doi.org/10.1007/s00221-015-4368-8

Ptak, Radek, and Julia Fellrath. 2013. "Spatial Neglect and the Neural Coding of Attentional Priority." *Neuroscience & Biobehavioral Reviews* 37 (4): 705–722. https://doi.org/10.1016/j.neubiorev.2013.01.026

Rossetti, Angela, Daniele Romano, Nadia Bolognini, and Angelo Maravita. 2015. "Dynamic Expansion of Alert Responses to Incoming Painful Stimuli Following Tool Use." *Neuropsychologia* 70 (April): 486–494. https://doi.org/10.1016/j.neuropsychologia.2015.01.019

Schwitzgebel, Eric. 2008. "The Unreliability of Naive Introspection." *Philosophical Review* 117 (2): 245–273.

Shomstein, Sarah. 2012. "Object-Based Attention: Strategy versus Automaticity." *Wiley Interdisciplinary Reviews: Cognitive Science* 3 (2): 163–169.

Siegel, Susanna. 2005. "Which Properties Are Represented in Perception." In *Perceptual Experience*, edited by Tamar S. Gendler and John Hawthorne, 481–503. Oxford: Oxford University Press.

Simons, D. J., and C. F. Chabris. 1999. "Gorillas in Our Midst: Sustained Inattentional Blindness for Dynamic Events." *Perception* 28 (9): 1059–1074. https://doi.org/10.1068/p281059

Spener, Maja. 2015. "Calibrating Introspection." *Philosophical Issues* 25 (1): 300–321.

Wu, Wayne, forthcoming. "Mineness and Introspective Data." In *Mineness*, edited by Marie Guillot, and Manuel García-Carpintero. Oxford: Oxford University Press.

———. 2008. "Visual Attention, Conceptual Content, and Doing It Right." *Mind* 117 (468): 1003–1033.

———. 2013. "Mental Action and the Threat of Automaticity." In *Decomposing the Will*, edited by Andy Clark, Julian Kiverstein, and Tillman Vierkant, 244–261. Oxford: Oxford University Press.

———. 2014. *Attention*. Abingdon, UK: Routledge.

———. 2017. "Shaking Up the Mind's Ground Floor: The Cognitive Penetration of Visual Attention." *Journal of Philosophy* 114 (1): 5–32.

———. n.d. "Action Always Involves Attention." *Analysis*. https://doi.org/10.1093/analys/any080

Zeimbekis, John, and Athanassios Raftopoulos. 2015. *The Cognitive Penetrability of Perception: New Philosophical Perspectives*. Oxford, Oxford University Press.

Zelinsky, Gregory J., and James W. Bisley. 2015. "The What, Where, and Why of Priority Maps and Their Interactions with Visual Working Memory." *Annals of the New York Academy of Sciences* 1339 (March): 154–164. https://doi.org/10.1111/nyas.12606.

PART II
SPACE AND MAP

8

Do we represent peripersonal space?

Colin Klein

8.1 Introduction

The space near our bodies is important. It is where we act. It buffers us from threats. We spend a lot of time trying to keep people out of this space, and letting them into it is a sign of trust and intimacy.

There is now considerable evidence that the brain contains neurons dedicated to the representation of this *peripersonal space* (PPS). Graziano and colleagues (1994) showed that the precentral gyrus of macaques contained neurons with bimodal visual and tactile receptive fields. Similar neurons are found in the ventral intraparietal area. These neurons respond to both touch on a part of the body and to a visual region extending out from that body part. The visual receptive field is anchored to the relevant body part and moves as the body part moves. Some of these neurons are also responsive to nearby sounds, regardless of absolute sound intensity or whether the stimulus is visible (Graziano, Hu, and Gross 1997; Graziano, Reiss, and Gross 1999). There are more, and more densely overlapping, receptive fields closer to the body (Graziano and Cooke 2006, Figure 7.4).

Further, stimulation of the same neurons using biologically realistic parameters can evoke complex defensive behaviours (Graziano, Taylor, and Moore 2002; Graziano et al. 2002; Graziano 2006). Stimulation of a region containing neurons with face-centered bimodal receptive fields, for example, evoked a complex defensive response involving head motion and a guarding response with the hand and arm, all directed towards the portion of space to which those responded (Graziano, Taylor, and Moore 2002). Although the organization of the precentral gyrus seems to involve a variety of different action types, the polysensory zone seems to be especially concerned with defensive motions (Graziano et al. 2002). Reviewing the first two sets of findings, Graziano and Cooke (2006, p. 846) thus suggested that 'a major function of these cortical areas is to maintain a margin of safety around the body and to coordinate actions that defend the body surface.'

Further research has deepened our appreciation of these neural systems (see Bufacchi and Iannetti, 2018, for a recent review). Yet this research has also raised significant empirical and conceptual questions about the nature of the PPS representation. This debate arises in part because the concept of representation itself remains the subject of considerable philosophical debate (Shea 2018).

In what follows, I will use PPS to tackle some of these outstanding questions about representation—both about PPS in particular and about representation in general. The strategy will be oblique. I will start with the problem that PPS must solve: transformation between different coordinate systems. I will then show how the computational demands on coordinate transformation differ in important ways depending on the bases used for the two

Colin Klein, *Do we represent peripersonal space?* In: *The World at Our Fingertips*. Edited by Frédérique de Vignemont, Andrea Serino, Hong Yu Wong, and Alessandro Farnè, Oxford University Press (2021). © Oxford University Press.
DOI: 10.1093/oso/9780198851738.003.0008.

representational spaces being linked. I argue that there are good reasons to think that PPS uses what I will call a *functional* basis (rather than a *cartographic* one) not the least of which is that it solves certain outstanding empirical problems. Finally, I will use this result to loop back around to questions of representation more broadly.

8.2 Coordinate transformation and the choice of basis

Brozzoli et al. make a crucial observation about our actions in PPS:

> ... in order to interact successfully with the objects in the surrounding of our body, it is neces-
> sary to represent the position of the target object relative to the observer's body or body parts.
> Given that our hands can move simultaneously with and independently from our eyes, the
> brain needs to integrate and constantly update information arising in an eye-centered refer-
> ence frame with information about the current position of the hand relative to the body and to
> nearby potential target objects. The perihand space representation provides an effective mech-
> anism to support such a fundamental function. (2014, 130)
>
>   Reproduced from Brozzoli, Claudio, H Henrik Ehrsson, and Alessandro Farnè. 2014.
>   "Multisensory Representation of the Space Near the Hand: From Perception to
>   Action and Interindividual Interactions." The Neuroscientist 20 (2): 122–35.

In other words, the primary problem that needs to be solved to enable effective action in PPs is that of *coordinate transformation* between different representational spaces. Theories of PPS, which otherwise disagree, converge on this as something that PPS representations must do.

While I will mostly be concerned with the representation *of* space, it is important to note that the notion of a 'representational space' is far more general. Any determinable property which has n different determination dimensions can be represented as an n-dimensional space, with the particular value taken by that property corresponding to a point in that space (Funkhouser 2006). If the dimensions that span this space additionally represent degrees of similarity, then overall more similar instances will appear closer together in that space. (Things can be similar in a variety of different ways, of course, and these different ways correspond to spaces with dif-ferent bases.) So for example, neuroscientists regularly talk of colour spaces in perception, of feature spaces for recognizing objects, and motor spaces for action. None of these are concerned with representing space per se.

The coordinate transformation problem has four related but distinct parts (McCloskey 2009; Grush 2007, footnote 3). First, what corresponds to the origin in one space may not coincide with the origin in a different space. This is the natural reading of 'an X-centred reference frame.' So, for example, an eye-centred reference frame might have its origin at the eye, while a hand-centred reference frame would have it at the hand. Events at the same location would have dif-ferent coordinates in each frame.

Transformation of origins is a relatively trivial problem when the coordinate systems stand in a stable relationship. The second, more serious, problem noted by Brozzoli et al. (2014) stems from the fact that the different origins can also move relative to one another. If the hand moves, the mapping from eye-centred to hand-centred coordinate space must also shift. Further, cal-culation of the relationship between these spaces cannot (in general) be derived solely from information within the two coordinate spaces. Coordinate transformation thus requires inte-grating additional information, like proprioceptive inputs about the location of the hand relative to the body.

The third, and more pressing, problem is that different coordinate systems can also use different bases. Formally, a basis is a mathematical notion: a minimal set of linearly independent vectors that span a given representational space.[1] Informally, the basis of a representational system can be thought of as the axes that define the space we are using to represent the world—its coordinate system, if you like. The same set of properties can be represented in different but equivalent ways, each of which can be thought of as corresponding to a representational space with a different basis.

So, for example, a retinotopic representation might use a two-dimensional polar representation, representing stimuli in terms of angle and distance from the fovea. An amodal allocentric representation might use something like a cartesian coordinate system with the origin centred on the head. An arm-centred representation might represent a location in space in terms of a multi-dimensional space with each axis representing a joint angle. Each of these represents a perfectly valid way to map out points in space. However, translating between these different bases can require complex and *nonlinear* mappings.

It is nonlinearity that makes this third problem difficult. Optimization of linear functions is a well-studied problem in both mathematical and neural contexts. Nonlinear functions, in contrast, are comparatively difficult to learn. Linearity is thus simpler from a developmental and a theoretical point of view. Of course, the coordinate transformations *must* be nonlinear in some sense. But all things considered, the fewer the nonlinearities one needs, the better.

The idea of a motor space is especially relevant for thinking about PPS, since at least some of the point of PPS is to guide motor actions in order to perform successful defensive motions. In present terminology, that gives rise to the fourth problem: however PPS is represented, there is also a prima facie coordinate transformation problem between PPS and motor space. Motor space is probably not spatially organized: the coordinate axes are best understood either as involving parameters like joint angles and muscle contractions or else as more complex combinations of basic actions (Graziano 2016). Whichever way PPS is represented, then, there is an additional mapping required to get to the representational space necessary for motor output.

8.3 Two strategies for representing space

Mathematically speaking, the choice of basis is usually irrelevant: there are many ways to represent the same space, each of which is sufficient for the job. *Computationally* speaking, the situation is quite different. How we represent space affects the complexity of the computations performed with that representation. I will focus on representations of spatially organized information, but the points here are very general.

As an illustration, suppose that we are tasked with coordinating cartographic data that is given using different map projections. Each projection can be thought of as using a different basis, and corresponds to a different way for projecting a globe to a map.

Moving directly from one map projection to another is often mathematically complex. Further, the transformation that takes one map projection to another is typically irrelevant to any other pair of projections. As illustrated in Figure 8.1 a we could simplify our task by translating the information in each map back to a point on the globe that the map is meant to represent. This strategy not only simplifies the math: it is flexible—we can work with any

[1] I also assume that any particular representation has a privileged basis that serves as its coordinate system, which is not required mathematically but would seem to be necessary for any concrete instantiation of that representation. This can complicate empirical inference: see Goddard et al. (2018) for an extended discussion of these issues as they arise in single-cell recordings.

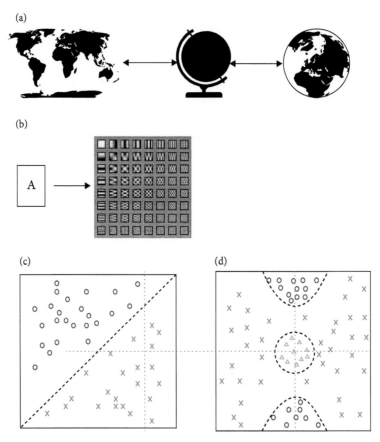

Figure 8.1 (a) A globe represents a cartographic basis that simplifies movement between different map projections. (b) The Discrete Cosine Transform 9 representation of a simple letter. (c) A decision problem with a linear decision boundary. (d) Two decision problems with nonlinear decision boundaries: neither the noughts nor the triangles can be distinguished from the crosses with a linear cut.

projection that has a well-defined forward and inverse mapping to the globe. Adding a new projection only requires figuring out the mapping to and from the globe, rather than to and from every other projection we care about. The intermediate representation is also fairly easy for us to understand and work with: there is a simple isomorphism between a globe and the world it represents.

Call this the strategy of building a *cartographic basis*. A cartographic basis has (roughly) as many dimensions as the actual space we wish to represent. Each dimension corresponds in a relatively straightforward way to a single spatial dimension in the real world.

Contrast cartographic bases with a less familiar, but widely used, strategy for representing spatially organized information. A two-dimensional Fourier transform takes spatially organized information and represents it in the frequency domain instead. Variants of this strategy are widely used. The JPEG standard for encoding pictures (ISO/IEC 10918-1:1994), for example, uses a Discrete Cosine Transform (DCT). As shown in Figure 8.1b, the encoding step of a JPEG image represents each 8 × 8 block of pixels as a combination of weights on each of the 64 basis functions. The inverse of this process can be used to recover the original image.

Call this second strategy that of finding a *functional basis*. Spatial information is represented as points in a high-dimensional space, the bases of which correspond to nonlinear functions of the original input.[2] The basis functions themselves are chosen for their useful mathematical properties, rather than for any obvious correspondence to the world. The dimensionality of this space is determined by the number of basis functions needed to span the original space with a desired degree of accuracy, rather than by the dimensionality of the target itself.

A functional basis is less intuitive and more complicated. Why go to the trouble? Because, in a functional basis, certain algorithms that would be difficult to perform in a cartographic basis become trivial.

The functional basis for the JPEG standard images is chosen to facilitate compression: the high-frequency components of a picture can be dropped without much difference in image quality (at least to the human eye). Compression thus becomes a simple matter of bounding the representation. In a functional basis like the DCT, simple local operations on the representation also affect global properties of the output. Operations like noise removal, orientation detection, or smoothing can thus become relatively trivial.

Of course, there are tradeoffs involved: simple editing—say adding a circle around a point—can be surprisingly hard in a functional basis. The choice of basis thus involves tradeoffs that are determined by what you want to do with the resulting representation. There is always a close relationship between the way data is represented and the efficiency of algorithms performed upon it (Wirth 1976, p. 2).

A further advantageous application of functional bases is worth emphasizing. A large number of decision-making problems in machine learning involve finding a linear function that lets us distinguish different classes. Just as linear projections are preferable to nonlinear ones, linear separability is desirable for several reasons. Determining the best linear separator between points is a well-studied and relatively simple procedure: it can be done analytically in many cases, and is rapidly learned by a wide variety of neural networks. Nonlinear boundaries are more difficult to learn, especially in higher dimensions.

Sometimes the data is cooperative, as in Figure 8.1c. That is, sometimes there is linear separation when we represent the data in a 'natural' basis where the dimensionality of the representation corresponds to the dimensionality of the problem space. This is often not the case. Figure 8.1d shows a case where the separating boundaries for two classes from a third are nonlinear. No straight line drawn through the space will separate the crosses and the noughts, even approximately.

One possibility for dealing with data like this would be to learn a nonlinear function. This nonlinear function, however, will be no help if we move to a different problem (say, distinguishing noughts from triangles in Figure 8.1d). The problems that nonlinearity introduces will recur with each new class to be learned. Instead, we can avail ourselves of a sneaky trick widely used in machine learning.

Note first that the separating boundaries in Figure 8.1d are conic sections. Any conic section can be represented as $ax^2 + by^2 + cx + dy + e$. Suppose we project our inputs x and y to a new four-dimensional functional basis W, with axes corresponding to $\{x^2, y^2, x, y\}$.

---

[2] To ease exposition, I will sometimes treat the strategy as involving a proper function space and sometimes as involving a space in $\mathbb{R}$ into which some original set of data is projected, with each of the basis functions corresponding to one dimension of the target space.

Any conic section in our original space corresponds to a hyperplane in W. Thus in W the classes we care about become linearly separable. In other words, when faced with a nonlinear decision boundary to learn, we can sometimes push that nonlinearity into the space in which the problem is represented. Then linear boundaries of the new space correspond to nonlinear boundaries in the original space.

The bases of W were chosen solely for their mathematical properties. W is larger than it needs to be to solve any individual classification problem. However, having picked a single, slightly more complex representation, we can then solve innumerable different categorization problems by simply learning linear functions in W.

Along the same lines, note that the class boundaries can be modified in useful ways simply by biasing the corresponding linear functions from W. Suppose, for example, we wanted to stretch the class boundary of the triangles in Figure 8.1d so as to be more elliptical along the x-axis. We can keep W fixed and simply change the weight corresponding to the x^2 axis. As this comes at the readout stage, not in W itself, we could thus stretch the class boundary without affecting our projection from any other class.

Functional bases are thus powerful tools for representing the world. By systematically projecting input to a higher dimensional space in a nonlinear way, an indefinite variety of decision problems can be represented as linear maps from that space. In practice, the dimensionality of the spaces of interest will be much higher, and the most useful bases will be specific to the set of problems at hand.

8.4 Peripersonal space is represented by a functional basis

In the opening section, I canvassed some reasons to think that there are neurons specialized for representing PPS at all. But that does not really settle the question of the basis of that representation. Most work on PPS, at least until recently, has tacitly assumed a cartographic basis. So, for example, Bufacchi and Iannetti (2018; 2019)'s recent presentation in terms of fields surrounding the body is most naturally read as asserting that the internal representation of PPS is structured along broadly cartographic lines, with action values specified at each spatial location.

I think there is a reasonably strong argument that PPS is represented using a functional basis. The argument is primarily from computational parsimony. Given the variety of things that PPS representations are invoked to do, a cartographic basis is forced to proliferate PPS representations and their interactions. Each of these new representations introduces new learning challenges. A functional basis, by contrast, allows for a fixed and fully general basis that can be used across a variety of contexts, and which faces only linear learning problems. Absent reasons to believe otherwise, I think this is decent evidence for a functional basis.

To begin, the neural evidence clearly suggests that there are nonlinear interactions at the representation stage. Anderson and colleagues noted that successful visually directed action of external objects must involve combining information from three different coordinate frames: head, eye, and retina. They found neurons in the macaque parietal lobe that were best modelled as involving a multiplicative—i.e. nonlinear—interaction between retinal and head position (Andersen and Zipser 1988). Andersen et al. (1993) extended this work, showing similar nonlinear interactions with bodily position.

Anderson and colleagues took this as evidence of multiple, distributed spatial representations in the brain, corresponding to distinct coordinate systems. However, Pouget and Sejnowski (1997) showed that this data could be better accommodated by a model in which there is a single representation that uses nonlinear basis functions to represent the location of objects. The nonlinear transformations used are different from the ones considered in the previous section: products of gaussian functions of retinal position, and sigmoidal functions of retinal location (1997, p. 224). A suitable ensemble of these functions can combine and encode motor input in a way that is suitable for *linear* readout by the motor system. Further, in such a model, a stimulus 'is represented in multiple frames of reference simultaneously by the same neuronal pool' (p. 223). This explains, among other things, why hemineglect due to damage of this pool of neurons does not seem to be confined to a single frame of reference.

Of course, functional and cartographic bases must involve some nonlinearity: both require a nonlinear projection from earlier sensory areas to the representation of PPS. I take it that the significance of the neural evidence is twofold. First, it shows that there are neurons with the relevant sorts of properties to form a functional basis. Second, it shows that these neurons collectively form a basis suitable for overall functional representation.

Consider next the computations that would be needed to efficiently generate the actions for which PPS is posited in the first place. PPS is implicated in (among other things) tool use and defensive actions. PPS neuron receptive fields are plastic and can be altered.

A second set of evidence comes from the plasticity of PPS receptive fields. There is both neural and behavioural evidence that the zone of PPS is plastic and can be altered in response to both task and stimulus. Receptive fields in the ventral intraparietal area can be altered by spatial attention (Cook and Maunsell 2002). Bufacchi and Iannetti (2018) canvass a variety of ways in which defensive motions need to be context-sensitive. When faced with a hungry tiger, the decision to flee or to climb a tree depends crucially on the distance to both the tiger and to the nearest tree.

Yet this plasticity puts constraints on the underlying representation. As De Vignemont and Iannetti (2015) note, tool use and defensive actions give rise to fundamentally different kinds of action, with different ends, different trajectories, and different informational grain. Defensive actions are fast and often automatic, whereas skilled actions are often slow and conscious. The modulation of PPS also differentially affects defensive and tool-using actions. While there is some evidence that tool use can extend the defensive peripersonal field (Rossetti et al. 2015), the two kinds of actions would seem to require fundamentally different sorts of plasticity. Indeed, as Povinellia, Reaux, and Frey (2010) note:

> ... during most activities of daily living, tools and utensils are used to perform actions within our natural peripersonal space. Moreover, tools are frequently used in ways that we would never employ our hands. For instance, we will readily use a stick to stoke the hot embers of a campfire, or stir a pot of boiling soup with a wooden spoon. In these circumstances, the target of the actions may be located well within reach, but a tool is chosen as a substitute for the upper limb in order to avoid harm. These examples suggest that we maintain separate non-isomorphic representations of the hand vs. tool as concrete entities even when using handheld tools within our normal peripersonal space. (pp. 243–4)

This suggests that the defensive PPS representation should remain unchanged even when a skilled PPS representation is extended by tools.

So far, so good. Now, either sort of basis can accommodate these behavioural facts. However, the malleability of PPS presents something of a challenge if one assumes that PPS is represented by a cartographic basis. On such accounts, the change in PPS is typically modelled as deriving from a bias or distortion of the underlying representation itself. Anxiety (say) warps the representation of PPS in subtle ways, with the result that we act as if there were larger boundaries around objects.

That distortions of PPS are effected by distortion of the underlying representation is, I suggest, the natural reading when using a cartographic basis. The input and output transformations are complex and nonlinear, and subtle variation of these mappings would be difficult to manage. Simply changing the space—representing a spider as bigger than it actually is, for example—is the only way to allow the hard-won nonlinear mappings to stay stable while putting linear biases on parts of the represented space itself.

De Vignemont and Iannetti (2015) thus draw the natural conclusion: if PPS is altered by altering the underlying representation, then there cannot be a *single* PPS representation (see, similarly, Bufacchi and Iannetti 2018). The effect that different stimuli have on PPS appear to differ depending on whether we consider defensive or tool-use behaviours. This would create contradictory demands on modification of the representation of PPS. Hence, they argue, there must be at least two distinct representations of PPS, covering different use cases.

Yet proliferating PPS cannot stop there. There are a great number of circumstances in which the line between tool use and self-protection breaks down. Many tools, especially when working close to the body, can injure. Flint knapping, for example, involves working with sharp instruments close to the body, and comes with well-attested possibilities of serious injury.[3] Success in such situations requires close coordination between defence and tool use. Similarly, we do actually protect tools in many cases: a butcher's knife will chip and break on bone, a poorly handled marshmallow stick will catch fire. This may be especially pressing with premodern, hand-crafted tools.

These problems are compounded when there is cooperative action with tools, such as in hunting: the polite hunter avoids spearing himself *and* his companions. Indeed, one should probably treat the standard cases—objects unexpectedly threatening the body, and uncomplicated tool use in constrained situations—as the limit cases of more realistic, complex situations. Finally, there are a number of situations that do not seem to fit neatly into the defensive/tool use dichotomy. Consider walking through crowds, or trying to punch someone nearby: sometimes the body is both the tool and the thing being protected.

The cartographic strategy, therefore, must proliferate representations and interactions in order to deal with the complexity of PPS behaviour. Again, this is a direct computational consequence of the choice of basis. Since (*ex hypothesi*) the mapping from sensory organs to PPS and from PPS to motor space are both nonlinear, it is difficult to see how consistent alterations to either could easily be made. Conversely, since representations in a cartographic basis bear a relatively simple relationship to external space, it is easy to see how alterations of the PPS representation would fit the bill. This means that the only convenient point for simple alteration is at the representation of PPS itself—but then we must proliferate representations, because there are too many different ways in which PPS can be altered.

Contrast this with the use of a functional basis. Figure 8.2 shows the overall picture. Different coordinate frames are projected to a single, stable high-dimensional representation. This can then be projected linearly into a variety of different coordinate frames (or into the same frame in different ways) by simple linear readouts of the functional space. A single representation can be used in a variety of simple ways. Further, by assumption, linear transformations are comparatively easy to learn, so the proliferation of *readouts* for different tasks presents no deep difficulties.

---

[3] See Lycett, von Cramon-Taubadel, and Eren (2016) for ethnographic references, and Whittaker (1994, chapter 5) for some vivid modern examples.

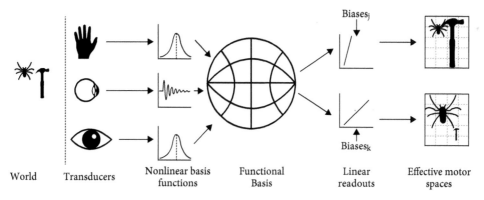

Figure 8.2 A schematic picture of the functional basis model of peripersonal space.

Similarly, linear readouts from a functional basis provide an easy method for biasing motor output via simple linear biases on the readout functions themselves. And because there can be multiple independent readout functions from the same functional space, different demands can give rise to distinct biases on this representation. Indeed, although I propose the model to account for PPS activity, there is no particular reason why it need be confined to PPS. Any nonlinear function from the same inputs to the same output might be profitably represented as linear readouts of this space, so long as the nonlinearities involved are in the span of the basis functions themselves.

The overall picture is one in which the core representation of PPS is a high-dimensional space that takes as inputs nonlinear functions of head position, limb position, eye position, and so on. The nonlinear basis functions that underly that transformation are chosen so that the readout end—here, I will assume the readout is done by various aspects of the motor system—is a linear function of the PPS space. These linear readouts can in turn be biased by different contextually sensitive inputs. Figure 8.2 puts everything together.

To sum up the argument: the choice of basis in which PPS is represented matters because it makes certain kinds of computational operations easier or harder. There is considerable evidence that PPS is represented by neurons with the right sorts of properties to underlie a functional basis. A cartographic basis has relatively restricted resources with which to accommodate shifts across different contexts. In addition, a cartographic basis would require the individual to learn two different, specific nonlinear mappings in developmental time. This adds additional complexity. By contrast, a functional basis requires learning only a generic nonlinear mapping into functional space, and then a series of linear mappings from that space. Thus a functional basis ends up apparently simpler and more parsimonious across a variety of fronts. Of course, computational parsimony arguments for empirical findings are not conclusive, but, all things being equal, I suggest that parsimony provides a useful set of constraints on available theories.

8.5 Do we represent peripersonal space?

I conclude by returning to the question of representation. I have spoken so far as if it were obvious that we represent PPS. Yet, even if we keep the empirical evidence fixed, there remain several interesting questions that turn on broader points about the nature of representation itself. The question of whether the brain represents PPS admits of three distinct readings, each with different emphasis, each important. I consider each in turn.

8.5.1 Peripersonal representation

First, we might want to know whether the brain really represents *peripersonal* space. That is a question about whether the representations we use are specially tuned to PPS itself. For all I have said above, this is not obvious. It might be that we simply represent space and threats and tools, and the space near our body happens to be a place where these interests coincide. The differential representation on the space close to the body might then just be a side effect of the fact that most relevant action categories happen nearby.

Indeed, this is a possibility that has been explicitly raised in the recent literature. In a recent review, Bufacchi and Iannetti (2018) suggest that PPS is represented by a series of 'fields', which (like their analogues in physics) are defined at every point in space. Noel and Serrino (2019) protest that this is no longer a theory of PPS as such, because it does not confine itself to the space near the body. In response, Bufacchi and Iannetti (2019) appear to concede the point, arguing instead that PPS representations merely serve to protect the body, which might require tracking distant objects. But that would appear to stretch the definition of 'peripersonal space' to cover nearly any space. Conversely, Noel and Serrino (2019)'s argument appears to rely on the claim that certain action values are 'explicitly hard-coded in PPS neurons'. This strikes me as inconsistent with the previously canvassed arguments for behavioural and neural plasticity.

I think a focus on representational basis suggests a useful—and empirically testable— intermediate position. The question, it seems to me, turns on what PPS representations track. In a fully general case, where the functional basis covers all space with equal fidelity, there would be a good argument that PPS is not represented as such. Conversely, if the functional basis covered *only* PPS—that is, if it only enabled activity in nearby space—then we would have good evidence that PPS as such is represented. We have good reason to think that the latter is not the case, because of the extension of PPS by reaching tools.

Instead, an intermediate possibility is likely to be instantiated. Pouget and Sejnowski (1997, p. 233) note that basis function representations with multiple input dimensions and uniform coverage can get large rather quickly. The power of functional bases comes from their increased dimensionality, but that high dimensionality comes at a cost as well. One possible solution they suggest is that the parietal cortex might 'selectively span the input space to achieve greater efficiency'.

There is evidence that this is what occurs with PPS. As Graziano and Cooke (2006) note, the space near the body appears to be covered by more, and more densely overlapping, receptive fields (see their Figure 7.4). Space further from the body is still represented by multimodal neurons, but with lower fidelity. Assuming this represents the organization of the underlying functional basis, this might represent a compromise between full coverage and inflexibility. Conversely, this would also set limits on the degree to which further and further portions of the bodily space could be incorporated into PPS, and the accuracy with which the resulting incorporation might be achieved.

Here it strikes me that the decisive factor might involve experimentation with a wider variety of tools. Most experiments with tool use tend to focus on ethnologically realistic reaching tools that still operate near the body. Extremely long or large tools—telescoping tree pruners, or hydraulic excavators, for example—represent a novel empirical arena for testing the degree to which PPS can be extended far past the usual limits. The further PPS can be extended, the better an argument that there is no PPS *as such*, only regions of space where there are typically important actions to be done. Conversely, evidence of a spatially restricted area of high fidelity would suggest a specific and relatively inflexible tuning for near space, albeit one that can manage coarse and low-fidelity representation further afield.

8.5.2 Spatial representation

Following from the first question, we might ask a second, more general one. Supposing the brain represents something relevant to peripersonal defence (and the like), we might ask whether the brain represents peripersonal *space*. Our representation of PPS might be best understood as a representation of affordances for action in PPS (say) rather than as a representation of the space itself. Indeed, one thing that makes questions about PPS so interesting is that it is not at all clear that there is a distinctive phenomenology of PPS. That is, what is most salient in PPS (at least when I introspect) is usually not space itself but objects, tools, threats, and the like.

Focus on representational bases lets us reformulate this question in a useful way. There is a clear sense in which a cartographic basis represents space. But it is not obvious whether a functional basis can be said to represent space as such. The constraints on a functional basis are primarily that of mathematical convenience, rather than similarity to external space. What is more, because one of the goals of a functional basis is to allow for simple linear representations of output, one might claim that the point of a functional basis is to represent stimuli *in terms of* behaviours that might be performed on them.

Grush, for example, reads Pouget's model as having this sort of structure (Grush 2007; 2009). He claims that 'The basis functions' entire purpose is to support a linear decoding via a motor-behavior-specific set of coefficients to produce a bodily action that is directed on that stimulus—such as grasping or foveating the seen object'; the objects in question 'are objects in the sense of a potential focus of perception and action', and the functional basis as a whole underlies the 'capacity to represent an environment of actionable objects' (Grush 2009, pp. 342–43).

While I can see the attraction of such a view, I think it is mistaken. It is true that PPS is very often used for detecting and acting upon affordances. So a functional basis *plus* a readout function might be considered as something like a representation of affordances. But, as I have set it up, a functional representation of PPS is also a *decoupled* representation in the sense of Sterelny (2003). It can be used for a variety of different readout functions in different contexts. Further, since it forms a complete representational space, it is fully general. Together, this means that the representation of PPS is useful for an indefinite number of tasks. And what seems common to all of them is the concern with spatial location. Thus even though there's no obvious isomorphism between a PPS representation with a functional basis and space itself, the use of that representation means that it is best understood as spatial.

8.5.3 Representation as such

Third and finally, we might ask whether the brain *represents* PPS. That is both a philosophical question about the nature of representation and an empirical question about how the brain tracks PPS. What the brain does with PPS might fall short of fully representational: it might be, for example, that our facility with PPS is best considered as a species of online skill, and that online skill is itself nonrepresentational. This is the sort of view that might be championed by non-representationalist strands in philosophy of mind (Chemero 2009). I think this is too hasty, and ultimately depends on a false dichotomy. Again, focus on the particular properties of functional bases is helpful.

The use of a functional basis for PPS was motivated in part by the problem of coordinate transformation. Yet coordinate transformation itself does not seem like it necessarily involves representation as such. Broadly speaking, it seems like coordinate transformation is more like a computational function than a representation per se. Webb and Wystrach (2016, p. 29), discussing place learning in cockroaches and crickets, contrasts learning strategies that are sufficient to do simple coordinate transformations yet that fall short of 'internal representation of the spatial layout' of an arena.[4] O'Brien and Opie (2009) distinguish digital representations, which stand in a completely arbitrary relationship to their target, and analogue representations, which stand in a resemblance relationship to their target. Functional bases do not seem like they fit either bill, which again suggests that representation might be an inappropriate category.

Yet this argument relies on an overly restrictive notion of representation as something that is wholly explicit and always present to stand in for the original stimulus, whether or not that is being used at the time. But the primary function of a representation in more complex tasks is always transformative: the goal is to take information and re-present it in a form that is more amenable to further use (Ritchie 2015). DiCarlo and colleagues (2007; 2012), for example, model visual object recognition as a series of progressive transformations of information in the retinal image in a way that makes latent information simply accessible.

As before, I suggest that the more basic feature of representations is that they are decoupled from the immediate stimulus (Sterelny 2003) but still provide the same systematic handle on the world that the original stimulus could provide. Systematicity is key. Mere mappings from a single stimulus to a single response are not representational, even if the mapping is complicated. It is systematicity that allows the same representation to be used for a variety of different responses in a variety of circumstances.[5]

It is clear that both cartographic and functional bases are decouplable. Both are equally concerned with combining information from the input end into a form that can be used in a variety of different ways. Indeed, the entire motivation behind moving to a functional basis was to facilitate decoupled use in distinct contexts. Of course, this does not settle the philosophical debate about the fundamental nature of representations, but it suggests that both sorts of bases are similar enough that we ought to feel comfortable treating both or neither as representations.

That said, I think an assumption that PPS has a cartographic basis tacitly suggests a picture on which PPS is *explicitly* represented: that is, that everything about local space must be present like a map somewhere in cortex. There is no doubt that explicit maps are useful for some cognitive tasks (Rescorla 2009). But as Bufacchi and Iannetti (2019) point out in the defence of their field theory, what is really important is being able to systematically recover and transform action values for specific points in space. Describing a space with a functional basis also makes it more clear that explicit representation is less important than decouplable, systematic representation for most tasks.

---

[4] More precisely, the contrast is with the strategy of learning a single exemplar of a panoramic view and then comparing it with the present view via gradient descent. For a stronger statement, see also Webb (2006). I think one can argue that even this actually tacitly represents a space (compare with the so-called 'kernel trick' in machine learning that allows computation of a gradient in space without requiring an intermediate transformation to that space), but that is stronger than needed for the moment.

[5] For a full presentation of this argument and its ramifications, see Klein and Clutton, forthcoming.

It strikes me that the focus on PPS and coordinate transformation would be a good place to help with this, because there are a great number of cognitive tasks that can plausibly be modelled as species of complex coordinate transformations. There is a well-known body of work on the more general issue of egocentric to allocentric transformations of space, of course (Klatzky 1998). Effective action requires intertranslation between representations of outcomes and representation of motor actions (Butterfill and Sinigaglia 2014). Even some apparently higher-level judgements, such as stereotype-congruency, might be usefully modelled in terms of biases on underlying transformations, rather than explicit representation of stereotypes as such.[6] Focus on PPS, and on the possibility of functional bases for its representation, opens up a host of options that go beyond a simplistic distinction between transformation and representation.

8.6 Conclusion

I have argued that we represent PPS, but that the structure of that representation is best understood as involving a functional rather than a cartographic basis. We represent PPS in the sense that we have an efficient means for coordinate transformation that allows for the efficient learning and execution of PPS-related behaviours.

I take it that there is a broader lesson to be drawn. Debates over whether we represent PPS are ultimately questions about the number and taxonomy of processes involved in dealing with our behaviour near our bodies. That is, they are questions about the *cognitive ontology* that we need to posit in order to make sense of observed behaviour (Janssen, Klein, and Slors 2017). Thinking about PPS shows that questions about the number of processes often cannot be separated from the question of the format of the representations upon which the processes operate.

Acknowledgements

Thanks to Dan Burnston, Peter Clutton, Carl Craver, Frédérique de Vignemont, Stephen Gadsby, Thor Grünbaum, Julia Haas, Ben Henke, Gabbrielle Johnson, Tony Licata, Ross Pain, two anonymous referees, and audiences at the Australian National University, Institut Jean Nicod, and Washington University in St. Louis for helpful comments and feedback. Work on this paper was supported by Australian Research Council grant FT140100422.

References

Andersen, Richard A, and David Zipser. 1988. "The Role of the Posterior Parietal Cortex in Coordinate Transformations for Visual–motor Integration." *Canadian Journal of Physiology and Pharmacology* 66 (4): 488–501.

Andersen, Richard A, Lawrence H Snyder, Chiang-Shan Li, and Brigitte Stricanne. 1993. "Coordinate Transformations in the Representation of Spatial Information." *Current Opinion in Neurobiology* 3 (2): 171–176.

---

[6] This argument is made forcefully by Gabbrielle M. Johnson, "The Structure of Bias", *Mind*, https://doi.org/10.1093/mind/fzaa011

Brozzoli, Claudio, H Henrik Ehrsson, and Alessandro Farnè. 2014. "Multisensory Representation of the Space Near the Hand: From Perception to Action and Interindividual Interactions." *The Neuroscientist* 20 (2): 122–135.

Bufacchi, Rory J, and Gian Domenico Iannetti. 2018. "An Action Field Theory of Peripersonal Space." *Trends in Cognitive Sciences* 22 (12): 1076–1090.

———. 2019. "The Value of Actions, in Time and Space." *Trends in Cognitive Sciences* 23 (4): 270–271.

Butterfill, Stephen A, and Corrado Sinigaglia. 2014. "Intention and Motor Representation in Purposive Action." *Philosophy and Phenomenological Research* 88 (1): 119–145.

Chemero, A. 2009. *Radical Embodied Cognitive Science*. Cambridge: The MIT Press.

Cook, Erik P, and John HR Maunsell. 2002. "Attentional Modulation of Behavioral Performance and Neuronal Responses in Middle Temporal and Ventral Intraparietal Areas of Macaque Monkey." *Journal of Neuroscience* 22 (5): 1994–2004.

de Vignemont, F, and GD Iannetti. 2015. "How Many Peripersonal Spaces?" *Neuropsychologia* 70: 327–334.

DiCarlo, James J, and David D Cox. 2007. "Untangling Invariant Object Recognition." *Trends in Cognitive Sciences* 11 (8): 333–341.

DiCarlo, James J, Davide Zoccolan, and Nicole C Rust. 2012. "How Does the Brain Solve Visual Object Recognition?" *Neuron* 73 (3): 415–434.

Funkhouser, Eric. 2006. "The Determinable-Determinate Relation." *Nous* 40 (3): 548–569.

Goddard, Erin, Colin Klein, Samuel G Solomon, Hinze Hogendoorn, and Thomas A Carlson. 2018. "Interpreting the Dimensions of Neural Feature Representations Revealed by Dimensionality Reduction." *Neuroimage* 180: 41–67.

Graziano, Michael SA, 2006. "The Organization of Behavioral Repertoire in Motor Cortex." *Annual Review of Neuroscience* 29: 105–134.

Graziano, Michael SA, 2016. "Ethological Action Maps: A Paradigm Shift for the Motor Cortex." *Trends in Cognitive Sciences* 20 (2): 121–132.

Graziano, Michael SA, GS Yap, and CG Gross, 1994. "Coding of Visual Space by Premotor Neurons." *Science* 1054.

Graziano, Michael SA, and Dylan F Cooke. 2006. "Parieto-Frontal Interactions, Personal Space, and Defensive Behavior." *Neuropsychologia* 44 (6): 845–859.

Graziano, Michael SA, Xin Tian Hu, and Charles G Gross. 1997. "Coding the Locations of Objects in the Dark." *Science* 277 (5323): 239–241.

Graziano, Michael SA, Lina AJ Reiss, and Charles G Gross. 1999. "A Neuronal Representation of the Location of Nearby Sounds." *Nature* 397 (6718): 428.

Graziano, Michael SA, Charlotte SR Taylor, and Tirin Moore. 2002. "Complex Movements Evoked by Microstimulation of Precentral Cortex." *Neuron* 34 (5): 841–851.

Graziano, Michael SA, Charlotte SR Taylor, Tirin Moore, and Dylan F Cooke. 2002. "The Cortical Control of Movement Revisited." *Neuron* 36 (3): 349–362.

Grush, Rick. 2007. "Skill Theory V2.0: Dispositions, Emulation, and Spatial Perception." *Synthese* 159: 389–416.

———. 2009. "Space, Time, and Objects." In *The Oxford Handbook of Philosophy and Neuroscience*. Oxford: Oxford University Press.

Janssen, Annelli, Colin Klein, and Marc Slors. 2017. "What Is a Cognitive Ontology, Anyway?" *Philosophical Exploration* 20 (2): 123–128.

Klatzky, Roberta L. 1998. "Allocentric and Egocentric Spatial Representations: Definitions, Distinctions, and Interconnections." In *Spatial Cognition*, 1–17. Dordrecht: Springer.

Klein, Colin, and Peter Clutton (forthcoming). "What is the Job of the Job Description Challenge? A Case Study from Body Representation." In *Neural Mechanisms: New Challenges in Philosophy of Neuroscience*, eds. Fabrizio Calzavarini, and Marco Viola. Dordrecht: Springer.

Lycett, Stephen J, Noreen von Cramon-Taubadel, and Metin I Eren. 2016. "Levallois: Potential Implications for Learning and Cultural Transmission Capacities." *Lithic Technology* 41(1). 19–38.

McCloskey, Michael. 2009. *Visual Reflections: A Perceptual Deficit and Its Implications.* Oxford: Oxford University Press.

Noel, Jean-Paul, and Andrea Serino. 2019. "High Action Values Occur Near Our Body." *Trends in Cognitive Sciences* 23(4): 269–270.

O'Brien, Gerard, and Jon Opie. 2009. "The Role of Representation in Computation." *Cognitive Processing* 10 (1): 53–62.

Pouget, Alexandre, and Terrence J Sejnowski. 1997. "Spatial Transformations in the Parietal Cortex Using Basis Functions." *Journal of Cognitive Neuroscience* 9 (2): 222–237.

Povinellia, Daniel J., James E. Reaux, and Scott H. Frey. 2010. "Chimpanzees' Context-Dependent Tool Use Provides Evidence for Separable Representations of Hand and Tool Even During Active Use Within Peripersonal Space." *Neuropsychologia* 48: 243–247.

Rescorla, M. 2009. "Cognitive Maps and the Language of Thought." *British Journal for the Philosophy of Science* 60 (2): 377–407.

Ritchie, J Brendan. 2015. "Representational Content and the Science of Vision." PhD thesis, University of Maryland.

Rossetti, Angela, Daniele Romano, Nadia Bolognini, and Angelo Maravita. 2015. "Dynamic Expansion of Alert Responses to Incoming Painful Stimuli Following Tool Use." *Neuropsychologia* 70: 486–494.

Shea, Nicholas. 2018. *Representation in Cognitive Science.* Oxford: Oxford University Press.

Sterelny, Kim. 2003. *Thought in a Hostile World: The Evolution of Human Cognition.* Malden, Massachusetts: Blackwell Publishers.

Webb, Barbara. 2006. "Transformation, Encoding and Representation." *Current Biology* 16 (6): R184–R185.

Webb, Barbara, and Antoine Wystrach. 2016. "Neural Mechanisms of Insect Navigation." *Current Opinion in Insect Science* 15: 27–39.

Whittaker, John C. 1994. *Flintknapping: Making and Understanding Stone Tools.* Austin, TX: University of Texas Press.

Wirth, Niklaus. 1976. *Algorithms + Data Structures = Programs.* Upper Saddle River, NJ: Prentice Hall.

9

What do 'peripersonal space measures' really reflect? The action field perspective

R.J. Bufacchi and G.D. Iannetti

9.1 Introduction

9.1.1 How has the concept of peripersonal space appeared?

Interactions occurring between an agent and objects in the space near the body have been studied in a wide range of disciplines, including ethology, neurophysiology, social science, architecture, and philosophy (Hediger, 1955; Hall, 1969; Remland et al., 1995; Høgh-Olesen, 2008; de Vignemont, 2018). Such studies have shown that many behavioural responses are increased when stimuli occur near the body. This phenomenon makes evolutionary sense: a predator within striking distance is more relevant than one farther away. Neuroscientific studies in both humans and non-human primates have found neurophysiological measures that are similarly modulated by proximity to the body, suggesting a foundation for the behavioural modulations, and leading to the concept of peripersonal space (PPS).

9.1.2 But what is precisely meant when referring to peripersonal space?

What is precisely meant when referring to peripersonal space, and how does the modulation of biological and behavioural responses on the basis of stimulus position in egocentric coordinates[1] fit into current systems-level understanding of brain function? It is our opinion that these seemingly naïve questions have not yet been satisfactorily answered. A clear theoretical framework is lacking, as demonstrated by the great deal of terminological and conceptual confusion among researchers interested in the field of PPS. Three categories of PPS definitions are commonly used: (a) the portion of space within a given Euclidian distance of the body; (b) the space within which certain physiological or behavioural responses are larger when the stimuli eliciting them occur near the body; (c) the mechanisms through which the brain encodes the space near the body. To avoid confusion in this article, we use the term 'near space' to refer to definition (a), and we use the term 'PPS measures' to refer to those measures described in definition (b). We choose this specific approach, because we

---

[1] By the term 'egocentric reference frame', we mean any coordinate system that has its origin set at some part of the body.

R.J. Bufacchi and G.D. Iannetti, *What do 'peripersonal space measures' really reflect? The action field perspective* In: *The World at Our Fingertips.* Edited by Frédérique de Vignemont, Andrea Serino, Hong Yu Wong, and Alessandro Farnè, Oxford University Press (2021).

have previously argued that major conceptual problems arise when definitions of PPS are used interchangeably (Bufacchi and Iannetti, 2018).

9.1.3 Using the term 'PPS' with multiple meanings causes conceptual issues

In brief, the use of the same word for three separate concepts has created the idea of an entity or mental construct called 'the PPS', which has aspects of each of the three definitions mentioned earlier. This has led to multiple loose notions about the nature of 'the' PPS. These notions bear substantial theoretical implications, while having little evidence backing them up. Three particularly problematic notions are those describing PPS as a (1) in-or-out (i.e. a sharp border between near space and far space; this is a consequence of spatially describing an all-or-nothing response), (2) distance-based (i.e. purely dependent on and indicative of distance to a body part), and (3) single space (i.e. a single entity, rather than a collection of a set of response types). While a sharp spatial boundary (notion 1) may be intuitively appealing, neurophysiological and behavioural data contradict the description of PPS as an in-or-out space. In fact, many PPS-related measures have graded or even reverse relationships to distance (Colby et al., 1993; Duhamel et al., 1998; Longo and Lourenco, 2007; Van der Stoep et al., 2015; Bufacchi et al., 2016). Most scholarly works about PPS highlight stimulus position in egocentric coordinates as the most important feature that modulates PPS measures (notion 2). However, PPS responses are influenced by many other factors such as environmental landscape, pregnancy, stimulus temperature, action planning and execution, tool use, stimulus trajectory, and stimulus valence (Iriki et al., 1996; Brozzoli et al., 2010; Sambo et al., 2012b; Taffou and Viaud-Delmon, 2014; Cléry et al., 2015a; Cardini et al., 2019; Lohmann et al., 2019; Ruggiero et al., 2019. See Section 9.3.3 and Figure 9.2B for more details and examples). Finally, PPS is often presented as 'the' PPS (notion 3), implying it to be a single entity. This is difficult to reconcile with the fact that many different PPS-related responses exist (Cléry et al., 2015b; Van der Stoep et al., 2015), and each of them can be used to describe a different space. While loosely conceptualizing PPS as a single distance-based in-or-out zone may allow for efficient summary of results, it is increasingly clear that this simple framework has become a source of data misinterpretations and conceptual misunderstandings.

9.1.4 An alternative framework that solves many issues

Here we offer a perspective that derives from considering the functional significance of the many PPS measures in light of what we consider one of the most promising systems-level frameworks of action selection and behaviour: the affordance competition hypothesis (Cisek and Kalaska, 2010). In this hypothesis, multiple actions are constantly being readied simultaneously by competing neural populations. This allows neural firing in sensorimotor areas to be understood as defining action *relevance* (or action *value*).[2] It follows that PPS-related neurons and behavioural measures could also reflect the behavioural relevance of a stimulus

---

[2] In the language of reinforcement learning, the equivalent to 'action relevance' is the 'value' associated with a particular action, in a particular state (Bufacchi and Iannetti, 2019; Noel and Serino, 2019). For this reason, we use the terms 'relevance' and 'value' interchangeably in this chapter.

for a given action or set of actions. Following this reasoning, we propose that PPS measures *reflect the value of actions aiming to either create or avoid contact between objects and the body*. Indeed, given that contact-related actions necessarily depend on the distance between the stimulus and the body, the proximity-dependence of PPS measures arises naturally, but not necessarily as their most functionally important feature.

To describe the idea of the relevance of contact actions in a spatial manner, we have suggested using the concept of *fields* (Bufacchi and Iannetti, 2018): properties which have a magnitude for each point in space and time[3] (McMcmullin, 2002). In this way, rather than describe PPS simply as a singular in-or-out zone, the hypothesis allows us to spatially visualize PPS responses as *a set or a class* of *graded response fields*, each describing the behavioural value of *different contact-related actions*. This means that the term 'PPS' refers to a *type* of response field, rather than an actual portion of space, or construct of the brain. These concepts have important implications over previous notions. The concept of a *field* allows PPS measures to change gradually with distance, rather than to define an in-or-out space. The concept of *mutiple fields* reflects the fact that there are many different PPS measures showing different response profiles. The concept of *behavioural relevance to actions aiming to create or avoid* contact between objects and the body explains the functional significance of the values composing the PPS field of each action, and the fact that factors other than proximity affect PPS measures. We believe this framework can explain seemingly anomalous empirical observations, and resolve some of the definitional and conceptual issues affecting the field.

Other PPS frameworks can similarly solve some of the conceptual issues presented here, and lead to the concept of multiple graded fields: multisensory integration and impact prediction. In the final section of this chapter, we discuss the advantages and shortcomings of these frameworks, and examine how they might link with contact-related action relevance.

9.2 The affordance competition hypothesis

9.2.1 A short primer on the affordance competition hypothesis

Given that the reconceptualization of PPS we proposed earlier (Bufacchi and Iannetti, 2018) and further elaborate here arises from the interactive behaviour framework (also labelled the 'affordance competition hypothesis'; Fagg and Arbib 1998, p. 1292; Shedlan et al., 2008), we attempt to do it justice in a brief summary. It is important to stress that the affordance competition hypothesis is not our model for PPS. Rather, it provides a scaffolding for the perspective that we propose. Cisek and Kalaska recently refined this hypothesis, and we refer the reader to their review for a full description and more elaborate argument than the one we present here (Cisek and Kalaska, 2010).

The affordance competition hypothesis argues for a parallel representation of multiple possible actions in competing neural populations. These populations are constantly competing, even in the absence of environmental stimuli and, importantly, these same neural populations also enact the action or the actions that outcompete their alternatives (Figure 9.1). This view is based on a school of thought that has existed for more than a century, particularly in ethology (Hinde 1966). It has recently gained popularity in robotics and artificial intelligence (Sahin et al., 2007), where computational architectures based on

---

[3] We use the term 'field' with a different meaning from that of receptive field. When we refer to receptive fields later in this work, we explicitly call them 'receptive fields'.

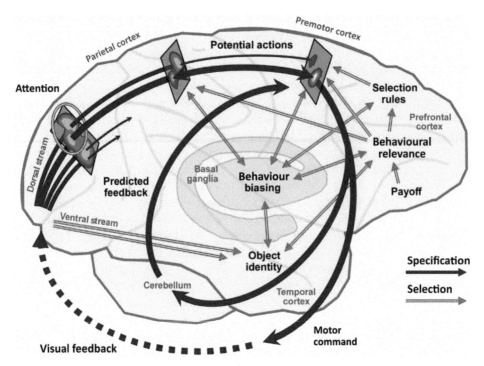

Figure 9.1 The interactive behaviour framework (Cisek and Kalaska, 2010), exemplified here for visually guided movements, postulates that neural populations in the parieto-premotor loop transform visual input into representations of potential actions. The strength of each action representation is also influenced by basal ganglia and prefrontal regions, so that it reflects the relevance of that action depending on the environmental circumstances. Reproduced, with permission, from (Cisek and Kalaska, 2010). Note that the use of visual input to the system in this figure is merely an example. The origin of sensory input can just as well be other sensory areas (Battaglia-Mayer and Caminiti, 2019).

Reproduced from cisek P., Kalaska JF. Neural mechanisms for interacting with a world full of action choices. Annu Rev Neurosci. 2010;33:269-98. doi:10.1146/annurev.neuro.051508.135409.

parallel action processing often prove more effective for adaptive interaction with the environment than traditional, serial architectures (Arkin, 1998).[4]

9.2.2 Issues with the traditional hierarchical perspective on brain function

The traditional ideas of hierarchical brain function—to which the affordance competition hypothesis has been in opposition for decades, and is progressively replacing—postulate

---

[4] The use of computational architectures based on parallel processing is reflected in the current rise of deep reinforcement learning, which operates on very similar principles and thereby has allowed computer systems to outperform humans across a range of board and computer games (Mnih et al., 2016; Silver et al., 2016). Not surprisingly, reinforcement learning has proven to have strong explanatory power for many different types of neural activities in rodents and humans (Niv, 2009; Ribas-Fernandes et al., 2011; Lee et al., 2012; Stachenfeld et al., 2017).

a serial, three-step perception-cognition-action information-processing pathway. Many have cogently argued that, while these three steps are conceptually attractive, there is little neurophysiological evidence for distinct modules that directly reflect them (Dewey, 1896; Treue, 2008; Ebbesen et al., 2018). In fact, 'functions that should be unified [under the traditional perception-cognition-action framework] appear distributed throughout the brain, whereas those that should be distinct appear to involve the same regions, or even the same cells' (Cisek and Kalaska, 2010). For example, in the case of perception, information about the same object exists in different visual processing sub-streams. Furthermore, virtually all neuronal responses to visual stimuli are modulated by attention, although to a different degree throughout visual hierarchal processing (Boynton, 2005). Thus, neural representations appear 'dominated by the behavioral relevance of the information, rather than designed to provide an accurate and complete description of [the visual world]' (Treue, 2008). In the same way that there does not seem to be a single, separate, stable perception module, there is no separate cognitive module either: neural indicators of future decisions, for example, can be found throughout sensorimotor circuits, even in subcortical structures such as the superior colliculus (Ebbesen et al., 2018). In fact, the same 'sensorimotor' neurons often reflect both decision-related variables first, and later the metrics of the action used to report the decision (Roitman and Shadlen, 2002; Cisek and Kalaska, 2005; Hoshi and Tanji, 2007). Therefore, a simple motor-output module, which only enacts the commands of a cognitive module, is physiologically implausible.

9.2.3 Evidence for the affordance competition hypothesis

It is evident that the neural processing mechanisms described earlier do not fit with a traditional serial view of brain function. Instead, they are to be expected in a nervous system whose main goal is to continuously choose the most useful action while navigating in a constantly changing world (Gibson, 1979). In such a system, the brain has pragmatic representations of potential actions that compete with each other to maximise the individual's well-being. When a representation outcompetes the others, the action encoded in that representation is immediately enacted. Thus, under this framework, action selection and action specification operate in a simultaneous and integrated manner. This type of functional organization fits the three types of observations described in the previous section. First, there should be many seeming 'representations' of the same visual object, each corresponding to a different action. Second, these representations should be highly affected by attention, given that attention is a prime and early mechanism for computing behavioural relevance (Allport, 1987; Neumann, 1990). Third, correlates of decisions and stimulus relevance should be visible in sensorimotor systems, because they are directly linked to the competition between action representations. Remarkably these three types of observations also characterise PPS measures (see Section 9.3.4).

9.2.4 Neural substrates of competing actions

At least when visually guided, competing actions have been suggested to be encoded in the dorsal stream of visual processing and the parieto-premotor circuit (Milner and Goodale, 1995; Caminiti et al., 2017). Structures such as the prefrontal cortex and basal ganglia are likely to bias this competition (Figure 9.1; Andersen and Buneo, 2002; Whitlock et al., 2008;

Cisek and Kalaska, 2010), while prefrontal cortex in turn likely receives information on object identity from temporal lobe through the ventral stream and subjective value from orbitofrontal cortex (Cisek and Kalaska, 2010). We believe it is conceptually revealing that, similarly to competing action representations, PPS-related neurons have been mainly found in the parieto-premotor loop (Cléry et al., 2015b), and that many behaviours that have been used as PPS measures have also been linked to neural activity occurring within this loop (Brozzoli et al., 2014).

9.3 Peripersonal space as a set of contact-related action fields

9.3.1 Proximity dependence of peripersonal space measures arises naturally for contact-related actions

Behavioural or neural responses whose magnitude depends on proximity to the body have traditionally been labelled 'PPS measures'. If one accepts the affordance competition hypothesis, such proximity dependence arises naturally in the case of a particular type of action, whose relevance will by definition depend on the distance of an object: the actions aiming to *create or avoid contact between objects and the body*. In the following sections, we expand on this idea by briefly describing some of the current interpretational issues regarding PPS, and then demonstrating how the affordance competition hypothesis can solve each of them.

9.3.2 Not an 'in-or-out' bubble, but a graded field

The first issue is that PPS is often referred to as an 'in-or-out space', even though this is not supported by empirical data (see Bufacchi and Iannetti, 2018, for an expanded argument). Why then has the notion of an in-or-out PPS been so persistent in the field?

9.3.2.1 Early reports of an in-or-out peripersonal space might have guided us in a wrong direction
We believe that this notion stems from oversimplifications used by researchers to convey their findings efficiently and concisely. Such binary descriptions are already present in seminal studies that eventually led to the concept of PPS: multiple early researchers seem to have opted to present their findings as if there were a only few *distinct* spaces at different distances from an animal, probably to ease the understanding of the reader (Brain, 1941; Hediger, 1950, 1955; Hall, 1969). Nonetheless, their empirical evidence did not support the existence of such distinct spaces. Their decision to present their findings as either-in-or-out zones, without referring to graded responses or fuzzy boundaries, might have implicitly set the tone, implying that near and far space have sharp boundaries and that they are separate constructs in the brain. In other words, these descriptions were simple labels to help observers of empirical results to classify data. However, such descriptions have later resulted in unwarranted implications about the mechanisms generating the data.

Examples of how early literature might have affected the tone of subsequent PPS studies can be found in studies of single neurons in macaques (Hyvarinen and Poranen, 1974; Mountcastle et al., 1975; Leinonen and Nyman, 1979; Rizzolatti et al., 1981; Gentilucci et al., 1983; Colby et al., 1993; Graziano and Gross, 1993), where a class of bimodal neurons were identified in both cortical and subcortical structures such as putamen, parietal, and premotor areas (do note the similarities to regions in which competing action representations are present under the affordance competition hypothesis!). These neurons respond to somatosensory stimuli, but also to visual or auditory stimuli presented in spatial proximity to the somatosensory receptive field. Such neurons were often described as demarcating zones of space, represented visually as bubbles with clear boundaries (Graziano and Cooke, 2006). These boundaries, however, were lines of arbitrary response magnitude (Graziano et al., 1997; Gentilucci et al., 1988): the actual data reported in these studies clearly illustrate that the response magnitudes were not simple step-like functions but instead gradual proximity functions (Rizzolatti et al., 1981; Gentilucci et al., 1988; Colby et al., 1993; Graziano et al., 1994; Fogassi et al., 1996; Duhamel et al., 1998); see also (Figure 9.2).

The psychophysical literature suffers analogous oversimplifications. Early psychophysical experiments would often test only two conditions: near versus far (Làdavas et al., 1998; McCourt and Garlinghouse, 2000; Farnè and Làdavas, 2002; Sambo and Forster, 2009; Sambo et al., 2012b; De Paepe et al., 2014). The desire to keep experimental design and analysis simple is understandable, but a binary experimental design can inadvertently imply a binary response pattern, and a consequent 'in-or-out' description of results. Even when empirical data showed a clearly graded fall-off and the authors explicitly stated that these changes were gradual, results have often been discussed implying a binary 'near space versus far space' distinction (Longo and Lourenco, 2006; Canzoneri et al., 2012; Sambo and Iannetti, 2013; Ferri et al., 2015; Longo et al., 2015; Serino et al., 2015; Bufacchi et al., 2016). Notably, we are not exempt from presenting results in such a simplified manner (Sambo and Iannetti, 2013; Bufacchi and Iannetti, 2016; Bufacchi et al., 2016, 2017). Such type of discussion can preclude a more complete conceptual understanding of PPS.

9.3.2.2 Action relevance allows for a graded description of 'peripersonal space measures'

Therefore, a better method of discussing PPS responses is warranted. The affordance competition hypothesis can provide some insight into a solution: the value, or relevance, of actions is not a binary quantity. Populations of neurons underlying competing contact-related actions should increase their response magnitude monotonically with the probability that the action they underlie is useful. For example, if a particular action may be useful in the near future, we expect measures that reflect the action's relevance to be at intermediate levels between when that action is enacted and when it bears no relevance to a situation whatsoever. Reasoning spatially, it follows that contact-action relevance should depend on proximity in a graded manner: as an object moves closer to the optimal position in which it can be touched or avoided through a specific action, the probability that that action needs to be enacted increases gradually. This action relevance perspective also allows for high action values even when the stimulus is beyond reaching distance (e.g. the relevance of grabbing a chocolate that you want to eat, at the other side of a shop). While this might seem to go against the traditional dogma that PPS responses only show larger magnitude when stimuli are in arm reaching distance, it in fact fits the existing data well: certain PPS neurons respond to stimuli presented at long distances from the body (Colby et al., 1993; Graziano et al., 1997), and during locomotion behavioural PPS measures are enhanced in response to stimuli beyond reaching distance (Noel et al., 2014; Berger et al., 2019).

9.3.2.3 The concept of field allows for spatial description of action relevance

A useful tool to illustrate the spatial distribution of such continuous action relevance values is the concept of a *field*. As anticipated earlier in the chapter, we use the term 'field' in the

Panal A: Heterogeneity of PPS fields

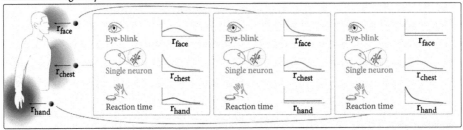

Panal B: Additional factors modulate PPS-related measures

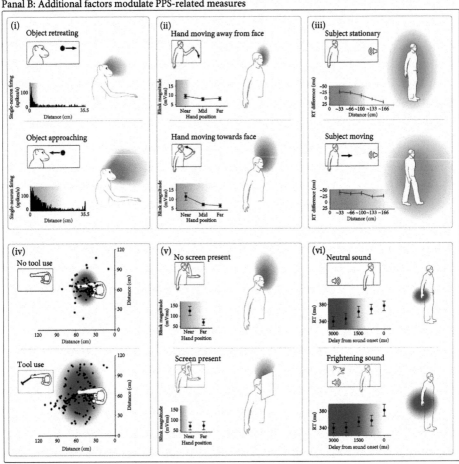

Figure 9.2 As detailed in the main text, we propose a reconceptualization of peripersonal space (PPS) as a set of fields reflecting the relevance of actions aimed at creating or avoiding contact between objects and the body. This figure illustrates the idea that there is not a single PPS, but that instead there are many PPS fields, all of which depend on proximity in a graded manner. Panel A shows the heterogeneity of PPS fields. Different PPS fields can be derived from the many types of biological measures that differently depend on spatial proximity. Here we show, as an example, the PPS fields derived from the modulation exerted by the proximity between a visual stimulus and the body on different biological measures: (1) the somatosensory-evoked eye blinking (green); (2) the response of a visuo-tactile single neuron with a somatosensory receptive field on the chest (red); and (3) the reaction times to somatosensory stimuli

Figure 9.2 Continued
delivered to the hand (blue). Note how the same visual stimulus in an identical position elicits different responses and defines PPS fields with different spatial features. Panel B: not just proximity: additional factors modulate PPS-related measures. Although the magnitude of PPS-related measures is commonly affected by proximity to a body part, many other factors affect these PPS measures as well. Such factors include various types of motion: motion of a visual stimulus (top left), stimulated limb (top middle), and the entire body (top right) can all cause expansion of the response fields. Factors independent of motion also affect PPS measures: tool use can expand response fields (bottom left), a protective screen can deform them (bottom middle), and frightening sounds can expand them (bottom right). Response fields are colour-coded by body part near which their magnitude is maximal: face (green), hand (blue), and trunk (red).
Adapted from (Graziano et al., 1997; Sambo et al., 2012a; Noel et al., 2014b, 2014a; Taffou and Viaud-Delmon, 2014; Serino et al., 2015; Bisio et al., 2017).

same sense it is used in modern physics—i.e. to express a quantity that has a magnitude for each point in space and time[5] (McMcmullin, 2002). Because this magnitude can change continuously in space, the concept of a field accommodates gradual changes with distance in three dimensions. Thus, this concept does not necessitate PPS to be an 'in-or-out' zone, and additionally allows PPS-related measures to change non-monotonically with proximity to the body. For example, it accommodates empirical evidence of multiple separated regions with local maxima: high relevance values around both the hand holding a tool and the tip of that tool (Holmes et al., 2007), and maximal firing rates in response to stimuli placed a few centimetres away from the body instead of adjacent to it (Rizzolatti et al., 1981). In other words, fields provide a good way of spatially describing the probabilistic relevance of actions, and thus also the activity of competing action representations prescribed by the affordance competition hypothesis.

9.3.3 More than one space: many different peripersonal space measures

9.3.3.1 Evidence against peripersonal space as a single entity
The affordance competition hypothesis can also help solve the second issue posed by traditional descriptions of PPS. As described earlier, many PPS-related measures exist[6]. The main feature that all these measures have in common is that their magnitude increases with the proximity of a stimulus to a certain body part. However, *what* this body part is, and *how*

---

[5] Importantly, we are not arguing that the values for every point in space and time are being individually stored in 'PPS regions' of the brain (Bufacchi and Iannetti, 2019; Noel and Serino, 2019). Rather, we argue that the regions underlying PPS-related measures can be seen as function approximators, which return a particular instance of a PPS field value, given the input of a particular position and time (i.e. state).

[6] These measures include the activity of single visuo-tactile and audio-tactile neurons in monkeys (Hyvarinen and Poranen, 1974; Rizzolatti et al., 1981; Colby et al., 1993; Graziano and Gross, 1993), cross-modal extinction in brain-damaged patients (Farnè and Làdavas, 2002), crossmodal congruency distraction effects (Spence et al., 2004), performance in line bisection tasks (McCourt and Garlinghouse, 2000), reaction times to tactile stimuli during simultaneous visual/auditory stimulation (Canzoneri et al., 2012; Serino et al., 2015; de Haan et al., 2016), temporal order judgements (De Paepe et al., 2014), reachability and semantic estimates (Bourgeois and Coello, 2012; Kalénine et al., 2016), spatial demonstratives (Caldano and Coventry, 2019), defensive reflexes (Sambo et al., 2012b; Versace et al., 2019), EEG responses (Sambo and Forster, 2009; Wamain et al., 2016), fMRI responses (Bremmer et al., 2001; Makin et al., 2007), and TMS-evoked responses in muscles (Serino et al., 2011).

the magnitude depends on proximity, both vary from measure to measure (Farnè et al., 2005; Serino et al., 2015): it is well known that different types of biological measures, differently dependent on body proximity, yield different PPS fields (Figure 9.2). Furthermore, as described in the next section, PPS measures are not only affected by proximity but also by other factors, whose modulatory effects on each measure vary.

Thus, given (1) the multitude of PPS-related response functions, and (2) their different sensitivity to factors different from body proximity, referring to 'the' PPS as if it were a single entity does not bring clarity to the discourse.

9.3.3.2 Action relevance allows for many different peripersonal space measures

Here, the affordance competition hypothesis helps us again. If a given PPS measure reflects the value of a particular action or set of actions, the spatial profile (or, to use the terminology we suggested above in Section 9.1.4, the *field*) of such value will depend on what the action is. For example, the blink reflex aims to avoid contact between a dangerous stimulus and the eye; it is thus behaviourally useful that its magnitude depends on the likelihood that an environmental stimulus hits the eye—a likelihood that, in turn, depends on the proximity between the stimulus and the face (although by no means only on the proximity—see the next section) (Sambo et al., 2012a; Bufacchi et al., 2016). Similarly, the stimulation of some multimodal VIP neurons elicits defensive postures aimed at preventing contact between an environmental stimulus and a particular body part (Cooke and Graziano, 2004). Hence, it is the variety of different possible actions that explains why PPS fields derived from, for example, the modulation of the magnitude of the defensive hand-blink reflex (HBR; Sambo & Iannetti 2013), and from the firing rates of certain VIP neurons, can be vastly different: HBR responses are increased when stimuli are near the face (Bufacchi et al., 2016), whereas a specific VIP response might be stronger when the stimulus is near, say, the forearm (Duhamel et al., 1998). In other words, the PPS field derived from a particular measure will be shaped by whatever actions are linked to that measure (Figure 9.2). Thus, the brain estimates the value of external events differently for the behaviour triggered by these events. This estimation broadly depends on the spatial location of the sensory event in egocentric coordinates, but the precise spatial relationship differs depending on the type of action required given a particular sensory event. As an aside, this perspective can resolve the apparent conflict between 'appetitive' and 'defensive' PPS (de Vignemont and Iannetti, 2015): both grasping and defensive actions have the common denominator of determining whether contact is created or avoided between an object and a body part, while occasionally being affected differently by variables such as object valence and movement.

9.3.3.3 Peripersonal space as a class of response fields

If one accepts this line of reasoning, the conclusion is that there is no such thing as 'the PPS', but instead only a set of responses, which can be understood spatially under the umbrella term of 'PPS fields'. Rather than forcing us to choose which (or which combination) of the many PPS measures best represent 'the' PPS, or 'what is near', the concept of a set of fields allows us to see the various PPS measures as separate instances of a wider class of responses with common functional significance. In this way, the interactive behaviour framework and the resulting postulate of several contact-related action relevance fields allow us to shift perspective and safely abandon the concept of a single PPS, for which there is currently a lack of empirical evidence.

9.3.4 More than proximity: many variables influence peripersonal space measures

The concept of contact-related action relevance also helps resolve the third issue posed by traditional descriptions of PPS: although proximity to the body or a body part is a common factor determining the spatial properties of PPS measures, it cannot be ignored that many other factors affect these measures. In this section, we will first briefly describe some of these other factors (more substantial summaries of these factors can be found in multiple recent reviews: Fogassi and Gallese, 2004; Dijkerman and Farnè, 2015; Bufacchi and Iannetti, 2018; Hunley and Lourenco, 2018), and show that such seeming anomalies are easily explained through contact-related action relevance.

9.3.4.1 Evidence of many non-proximity factors affecting PPS measures
We will call some of these other factors 'movement-related'. With this term, we mean parameters derived from the temporal derivative of stimulus position (Figure 9.2, Panel Bi, ii, and iii). For example, many bimodal 'PPS neurons' respond more than twice as much to stimuli moving in a preferred direction (Colby et al., 1993), and some even prefer receding stimuli (Duhamel et al., 1998). This means that, because of differential sensitivity to movement, two neurons with the same tactile receptive field could show different response magnitudes even when the responses are elicited by identical visual stimuli. Bimodal PPS neurons are also sensitive to large numbers of other movement-related parameters such as speed, direction of translation and rotation, and active and passive joint movement (Rizzolatti et al., 1981; Fogassi et al., 1996; Graziano et al., 1997). Human PPS-related responses can also be influenced by movement-related factors such as walking (Noel et al., 2014; Amemiya et al., 2017), gravitational cues (Bufacchi and Iannetti, 2016), vestibular cues (Pfeiffer et al., 2018), motion of body parts (Brozzoli et al., 2009, 2010; Wallwork et al., 2016; Bisio et al., 2017), stimulus direction (Serino et al., 2015), stimulus trajectory (Cléry et al., 2015a), and trajectory of environmental objects (Somervail et al., 2019).

We also identify another group of modulating factors, which we call 'non-movement-related'. Under this category, we include parameters that are neither derived from stimulus position, nor from its temporal derivatives (Figure 9.2, Panel Biv, v, and vi). These non-movement-related factors, such as stimulus size and semantic value, affect canonical PPS neurons in primates (e.g. some respond more strongly to snakes than to apples) (Graziano et al., 1997; Graziano, 2018). In humans, many types of high-level, non-movement-related factors also affect PPS measures: these include stimulus valence and semantics (Heed et al., 2010; Taffou and Viaud-Delmon, 2014; Ferri et al., 2015; de Haan et al., 2016), learning (Cléry et al., 2015b), environmental landscape (Sambo et al., 2012a; Somervail et al., 2019), postural valence (Biggio et al., 2019), pregnancy (Cardini et al., 2019), sensorimotor adaptation (Leclere et al., 2019), stimulus temperature (Ruggiero et al., 2019), tool use (Berti and Frassinetti, 2000), action planning (Lohmann et al., 2019), and action execution (Brozzoli et al., 2009).

9.3.4.2 Action relevance explains modulation by many non-proximity factors
Given the large number of both movement-related and non-movement-related factors determining response magnitudes of PPS measures, the concept of a distance-based, in-or-out space leads to problems in reasoning. For example, when PPS measures are altered by

stimulus movement, people have claimed that PPS or 'the near space' has changed in size. This has even been stated as 'far becomes near' (Berti and Frassinetti, 2000). However, this reasoning is not necessarily correct: 'When a stimulus is close, a measure's magnitude increases. Therefore, when a measure's magnitude is increased, a stimulus is close.' This reverse inference is not logically sound, because other measures than proximity can cause the observation that the measure's magnitude was increased (Poldrack, 2012). In other words, the manner in which PPS is currently often discussed has staked a larger importance on the factor proximity than on the factor movement direction, only because of the assumption that the measure denotes a space. In reality, there is no reason to think that, from a functional perspective, stimulus proximity is more important to PPS measures than any of the other factors to which these measures are sensitive.

Contact-related action relevance provides a functional perspective in which proximity is one, but not the most important or only, factor. Consider how varying stimulus movement might affect an animal's action choice. If the animal observes a rock rolling towards it with high speed, the animal would need to deploy a motor repertoire vastly different from if the object were moving tangentially to it. This would be the case even if the rock began in the same location—i.e. at the same distance from the animal. Environmental factors might also affect action choices and thus PPS fields. If an animal is faced with a predator, the animal might behave very differently if there were a large tree to climb, compared with a small cave to scurry into (Evans et al., 2019; see also Figure 4 in Bufacchi and Iannetti, 2018). Thus, the PPS field derived from the same measure might be different depending on the contextual change represented by the presence of the tree or the cave. Accordingly, various studies have shown that PPS measures are affected by environmental objects. For example, the HBR is affected by the presence of a screen (Sambo et al., 2012a) and by moving environmental objects (Somervail et al., 2019), reaction times to multimodal stimuli are affected by the presence of a transparent screen (Costantini et al., 2010), and the receptive field of various visuotactile neurons changes when a barrier is placed in between the monkey and the stimulus (Bonini et al., 2014; Maranesi et al., 2014).

The proposed reconceptualization also allows us to understand why and how PPS fields might interact within and between individuals (Pellencin et al., 2018): conspecifics are also features of the environment. For example, when considering a given individual surrounded by other agents, the relevance of the actions available to that individual depends not only on the position of other agents, but also on the set of actions available to them (i.e. their various contact-related PPS fields). Thus, social influences on PPS measures can be explained by postulating that each individual uses his/her own coding of others' contact-related action relevance as input to generate his/her own PPS fields. In fact, as already suggested by some (Iachini et al., 2016; Quesque et al., 2017; Hunley and Lourenco, 2018; Coello and Iachini, this volume), the minimum distance at which an individual is comfortable with being approached by others, often referred to as 'personal space', can then also be understood as a PPS action-relevance field: that distance is likely dictated by the relevance of contact-actions available to both the individual and the others (Cartaud et al., 2018).

9.3.4.3 A similar unjustified primacy of space for hippocampal place cells

Interestingly, the same issue of the unjustified primacy of space might apply to the functional interpretation of place and grid cells. Classically, these cells have been considered to code the *spatial* position of an animal relative to the environment, due to the strong correlation between their firing and the animal's location (O'Keefe and Dostrovsky, 1971; Mullerl et al., 1989). However, recent evidence that these cells are not solely sensitive to spatial information

(Garvert et al., 2017) has led to the novel idea that place and grid cells do not represent the animal's state in space specifically, but instead code its current general state as a function of future possible states (Stachenfeld et al., 2017; Chen et al., 2019). This is particularly relevant for the current discourse on PPS, not only because it provides evidence that a greater understanding of the system might be gained by revoking the primacy of spatial location (Bufacchi and Iannetti, 2018), but also because the hippocampus and parieto-premotor loop likely work tightly together to plan and coordinate actions (Whitlock et al., 2008).

9.4 The link between contact-related action relevance, impact prediction, and multisensory integration

So far we have argued that the affordance competition hypothesis leads to an understanding of PPS as a set of contact-related action relevance fields, and that this understanding solves common problems in interpreting PPS measures.[7] Admittedly, a few other PPS frameworks can lead to the conclusion that PPS should be not be conceived as a single, proximity-dependent entity with sharp boundaries. In this section, we will discuss those frameworks and their relationship with contact-related action relevance, and hope to show that contact action relevance can explain and incorporate those other frameworks. To do so, we will use some simple mathematical notations.

9.4.1 Contact-related action relevance formalized

Borrowing terminology from the field of reinforcement learning, we write as Q the value of a particular contact-related action a (out of the set of all possible actions A), in a particular state s (out of the set of all possible states S): $Q(a, s)$ (Bufacchi and Iannetti, 2019; Noel and Serino, 2019).[8] Simplifying even further some of the assumptions we have made in writing

---

[7] We are not the first to suggest a motoric interpretation of PPS, but there are several differences between our perspective and others. First, other motor conceptualizations often imply that the firing of a particular 'PPS neuron' represents an action directed towards or away from a specific spatial location in external space (Rizzolatti et al., 1997; Fogassi and Gallese, 2004; Graziano and Cooke, 2006). In contrast, we postulate that the firing represents the value of a specific action, meaning that, for example, movements tangential to a specific location would be included under the current perspective as well.

Other motor conceptualizations also imply that stimulus position in egocentric coordinates and the level of neuronal firing reflecting a motor programme are obligatorily linked. For example, when a stimulus is in a certain position, a particular motor plan will always be activated to a certain degree. Thus, these conceptualizations allows one to flip between a motoric and a spatial interpretation of neural firing. Our view, in contrast, does not imply an obligatory relationship between stimulus position and neuronal firing.

[8] Adapted from (Bufacchi and Iannetti, 2019): the goal of reinforcement learning is to allow an agent to optimally interact with its environment. There are four core concepts in reinforcement learning: states S, actions A, rewards R, and values V or Q. States describe the configuration of the agent–environment system. Actions are the options available to the agent, and cause it to enter another state in the next time-step (this next state can be identical to the previous state). Each action–state pair is associated with a reward, which can be positive, negative, or zero. The objective of the agent is to maximize rewards by estimating at each time-step the value of each possible action. To estimate these action values Q, the agent makes assumptions about what future states can be visited (i.e. physical laws) and will be visited (i.e. the policy of the agent). Based on those (learned) assumptions, expected future rewards are summed and weighted, typically in a manner inversely proportional to time, such that predicted rewards in the near future weigh more heavily than those in the far future. Thus, when the agent finds itself in a particular state s, each of its available actions is associated with one value: the discounted cumulative expectation of future rewards $Q(s,a) = E[R \mid s,a]$.

(i.e. assuming linearity), we can say that the strength of a PPS-related measure M should be proportional to a weighted sum of the values of several contact-related actions C:

$$1)\ M \propto \sum_{a \in C} w_a Q(a, s)$$

where w_a is the weight of a given action. Note that the weights would be unique to each PPS measure, and that many of them, or even all but one, could be 0 for most actions a and measures M. Note also that we have alluded to the set of all contact-related actions C, which we use to understand and classify PPS measures but, importantly, is not a distinction that we feel the brain would make explicitly. While the meaning of contact-related actions is intuitively obvious, defining them formally is harder. In the current description, C is a subset of all actions A. Actions in subset C alter the probability P of a tactile event t_i occurring in the future, where i is the index of a small section of the body's surface. In other words, we can say that, for each contact-related action, there must exist a part of the body surface for which performing the action in the present τ_0 increases or decreases the future cumulative probability of contact with the external stimulus. In other words,

$$2)\ \forall a \in C, \exists i, \sum_{\tau_{0+1}}^{\tau_\infty} P(t_i \mid \tau, A_{\tau_0} = a) \neq \sum_{\tau_{0+1}}^{\tau_\infty} P(t_i \mid \tau)$$

A final point to note is that equation 1 is relevant for the present time (τ_0): a PPS measure M should be proportional to action values at the current time point. However, these action values will by definition depend on the model the agent has of the future (Bufacchi and Iannetti, 2019): the value of an action is the cumulative expectation of future rewards.

In the next two sections we briefly describe two other PPS frameworks, and in Section 9.4.4 we argue that action-relevance framework is the most parsimonious.

9.4.2 Impact prediction

Having established a mathematical formalism of the contact-related action value framework we propose, we now attempt to look at other PPS frameworks through a similar lens. One such popular framework states that PPS measures reflect impact prediction (Noel et al., 2018a; Dijkerman and Medendorp, this volume). In more formal language, PPS measures M are outputs from a system calculating the probability that a stimulus will impact a specific part of the body within a specific time window $[\tau_1 \ldots \tau_n]$. Thus, using the notation used above, we could write:

$$3)\ M \propto \sum_{\tau_{0+1}}^{\tau_n} \sum_i w_i P(t_i \mid \tau, s)$$

Support for such a framework comes from observations that certain PPS measures, such as tactile detection and reaction speed, are maximally enhanced when tactile stimuli are delivered at the predicted time of impact of a looming visual or auditory stimulus (Canzoneri et al., 2012; Kandula et al., 2014; Cléry et al., 2015a). These effects are especially strong when the tactile stimulus is also spatially congruent—i.e. it is delivered at the location toward which the visual stimulus looms (Kandula et al., 2014; Cléry et al., 2015a). If PPS measures reflect impact prediction, (1) we can still construct multiple response fields, one for each location of impact, and (2) these 'hit probability' fields would still be affected by movement-related variables, such as walking and stimulus velocity (Fogassi et al., 1996; Noel et al., 2014). It is less clear why (3) those impact-predicting PPS fields would gradually fall off with distance, but a possible explanation for this might be the fact that the sensory system does not have full access to all the information from a noisy environment.

We have in fact previously proposed a mathematical model in which the magnitude of PPS measures are proportional to the probability that a part of the body is hit by a stimulus, given some uncertainty in the stimulus' movement direction (Bufacchi et al., 2016). Specifically, in the case of the hand-blink reflex, that model corresponded to equation 3 under the following assumptions: that the trajectory of the environmental object is linear and randomly distributed; that $n = \infty$; that $w_i = 0$ for all surface elements not on the face; and that $w_i = 1$ for all surface elements that are on the face. We showed that this model can account for hand-blink reflex magnitude across a large three-dimensional region of space, and can also explain why gravitational cues affect it: when a stimulus is above the body, the stimulus is more likely to fall down and hit the body, resulting in a greater hit probability and hence blink magnitude (Bufacchi and Iannetti, 2016). Similarly, this model successfully explains the increase in blink magnitude when moving stimuli different from that eliciting the blink are presented near the blink-eliciting stimulus: these other stimuli might alter the assumed distribution of directions in which the blink-eliciting stimulus can move (Somervail et al., 2019).

However, as explained in detail in Section 9.4.4.2, both the impact prediction framework and the multisensory integration framework that we describe in the next paragraph do not satisfactorily explain the dependence of PPS measures on non-movement factors.

9.4.3 Multisensory integration

Another, related perspective on PPS holds that PPS exclusively reflects multisensory integration between touch and another sense (Bernasconi et al., 2018; Noel et al., 2018b). Indeed, many traditional PPS measures are sampled in response to the simultaneous presentation of stimuli belonging to touch and to another sensory modality—e.g. reaction times to tactile stimuli while auditory or visual stimuli approach (Canzoneri et al., 2012), and audio-tactile temporal order judgements (Kitagawa et al., 2005). The body proximity-dependence of these types of multimodal integration can be at least partially explained by the fact that multisensory events involving touch must by definition happen near the body (van der Stoep et al., 2016). Although frequently mentioned, multisensory integration is often only loosely alluded to as a PPS framework, without stating exactly how it would explain PPS findings. There is one recent attempt at a formal model relating multisensory integration to PPS (Noel et al., 2018b). This model assumes that a single visual and a single tactile stimulus are presented simultaneously, and that this leads to the observation of a multisensory percept. The model is formulated in terms of the external location of that multisensory event. The multisensory integration model we suggest here is inspired by that model, and attempts to generalize the multisensory model beyond the assumptions it makes, in order to link it formally to action relevance. Considering the case of vision and touch, a PPS measure might be proportional to the probability of any visuo-tactile event (t_i, v_j) occurring in the present.

$$4)\ M \propto \sum_i \sum_j w_i w_j P\left(t_i | \tau_0, v_j\right) = \sum_j w_j P\left(v_j\right) \sum_i w_i P\left(t_i | \tau_0, v_j\right)$$

where v_j is the jth visual event (and t_i, as mentioned earlier, is the ith tactile event). Importantly, whether using the formal model provided here or a more loose conceptualization, multisensory integration (1) would depend on the location of tactile events, thus allowing for the existence of multiple PPS fields, (2) would have a gradual fall off from the body, as opposed to a binary in-or-out function, and (3) might depend on stimulus movement and other movement-related factors (see earlier Section 9.3).

As can be seen by comparing equations 3 and 4, multisensory integration involving touch is conceptually very similar to impact prediction, given that impact prediction also involves integrating audio/visual events (i.e. the state s) with tactile events (Noel et al., 2018b). One clear difference between those equations, is that equation 4 (i.e. the multisensory framework) concerns the present τ_0, while equation 3 (i.e. the impact prediction framework) concerns the future $[\tau_1 \ldots \tau_n]$. We could then speculate that a system performing multisensory integration (equation 4) might, in the absence of touch, be performing future tactile prediction, and that a tactile prediction network might perform multisensory integration when stimuli are presented simultaneously. In fact, it has already been suggested that visuo-tactile and audio-tactile integration could rely on the same neural mechanisms as impact prediction (Fotowat and Gabbiani, 2011; Cléry et al., 2017; Noel et al., 2018a).

Another difference that becomes immediately clear from equation 4 is that, if there is no concomitant stimulation in both modalities (e.g. if no visual event occurs, meaning that $P(v_j) = 0$ for all j), PPS measures should not change, or be at baseline. However, it is important to note the existence of many PPS-related measures that (a) do not require simultaneous stimulation in multiple modalities (i.e. a single modality is enough to elicit a proximity-dependent response) (Mountcastle et al., 1975; Graziano et al., 1994; Bernasconi et al., 2018), or (b) even only exist in response to stimulation in a single modality (Longo and Lourenco, 2006; Sambo et al., 2012b; Blini et al., 2018). This makes the account of PPS responses as strictly multisensory unlikely, unless one a priori defines PPS measures as having to be multimodal. Rather, the existence of unimodal (or not strictly multimodal) PPS measures implies that the multisensory integration theory could be seen as a subset, or a resultant property of the impact prediction theory (Blini et al., this volume).[9]

9.4.4 Integrating impact prediction and multisensory frameworks with action relevance

9.4.4.1 Hit probability in different peripersonal space frameworks

Having established the similarities between impact prediction and multisensory integration, in this section we examine how contact-action relevance relates to them. Compare equations

---

[9] Blini et al. (2018) found that when visual stimuli are nearer the body, participants are better at discriminating between spheres and cubes than when the same stimuli are far away. Considering the motoric theory of perception (Berthoz, 1997), subjects could use the information from neurons that specify the relevance of potential actions to discriminate objects. Thus, when objects are near, they afford more types of contact-related actions than when they are far. Furthermore, the ball and the types of actions afforded by the cube and ball differ more when they are near than when they are far.

1 and 2 on the one hand with equations 3 and 4 on the other hand: all frameworks contain a term specifying the probability that a tactile event t_i occurs at the ith element of the body surface. Besides the exact variables defining this probability term, the important difference between PPS frameworks is what this term represents. In the hit probability and impact prediction frameworks (equations 3 and 4), this probability is being explicitly calculated in the brain, and PPS neurons output the result of this calculation. In other words, in the hit probability and multisensory frameworks, the *raison d'être* of PPS measures (and of PPS neurons in particular) is to calculate the probability of occurrence of tactile events—information that is then fed to other systems. In contrast, in the action relevance framework that we propose, hit probability is a lens used to understand neural responses, but is not calculated explicitly in a cortical 'hit probability module'. We say that certain actions inevitably alter hit probability (as an objective measure), and that the relevance of these actions obeys certain laws, which will result in the response profiles of the 'PPS measures' that we know and love. Thus, in the action relevance framework, the only necessary assumption is that brains have evolved to control movement, and not that brains explicitly calculate hit probability in certain neurons or systems.

9.4.4.2 Non-movement factors are only explained by the action relevance framework

Another important aspect of the hit probability and multisensory integration frameworks, in contrast to the action relevance framework, is that the former are purely predictive functions, specifying whether an event occurs or not (equations 3 and 4), whereas the effect of any non-movement-related factor on hit probability is hard to explain, such as action planning (Patané et al., 2018), stimulus valence and semantics (Heed et al., 2010; Taffou and Viaud-Delmon, 2014; Ferri et al., 2015; de Haan et al., 2016) (see Section 9.3.3). As such, impact prediction and multisensory frameworks are blind to the semantics and valence of the object producing the impact. We realized this important aspect when attempting to model the effect of trigeminal neuralgia on the proximity-dependent modulation of the hand-blink reflex (Bufacchi et al., 2017). In this chronic pain condition, one half of the face is affected by abrupt intense painful sensations (Cruccu et al., 2016). These painful sensations can be triggered by very light innocuous tactile stimuli (e.g. a gentle breeze), suggesting that it might be more behaviourally relevant to better defend the affected side of the face. A movement-related variable was not sufficient to model the larger blink magnitude to stimuli ipsilateral to the painful side of the face. Instead, we had to add a non-movement parameter—that is, the side bias consequent to the ipsilaterality of the stimulus to the painful side of the face (Bufacchi et al., 2017). There are of course many more examples than those mentioned so far of behavioural relevance influencing PPS measures (see also Section 9.3.3). We could even go so far as to say that there isn't a single PPS measure that is not affected.

For example, take the PPS fields derived from reaction times to tactile stimuli on either the hand, the face, or the torso. The proximity of *looming* auditory stimuli shortens reaction times for the tactile stimuli delivered to all these locations. In contrast, the proximity of *receding* auditory stimuli only shortens reaction times when tactile stimuli are delivered to the hand (Serino et al., 2015). This also supports the action relevance framework provided here: many actions of the arm, face, and torso are relevant to avoiding a looming stimulus. However, there are far more actions relevant to a receding stimulus executed by the arm (e.g. reaching to grasp) than by the torso or the head. A second example is that after tool use cross-modal congruency is only strengthened near the tip of the tool, which is the behaviourally relevant part of the tool to create contact, while congruency is not strengthened around the middle of the tool (Holmes et al., 2007).

9.4.4.3 Impact prediction and multisensory integration as byproducts of action relevance

Thus, in contrast to impact prediction and multisensory integration, action relevance can explain the effects of the many above-mentioned non-movement and non-proximity factors: besides an object's location, its nature, its movement, as well as the motivational set of the agent might all affect the value of actions aiming to create or avoid contact with that object. Impact prediction and multisensory integration are only partially satisfactory explanations of empirical data because they are necessary yet partial aspects of action relevance (Fogassi and Gallese, 2004). This point is exemplified by the formalization we provide: equation 2 shows that certain types of actions by definition change the probability of a tactile event on the skin, and their relevance will thus necessarily correlate with hit probability. Indeed, as we demonstrated earlier with the hand-blink reflex, the relevance of a contact-related action clearly correlates with the probability that a stimulus might impact a body part (Bufacchi et al., 2016). The calculation of *contact* action relevance can therefore resemble, but not be limited to, calculating impact prediction. As such, and given the similarities between equations 3 and 4, a measure that reflects contact action relevance should also strongly correlate with both impact prediction and multisensory integration involving touch. In other words, if PPS measures do reflect contact action relevance, their strong relationships with impact prediction and tactile multisensory integration are *arising properties*, much like their dependence on proximity and the other factors discussed above (see Figure 9.3): these correlations might be considered epiphenomena of the need to act.

Action relevance also addresses a further question about impact prediction: at which timescale does impact prediction correlate best with PPS measures (i.e. what is τ_n?), and why? From the action relevance perspective, the response is simple: impact prediction should correlate best with PPS measures at timescales that allow performing actions that cause or prevent the impact being predicted. Without action relevance, this question remains unanswered.

Seeing impact prediction and multisensory integration as arising properties of action relevance does not strictly exclude the possibility that there are separate 'body proximity' or 'impact prediction' modules, whose output is fed into an 'action relevance' module. However,

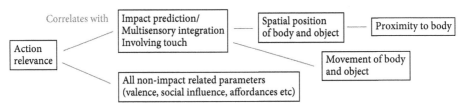

Figure 9.3 If a measure reflects the relevance of a contact-related action, and does so correctly, the same measure should also correlate with a number of factors that affect contact-action relevance. All these factors are known to affect peripersonal space (PPS) measures (Fogassi and Gallese, 2004; Dijkerman and Farnè, 2015; Bufacchi and Iannetti, 2018; Hunley and Lourenco, 2018). These factors include (but are not limited to) impact prediction and multisensory integration, which have been proposed as the explanations for the existence of PPS measures. However, given that some of the factors affecting PPS measures are not related to movement or position of stimuli relative to the body, action relevance is a more complete framework to understand the functional significance of 'PPS-related' measures.

to the best of our knowledge, there are no PPS measures that uniquely reflect the activity of any of these modules. Much like the general argument against serial processing (see Section 9.2), the existing PPS measures therefore do not support a serial computation of action relevance. Instead, if action relevance is factored in from the earliest computations (as predicted by the interactive behaviour framework [Cisek and Kalaska, 2010]), we should (and we do!) see PPS measures that always correlate to some degree with all three factors: body proximity, impact prediction, and other non-movement parameters. We therefore suggest that, to better understand the neural mechanisms behind the large number of PPS-related measures, we should attempt to test whether action relevance can be broken down into other types of distinguishing features that might better reflect individual PPS measures. That is where we believe the future challenges of PPS research lie.

References

Allport DA (1987) Selection for Action: Some Behavioral and Neurophysiological Considerations of Attention and Action. In Perspectives on Perception and Action, eds H Heuer, HF Sanders, pp 395–419. Hillsdale, NJ: Lawrence Erlbaum.

Amemiya T, Ikei Y, Hirota K, Kitazaki M (2017) Vibration on the soles of the feet evoking a sensation of walking expands peripersonal space. 2017 IEEE World Haptics Conference, 605–610.

Andersen RA, Buneo CA (2002) Intentional maps in posterior parietal cortex. *Annu Rev Neurosci* 25: 189–220. Available at: http://www.annualreviews.org/doi/10.1146/annurev.neuro.25.112701.142922

Arkin RC (1998) Behavior-based Robotics, 1st edn. Cambridge, MA: MIT Press.

Battaglia-Mayer A, Caminiti R (2019) Corticocortical systems underlying high-order motor control. *J. Neurosci* 39(23):4404–4421.

Berger M, Neumann P, Gail A (2019) Peri-hand space expands beyond reach in the context of walk-and- reach movements. *Sci Rep*:1–12. Available at: http://dx.doi.org/10.1038/s41598-019-39520-8

Bernasconi F, Noel J-P, Park HD, Faivre N, Seeck M, Spinelli L, Schaller K, Blanke O, Serino A (2018) Audio-tactile and peripersonal space processing around the trunk in human parietal and temporal cortex: an intracranial EEG study. *Cereb Cortex*:1–13.

Berthoz A. (1997) Le Sens du Mouvement. Paris: Editions Odile Jacob.

Berti A, Frassinetti F. (2000) When far becomes near: remapping of space by tool use. *J. Cogn Neurosci* 12:415–420.

Biggio M, Bisio A, Ruggeri P, Bove M. (2019) Defensive peripersonal space is modified by a learnt protective posture. *Sci Rep* 9(1):6739.

Bisio A, Garbarini F, Biggio M, Fossataro C, Ruggeri P, Bove M (2017) Dynamic shaping of the defensive peripersonal space through predictive motor mechanisms: when the 'near' becomes 'far'. *J. Neurosci* 37:2415–2424. Available at: http://www.jneurosci.org/content/37/9/2415.abstract

Blini E, Desoche C, Salemme R, Kabil A, Hadj-Bouziane F, Farnè A (2018) Mind the depth: visual perception of shapes is better in peripersonal space. *Psychol. Sci* 29(11):1868–1877.

Bonini L, Maranesi M, Livi A, Fogassi L, Rizzolatti G (2014) Space-dependent representation of objects and other's action in monkey ventral premotor grasping neurons. *J. Neurosci* 34:4108–4119. Available at: http://www.jneurosci.org/cgi/doi/10.1523/JNEUROSCI.4187-13.2014

Bourgeois J, Coello Y (2012) Effect of visuomotor calibration and uncertainty on the perception of peripersonal space. *Atten Percep Psychophys* 74:1268–1283.

Boynton GM (2005) Attention and visual perception. *Curr Opin Neurobiol* 15:465–469.

Brain WR (1941) Visual disorientation with special reference to lesions of the right cerebral hemisphere. *Brain* 64:244–272. Available at: http://books.google.com/books?hl=en&lr=&id=JHzt R7ITiOgC&oi=fnd&pg=PA1247&dq=Visual+disorientation+with+special+reference+to+l esions+of+the+right+cerebral+hemisphere.&ots=k5xAjk5UiA&sig=YkeWGV9tD4h_2BN-lb1J7bfRs_c

Bremmer F, Schlack A, Shah NJ, Zafiris O, Kubischik M, Hoffmann K-P, Zilles K, Fink GR (2001) Polymodal motion processing in posterior parietal and premotor cortex. *Neuron* 29:287–296.

Brozzoli C, Cardinali L, Pavani F, Farnè A (2010) Action-specific remapping of peripersonal space. *Neuropsychologia* 48:796–802.

Brozzoli C, Ehrsson HH, Farnè A (2014) Multisensory representation of the space near the hand: from perception to action and interindividual interactions. *Neuroscientist* 20:122–135.

Brozzoli C, Pavani F, Urquizar C, Cardinali L, Farnè A (2009) Grasping actions remap peripersonal space. *NeuroReport* 20:913–917. Available at: http://content.wkhealth.com/linkback/openurl?sid= WKPTLP:landingpage&an=00001756-200907010-00004

Bufacchi RJ, Iannetti GD (2016) Gravitational cues modulate the shape of defensive peripersonal space. *Curr Biol* 26:R1133–R1134. Available at: http://dx.doi.org/10.1016/j.cub.2016.09.025

Bufacchi RJ, Iannetti GD (2018) An action field theory of peripersonal space. *Trends Cogn Sci* 22:1076–1090. Available at: https://linkinghub.elsevier.com/retrieve/pii/S1364661318302225

Bufacchi RJ, Iannetti GD (2019) The value of actions, in time and space. *Trends Cogn Sci* 23(4):270–271.

Bufacchi RJ, Liang M, Griffin LD, Iannetti GD (2016) A geometric model of defensive peripersonal space. *J. Neurophysiol* 115:218–225.

Bufacchi RJ, Sambo CF, Di Stefano G, Cruccu G, Iannetti GD (2017) Pain outside the body: defensive peripersonal space deformation in trigeminal neuralgia. *Sci Rep* 7:12487. Available at: http://www.nature.com/articles/s41598-017-12466-5

Caldano M, Coventry KR (2019) Spatial demonstratives and perceptual space: to reach or not to reach? *Cognition* 191.

Caminiti R, Borra E, Visco-Comandini F, Battaglia-Mayer A, Averbeck BB, Luppino G (2017) Computational architecture of the parieto-frontal network underlying cognitive-motor control in monkeys. *eNeuro* 4:ENEURO.0306-16.2017.

Canzoneri E, Magosso E, Serino A (2012) Dynamic sounds capture the boundaries of peripersonal space representation in humans. *PLoS ONE* 7:e44306. Available at: http://www.pubmedcentral.nih.gov/articlerender.fcgi?artid=3460958&tool=pmcentrez&rendertype=abstract

Cardini F, Fatemi-Ghomi N, Gajewska-Knapik K, Gooch V, Aspell JE (2019) Enlarged representation of peripersonal space in pregnancy. Sci Rep 9.

Cartaud A, Ruggiero G, Ott L, Iachini T, Coello Y (2018) Physiological response to facial expressions in peripersonal space determines interpersonal distance in a social interaction context. *Front. Psychol* 9:1–11.

Chen G, Lu Y, King JA, Cacucci F, Burgess N (2019) Differential influences of environment and self-motion on place and grid cell firing. *Nat Comm* 10, 630. Available at: https://doi.org/10.1038/s41467-019-08550-1

Cisek P, Kalaska JF (2005) Neural correlates of reaching decisions in dorsal premotor cortex: specification of multiple direction choices and final selection of action. *Neuron* 45:801–814.

Cisek P, Kalaska JF (2010) Neural mechanisms for interacting with a world full of action choices. *Annu Rev Neurosci* 33:269–298. Available at: http://www.annualreviews.org/doi/10.1146/annurev.neuro.051508.135409

Cléry J, Guipponi O, Odouard S, Pinède S, Wardak C, Ben Hamed S (2017) The prediction of impact of a looming stimulus onto the body is subserved by multisensory integration mechanisms. *J. Neurosci* 37:10656–10670. Available at: http://www.jneurosci.org/lookup/doi/10.1523/JNEUROSCI.0610-17.2017

Cléry J, Guipponi O, Odouard S, Wardak C, Hamed S Ben (2015a) Impact prediction by looming visual stimuli enhances tactile detection. *J. Neurosci* 35:4179–4189.

Cléry J, Guipponi O, Wardak C, Ben Hamed S (2015b) Neuronal bases of peripersonal and extrapersonal spaces, their plasticity and their dynamics: knowns and unknowns. *Neuropsychologia* 70:313–326. Available at: https://www.ncbi.nlm.nih.gov/pubmed/25447371

Colby CL, Duhamel JR, Goldberg ME (1993) Ventral intraparietal area of the macaque: anatomic location and visual response properties. *J. Neurophysiol* 69:902–914.

Cooke DF, Graziano MS (2004) Super-flinchers and nerves of steel: defensive movements altered by chemical manipulation of a cortical motor area. *Neuron* 43:585–593. Available at: http://www.ncbi.nlm.nih.gov/pubmed/15312656

Costantini M, Ambrosini E, Tieri G, Sinigaglia C, Committeri G (2010) Where does an object trigger an action? An investigation about affordances in space. *Exp Brain Res* 207:95–103.

Cruccu G, Finnerup NB, Jensen TS, Scholz J, Sindou M, Svensson P, Treede RD, Zakrzewska JM, Nurmikko T (2016) Trigeminal neuralgia: new classification and diagnostic grading for practice and research. *Neurology* 87:220–228.

de Haan AM, Smit M, Van der Stigchel S, Dijkerman HC (2016) Approaching threat modulates visuotactile interactions in peripersonal space. *Exp Brain Res* 234:1875–1884.

De Paepe AL, Crombez G, Spence C, Legrain V (2014) Mapping nociceptive stimuli in a peripersonal frame of reference: evidence from a temporal order judgment task. *Neuropsychologia* 56:219–228. Available at: http://www.ncbi.nlm.nih.gov/pubmed/24486423

de Vignemont F (2018) Mind the Body: An Exploration of Bodily Self-Awareness. Oxford: Oxford University Press. Available at: https://global.oup.com/academic/product/mind-the-body-9780198735885?cc=gb&lang=en&

de Vignemont F, Iannetti GD (2015) How many peripersonal spaces? *Neuropsychologia* 70:327–334. Available at: http://dx.doi.org/10.1016/j.neuropsychologia.2014.11.018

Dewey J (1896) The reflex arc concept in psychology. Psychological Review 3: 357–370.

Dijkerman HC, Farnè A (2015) Sensorimotor and social aspects of peripersonal space. *Neuropsychologia* 70:309–312. Available at: http://linkinghub.elsevier.com/retrieve/pii/S0028393215001104

Duhamel JR, Colby CL, Goldberg ME (1998) Ventral intraparietal area of the macaque: congruent visual and somatic response properties. *J. Neurophysiol* 79:126–136.

Ebbesen CL, Insanally MN, Kopec CD, Murakami M, Saiki A, Erlich JC (2018) More than just a 'motor': recent surprises from the frontal cortex. *J. Neurosci* 38:9402–9413. Available at: http://www.jneurosci.org/lookup/doi/10.1523/JNEUROSCI.1671-18.2018

Evans DA, Stempel AV, Vale R, Branco T (2019) Cognitive control of escape behaviour. *Trends Cogn Sci* 23:334–348. Available at: https://doi.org/10.1016/j.tics.2019.01.012

Fagg AH, Arbib MA (1998) Modeling parietal-premotor interactions in primate control of grasping. *Neural Netw* 11(7–8):1277–1303.

Farnè A, Demattè ML, Làdavas E (2005) Neuropsychological evidence of modular organization of the near peripersonal space. *Neurology* 65:1754–1758. Available at: http://www.ncbi.nlm.nih.gov/pubmed/16344519

Farnè A, Làdavas E (2002) Auditory peripersonal space in humans. *J. Cogn Neurosci* 14:1030–1043.

Ferri F, Tajadura-Jiménez A, Väljamäe A, Vastano R, Costantini M (2015) Emotion-inducing approaching sounds shape the boundaries of multisensory peripersonal space. *Neuropsychologia* 70:468–475. Available at: http://linkinghub.elsevier.com/retrieve/pii/S0028393215001062

Fogassi L, Gallese V. (2004) Action as a Binding Key to Multisensory Integration. In: The Handbook of Multisensory Processes, pp. 425–441. Cambridge MA: MIT Press. Available at: https://www.academia.edu/2207407/Action_as_a_binding_key_to_multisensory_integration

Fogassi L, Gallese V, Fadiga L, Luppino G, Matelli M, Rizzolatti G (1996) Coding of peripersonal space in inferior premotor cortex (area F4). *J. Neurophysiol* 76:141–157. Available at: http://www.ncbi.nlm.nih.gov/pubmed/8836215

Fotowat H, Gabbiani F (2011) Collision detection as a model for sensory-motor integration. *Annu Rev Neurosci* 34:1–19.

Garvert MM, Dolan RJ, Behrens TEJ (2017) A map of abstract relational knowledge in the human hippocampal–entorhinal cortex. *Elife* 6:1–20.

Gentilucci M, Fogassi L, Luppino G, Matelli M, Camarda R, Rizzolatti G (1988) Functional organization of inferior area 6 in the macaque monkey I. Somatotopy and the control of proximal movements. *Exp Brain Res* 71:475–490.

Gentilucci M, Scandolara C, Pigarev IN, Rizzolatti G (1983) Visual responses in the postarcuate cortex (area 6) of the monkey that are independent of eye position. *Exp Brain Res* 50:464–468.

Gibson JJ (1979) The Ecological Approach to Visual Perception. Boston, MA: Houghton Mifflin.

Graziano M (2018) The Spaces Between Us: A Story of Neuroscience, Evolution, and Human Nature, 1st edn. New York: Oxford University Press.

Graziano MS, Cooke DF (2006) Parieto-frontal interactions, personal space, and defensive behavior. *Neuropsychologia* 44:845–859 Available at: http://www.ncbi.nlm.nih.gov/pubmed/16277998

Graziano MS, Gross CG (1993) A bimodal map of space: somatosensory receptive fields in the macaque putamen with corresponding visual receptive fields. *Exp Brain Res* 97:96–109.

Graziano MS, Hu XT, Gross CG (1997) Visuospatial properties of ventral premotor cortex. *J. Neurophysiol* 77:2268–2292.

Graziano MS, Yap GS, Gross CG (1994) Coding of visual space by premotor neurons. *Science* 266:1054–1057.

Hall E (1969) *The Hidden Dimension: Man's Use of Space in Public and in Private*. New York: Anchor Books.

Hediger H (1950) Wild animals in captivity. Buttersworth Scientific Publications. Available at: https://www.sciencedirect.com/book/9781483201115/wild-animals-in-captivity

Hediger H (1955) Studies of the psychology and behaviour of captive animals in zoos and circuses. Butterworths Scientific Publications.

Heed T, Habets B, Sebanz N, Knoblich G (2010) Others' actions reduce crossmodal integration in peripersonal space. *Curr Biol* 20:1345–1349.

Hinde RA. (1966) Animal Behavior: A Synthesis of Ethology and Comparative Psychology. New York: McGraw-Hill.

Høgh-Olesen H (2008) Human spatial behaviour: the spacing of people, objects and animals in six cross-cultural samples. *J. Cogn Cult* 8:245–280 Available at: http://booksandjournals.brillonline.com/content/journals/10.1163/156853708x358173

Holmes NP, Calvert GA, Spence C (2007) Tool use changes multisensory interactions in seconds: evidence from the crossmodal congruency task. *Exp Brain Res* 183:465–476.

Hoshi E, Tanji J (2007) Distinctions between dorsal and ventral premotor areas: anatomical connectivity and functional properties. *Curr Opin Neurobiol* 17:234–242.

Hunley SB, Lourenco SF (2018) What is peripersonal space? An examination of unresolved empirical issues and emerging findings. *Cogn. Sci.* 9.

Hyvarinen J, Poranen A (1974) Function of the parietal associative area 7 as revealed from cellular discharges in alert monkeys. *Brain* 97:673–692.

Iachini T, Coello Y, Frassinetti F, Senese VP, Galante F, Ruggiero G (2016) Peripersonal and interpersonal space in virtual and real environments: effects of gender and age. J Environ Psychol 45:154–164 Available at: http://dx.doi.org/10.1016/j.jenvp.2016.01.004

Iriki A, Tanaka M, Iwamura Y (1996) Coding of modified body schema during tool use by macaque postcentral neurones. *NeuroReport* 7:2325–2330.

Kalénine S, Wamain Y, Decroix J, Coello Y (2016) Conflict between object structural and functional affordances in peripersonal space. *Cognition* 155:1–7.

Kandula M, Hofman D, Dijkerman HC (2014) Visuo-tactile interactions are dependent on the predictive value of the visual stimulus. *Neuropsychologia* 70:358–366 Available at: http://linkinghub.elsevier.com/retrieve/pii/S002839321400462X

Kitagawa N, Zampini M, Spence C (2005) Audiotactile interactions in near and far space. *Exp Brain Res* 166:528–537.

Làdavas E, Zeloni G, Farnè A (1998) Visual peripersonal space centred on the face in humans. *Brain* 121:2317–2326.

Leclere NX, Sarlegna FR, Coello Y, Bourdin C (2019) Sensori-motor adaptation to novel limb dynamics influences the representation of peripersonal space. *Neuropsychologia*. Available at: https://doi.org/10.1016/j.neuropsychologia.2019.05.005

Lee D, Seo H, Jung MW (2012) Neural basis of reinforcement learning and decision making. *Annu Rev Neurosci* 35:287–308. Available at: http://www.annualreviews.org/doi/10.1146/annurev-neuro-062111-150512

Leinonen L, Nyman G (1979) II. Functional properties of cells in anterolateral part of area 7 associative face area of awake monkeys. *Exp Brain Res* 34:321–333.

Lohmann J, Belardinelli A, Butz MV. (2019) Hands ahead in mind and motion: active inference in peripersonal hand space. *Vision* 3:15.

Longo MR, Lourenco SF (2006) On the nature of near space: effects of tool use and the transition to far space. *Neuropsychologia* 44:977–981.

Longo MR, Lourenco SF (2007) Space perception and body morphology: extent of near space scales with arm length. *Exp Brain Res* 177:285–290.

Longo MR, Trippier S, Vagnoni E, Lourenco SF (2015) Right hemisphere control of visuospatial attention in near space. *Neuropsychologia* 70:350–357. Available at: http://linkinghub.elsevier.com/retrieve/pii/S0028393214003959

Makin TR, Holmes NP, Zohary E (2007) Is that near my hand? Multisensory representation of peripersonal space in human intraparietal sulcus. *J. Neurosci* 27:731–740. Available at: http://www.ncbi.nlm.nih.gov/pubmed/17251412

Maranesi M, Bonini L, Fogassi L (2014) Cortical processing of object affordances for self and others' action. *Front. Psychol* 5:1–10.

McCourt ME, Garlinghouse M (2000) Asymmetries of visuospatial attention are modulated by viewing distance and visual field elevation: pseudoneglect in peripersonal and extrapersonal space. *Cortex* 36:715–731.

McMcmullin E (2002) The origins of the field concept in physics. *Phys. Perspect* 4:13–39.

Milner AD, Goodale MA (1995) The Visual Brain in Action. Oxford: Oxford University Press.

Mnih V, Kavukcuoglu K, Silver D, Rusu AA, Veness J, Bellemare MG, Graves A, Riedmiller M, Fidjeland AK, Ostrovski1 G, Petersen S, Beattie C, Sadik A, Antonoglou I, King H, Kumaran D, Wierstra D, Legg S, Hassabis D (2016) Human-level control through deep reinforcement learning. *Nature* 518:529–533. Available at: http://dx.doi.org/10.1038/nature14236.

Mountcastle VB, Lynch JC, Georgopoulos A, Sakata H, Acuna C (1975) Posterior parietal association cortex of the monkey: command functions for operations within extrapersonal space. *J. Neurophysiol* 38:871–908. Available at: http://www.physiology.org/doi/10.1152/jn.1975.38.4.871

Mullerl RU, Kubie JL, Muller RU (1989) The firing of hippocampal place cells predicts the future position of freely moving rats. *J. Neurosci* 9:4101–4110. Available at: http://www.ncbi.nlm.nih.gov/pubmed/2592993

Neumann O (1990) Visual Attention and Action. In Relationships between Perception and Action: Current Approaches, eds O Neumann, W Prinz, pp. 227–267. Berlin, Heidelberg: Springer Berlin Heidelberg. Available at: https://doi.org/10.1007/978-3-642-75348-0_9.

Niv Y (2009) Reinforcement learning in the brain. *J. Math. Psychol.* 53:139–154.

Noel J-P, Blanke O, Serino A (2018a) From multisensory integration in peripersonal space to bodily self-consciousness: from statistical regularities to statistical inference. Ann N Y Acad Sci 1426:146–165. Available at: http://doi.wiley.com/10.1111/nyas.13867

Noel J-P, Samad M, Doxon A, Clark J, Keller S, Luca M Di (2018b) Peri-personal space as a prior in coupling visual and proprioceptive signals. *Sci Rep* 81(8):15819. Available at: https://www.nature.com/articles/s41598-018-33961-3.epdf?shared_access_token=brgB9BgjEAjLNCCIZzBaUtRgN0jAjWel9jnR3ZoTv0Oy3Xl9pn8BDfXISYpJuSyUkI0tmZCXvw6Of2ty0tbQ39aSQYbphhigbzC_KxMyggRgDIUigAXRxm_dBtEDFj1_3bVjzg1mjNP6g1g9yopyQS5ltCHghf7QLOPUHj-9Zbo%3D%0Aht

Noel J-P, Serino A (2019) High action values occur near the body. *Trends Cogn Sci* 23(4):269–270.

Noel JP, Grivaz P, Marmaroli P, Lissek H, Blanke O, Serino A (2014) Full body action remapping of peripersonal space: the case of walking. *Neuropsychologia* 70:375–384. Available at: http://linkinghub.elsevier.com/retrieve/pii/S0028393214002930

O'Keefe J, Dostrovsky J (1971) The hippocampus as a spatial map. Preliminary evidence from unit activity in the freely-moving rat. *Brain Res* 34:171–175.

Patané I, Cardinali L, Salemme R, Pavani F, Farnè A, Brozzoli C (2018) Action planning modulates peripersonal space. *J. Cogn Neurosci*:1–14. Available at: https://www.mitpressjournals.org/doi/abs/10.1162/jocn_a_01349.

Pellencin E, Paladino MP, Herbelin B, Serino A (2018) Social perception of others shapes one's own multisensory peripersonal space. *Cortex* 104:163–179.

Pfeiffer C, Noel J-P, Serino A, Blanke O (2018) Vestibular modulation of peripersonal space boundaries. *Eur. J. Neurosci* 47:800–811.

Poldrack RA (2012) Inferring mental states from neuroimaging data: from reverse inference to large-scale decoding. *Neuron* 72:692–697.

Quesque F, Ruggiero G, Mouta S, Santos J, Iachini T, Coello Y (2017) Keeping you at arm's length: modifying peripersonal space influences interpersonal distance. *Psychol. Res.* 81: 709–720.

Remland MS, Jones TS, Brinkman H (1995) Interpersonal distance, body orientation, and touch: effects of culture, gender, and age. *J. Soc. Psychol.* 135:281–297.

Ribas-Fernandes JJF, Solway A, Diuk C, McGuire JT, Barto AG, Niv Y, Botvinick MM (2011) A neural signature of hierarchical reinforcement learning. *Neuron* 71:370–379.

Rizzolatti G, Fadiga L, Fogassi L, Gallese V (1997) The space around us. *Science* 277:190–191. Available at: http://eutils.ncbi.nlm.nih.gov/entrez/eutils/elink.fcgi?dbfrom=pubmed&id= 9235632&retmode=ref&cmd=prlinks%5Cnpapers2://publication/uuid/C25FABDD- 52D9-45CA-A35B-0AC3B2CCFFEF

Rizzolatti G, Scandolara C, Matelli M, Gentilucci M (1981) Afferent properties of periarcuate neurons in macaque monkeys. II. Visual responses. *Behav. Brain Res* 2:147–163.

Roitman JD, Shadlen MN (2002) Response of neurons in the lateral intraparietal area during a combined visual discrimination reaction task. *J. Neurosci* 22:9475–9489.

Ruggiero G, Rapuano M, Iachini T (2019) Perceived temperature modulates peripersonal and interpersonal spaces differently in men and women. *J. Environ. Psychol* 63:52–59.

Şahin EC, Çakmak M, Doğar MR, Uğur E, Üçoluk G. (2007) To afford or not to afford: a new formalization of affordances toward affordance-based robot control. *Adapt. Behav.* 15(4):447–471.

Sambo CF, Forster B (2009) An ERP investigation on visuotactile interactions in peripersonal and extrapersonal space: evidence for the spatial rule. *J. Cogn Neurosci* 21:1550–1559. Available at: http://www.ncbi.nlm.nih.gov/pubmed/18767919

Sambo CF, Forster B, Williams SC, Iannetti GD (2012a) To blink or not to blink: fine cognitive tuning of the defensive peripersonal space. *J. Neurosci* 32:12921–12927. Available at: http://www.ncbi.nlm. nih.gov/pubmed/22973016

Sambo CF, Iannetti GD (2013) Better safe than sorry? The safety margin surrounding the body is increased by anxiety. *J. Neurosci* 33:14225–14230. Available at: http://www.ncbi.nlm.nih.gov/ pubmed/23986256

Sambo CF, Liang M, Cruccu G, Iannetti GD (2012b) Defensive peripersonal space: the blink reflex evoked by hand stimulation is increased when the hand is near the face. *J. Neurophysiol* 107:880– 889. Available at: http://www.ncbi.nlm.nih.gov/pubmed/22090460

Serino A, Canzoneri E, Avenanti A (2011) Fronto-parietal areas necessary for a multisensory representation of peripersonal space in humans: an rTMS study. *J. Cogn Neurosci* 23:2956–2967.

Serino A, Noel J-P, Galli G, Canzoneri E, Marmaroli P, Lissek H, Blanke O (2015) Body part-centered and full body-centered peripersonal space representations. *Sci Rep* 5:18603. Available at: http://dx. doi.org/10.1038/srep18603

Shadlen MN, Kiani R, Hanks TD, Churchland AK. (2008) Neurobiology of Decision Making: An Intentional Framework. In Better than Conscious? Decision Making, the Human Mind, and Implications for Institutions, eds C Engel, W Singer, pp. 71–101. Cambridge, MA: MIT Press.

Silver D, Huang A, Maddison C, et al. (2016) Mastering the game of Go with deep neural networks and tree search. *Nature* 529:484–489. Available at: http://dx.doi.org/10.1038/nature16961

Somervail R, Bufacchi RJ, Guo Y, Kilintari M, Novembre G, Swapp D, Steed A, Iannetti GD (2019) Movement of environmental threats modifies the relevance of the defensive eye-blink in a spatially-tuned manner. *Sci Rep* 9(1):3661.

Spence C, Pavani F, Driver J (2004) Spatial constraints on visual-tactile cross-modal distractor congruency effects. *Cogn. Affect. Behav. Neurosci.* 4:148–169.

Stachenfeld KL, Botvinick MM, Gershman SJ (2017) The hippocampus as a predictive map. *Nat Neurosci* 20:1643–1653.

Taffou M, Viaud-Delmon I (2014) Cynophobic fear adaptively extends peri-personal space. *Front. Psychiatry* 5:3–9.

Treue S (2008) Neural correlates of attention in primate visual cortex. *Trends Neurosci* 62:117. Available at: https://pubmed.ncbi.nlm.nih.gov/11311383/

Van der Stoep N, Nijboer TCW, Van der Stigchel S, Spence C (2015) Multisensory interactions in the depth plane in front and rear space: A review. *Neuropsychologia* 70:335–349. Available at: http://linkinghub.elsevier.com/retrieve/pii/S0028393214004618

Van der Stoep N, Serino A, Farnè A, Di Luca M, Spence C (2016) Depth: the forgotten dimension. *Multisens Res* 29:493–524. Available at: http://booksandjournals.brillonline.com/content/journals/10.1163/22134808-00002525

Versace V, Campostrini S, Sebastianelli L, Saltuari L, Kofler M (2019) Modulation of exteroceptive electromyographic responses in defensive peripersonal space. *J. Neurophysiol* 121(4):1111–1124. doi: 10.1152/jn.00554.2018.

Wallwork SB, Talbot K, Camfferman D, Moseley GL, Iannetti GD (2016) The blink reflex magnitude is continuously adjusted according to both current and predicted stimulus position with respect to the face. *Cortex* 81:168–175. Available at: http://www.sciencedirect.com/science/article/pii/S0010945216300727

Wamain Y, Gabrielli F, Coello Y (2016) EEG μ rhythm in virtual reality reveals that motor coding of visual objects in peripersonal space is task dependent. *Cortex* 74:20–30. Available at: http://linkinghub.elsevier.com/retrieve/pii/S0010945215003573

Whitlock JR, Sutherland RJ, Witter MP, Moser M-B, Moser EI (2008) Navigating from hippocampus to parietal cortex. *Proc. Natl. Acad. Sci. U.S.A.* 105:14755–14762. Available at: http://www.pnas.org/cgi/doi/10.1073/pnas.0804216105

10

Feeling the world as being here

Frédérique de Vignemont

10.1 Introduction

A vast array of experimental results have recently shown that there is something specific in the way we perceive the space immediately surrounding the body, also known as
'peripersonal space', by contrast with the perception of what lies farther away. As described
at length in this volume, the processing of peripersonal space has a distinctive sensorimotor
and multisensory signature. Research in cognitive neuroscience seems to indicate that we
process the space surrounding the body as a spatial buffer between self and world, both for
protecting one's body from immediate danger and for interacting with the environment.
However, we seem to have no conscious awareness of peripersonal space as being 'special'
in any sense. Instead, we are presented with a continuous visual field without a phenomenological boundary between what is close and what is far. At first sight, the mug that I see
next to my hand does not appear to me as different in any significant way from the mug on
the other side of the table. The former appears closer than the latter and it occupies a larger
area in the visual field. But the same could be said about the difference between the mug on
the opposite side of the table and the mug even farther away on the cupboard. Differences in
how things visually strike me are not specific to peripersonal space: they are merely a consequence of distance. The computational peculiarities of peripersonal perception thus seem to
have no consequences for phenomenology.

Here I shall propose that, when you see an object in the immediate surrounding of your
body, not only do you have a visual experience of the object (comparable to the experience
you can have of objects farther away), but you are also aware of the object as being *here* in
a primitive embodied way. This sense of here-ness can be conceived of as a specific type of
sense of presence. To better understand it, I shall turn to illusions in virtual reality (VR) and
to the feeling of disconnection in the psychiatric syndrome of depersonalization.

10.2 Preparing for impact

In what sense do our perceptual abilities differ between close and far space? To answer this
question, it suffices to acknowledge a simple fact: what is close to our body can be soon in
contact with it, either because the close-by object moves or because we move. A further basic
fact is that this contact may be welcome (when grasping an object), but not necessarily (when
colliding into an obstacle). Hence, there is an immediate significance of our surrounding,
which imposes clear rules for its perception. In brief, peripersonal perception must follow
the old Scout's motto: 'Be prepared' (*Semper paratus*). As I shall now describe in detail, one

Frédérique de Vignemont, *Feeling the world as being here* In: *The World at Our Fingertips*. Edited by Frédérique de
Vignemont, Andrea Serino, Hong Yu Wong, and Alessandro Farnè, Oxford University Press (2021). © Oxford University Press.
DOI: 10.1093/oso/9780198851738.003.0010.

must be prepared both at the sensory level (prepared to detect and process whatever may be soon on the skin) and at the motor level (prepared to react to it).

10.2.1 Seeing what is right under one's nose

In 1981, Rizzolatti and his colleagues recorded the activity of neurons in the ventral premotor areas in monkeys and found that some were activated not only by tactile stimuli on the body, but also by visual stimuli presented in the space close to the body, from a few centimetres from the body to approximately 30 cm. In humans too, it has been shown that the perception of visual stimuli presented in peripersonal space interferes with tactile perception. This multisensory signature of peripersonal perception is well illustrated by the cross-modal congruency effect. Participants are asked to localize a tactile stimulus applied on one finger, while trying to ignore visual distractors presented simultaneously at either congruent or incongruent positions. Crucially, incongruent visual distractors interfere with tactile localization (i.e. participants are both slower and less accurate), but only when the visual stimuli are close to the body (Spence et al., 2004). This visuo-tactile interaction can be explained by the fact that the sight of objects close to one's body generates expectation of a tactile event, which then influences the experience of the actual tactile stimulus.

The spatial relationship with your surroundings keeps changing: you see the tiger running towards you or the ball thrown at you, you navigate in rooms full of furniture and in streets full of people. In all cases, you need to be prepared for immediate impact and the prediction of impact entails a specific course of action. Consequently, peripersonal perception is especially sensitive to dynamic stimuli, some neurons responding more than twice as much to stimuli in movement relative to the subject (Colby et al., 1993). For example, it was found that participants detected a tactile stimulus on their hand earlier when they simultaneously perceived a dynamic sound, which gave the impression of a sound source approaching as long as it was heard at a limited distance from the hand (Canzoneri et al., 2012).

A further consequence of the anticipatory function of peripersonal perception is the type of spatial frame of reference that it exploits. The perceptual system predicts that the location of objects that are for now only close to the body will soon be *on the skin itself*. Consequently, peripersonal vision exploits the same frame of reference as touch. Tactile experiences are felt to be located in what we may call 'somatotopic coordinates' (or what others call 'skin-based coordinates' or 'bodily coordinates'). The location of pressure is encoded as occurring at a specific spot on the surface of the body: I feel touch on my hand, on my foot, or on my back. These somatotopic coordinates do not change when the sensing body part moves: I still feel touch on my hand no matter where my hand is. They are given within the frames of reference based on the representations of the body parts (de Vignemont, 2018). If my body representation has not been updated after amputation, then I would still feel sensation in the hand although it is no longer there. The somatotopic reference frames are common to all bodily sensations (i.e. touch, pain, thermal perception, proprioception), but it is normally not used by external sensory modalities, such as vision and audition. Clearly, I do not localize the bus stop that I see on the other side of the street in somatotopic coordinates. More precisely, I can experience it as being on the left or the right, and these coordinates are relative to some key body parts (such as eyes, head, and torso), but the visual system does not project the localization of the bus stop on the surface of my body. This is so because it does not anticipate its impact on me. By contrast, when I walk toward the bus stop, which is now 30 cm away from me, the visual system anticipates that my left arm might hit it. The location of the bus stop is then given in somatotopic coordinates, although I have access to it only through vision[1] (Graziano and Gross, 1993): 'a gelatinous medium surrounding the body, that deforms in a topology-preserving fashion whenever the head rotates or the limbs move' (p. 107). In his recent book, Graziano (2018) thus qualifies peripersonal space as a 'second skin'.

A last consequence of the anticipatory function of peripersonal perception is that impact prediction enables better processing of the anticipated event (Engel et al., 2001; Hyvärinen and Poranen 1974; Cléry et al., 2015). It has been shown that the detection of objects and the processing of their global spatial properties such as shape are facilitated when presented in close space (Dufour and Touzalin, 2008; Reed et al., 2006; Gozli et al., 2012; Kelly and Brockmole, 2014; Blini et al., 2018). For instance, Blini and colleagues (2018) found that one's visual capacities were better for objects presented in peripersonal space. Participants were presented with 3D shapes in an immersive VR environment either in close space or in far space, and the objects in far space appeared illusorily bigger than the objects in close space. Even so, participants were better in discriminating the shape of the objects for those that were presented next to their body.

Still one should not believe that the closer, the better. The fact is that peripersonal perception appears sometimes to be less efficient than the perception of far space. For instance, the processing of colour information and of fine-grained spatial properties is less reliable in close space than in far space (Gozli et al., 2012; Kelly and Brockmole, 2014). It has also been found that participants take more time at performing a visual search task when presented in close space. In one study, participants were asked to localize a target among distractors and they were slower when the visual display was close than far (Abrams et al., 2008). How could participants be bad at finding a target that was just under their nose?[2] I shall now argue that sensory processing of objects presented in peripersonal space is enhanced not for the sake of improving their recognition but instead for the sake of guiding the motor system for it to react in the best way.

10.2.2 Ready to react

To recapitulate, there are specific computational principles that characterize peripersonal perception, which differ from those that govern far perception. As I shall now argue, they can be explained by the specific relationship between peripersonal perception and action. Since its discovery by Rizzolatti and his colleagues, peripersonal perception has been ascribed a motor function (Rizzolatti et al., 1997, Brozzoli et al., 2012, Bufacchi and Iannetti, 2018). This motor function explains the properties of the seen object for which peripersonal perception is the most reliable: they are those that are useful for the selection of the type of movement to perform, what Grush (2007) calls type-selecting dispositions (for details, see de Vignemont, forthcoming a).

By definition, all bodily movements take place in peripersonal space. Hence, peripersonal perception cannot afford to be relatively detached from the motor system. What is seen

---

[1] A question that one may ask is whether peripersonal vision still qualifies as being visual since it uses a spatial frame that is normally used by touch and other types of bodily perception. Instead, one may claim that peripersonal perception is multimodal: neither visual, nor tactile. I shall leave this question open here.

[2] This shows that one cannot explain the advantage of peripersonal perception by attentional facilitation only. If whatever is close were given attentional priority just because it is close, then one should be better in every perceptual task and this is not the case (see Blini et al., this volume, for discussion).

there must be able to directly connect with what one does, and vice versa. This direct connection to action is made possible by the fact that peripersonal space is mainly represented in brain regions that are dedicated to action guidance (in the premotor areas and in the ventral section of the intraparietal sulcus). The tight link to bodily movements is also revealed by the fact that the practical knowledge of one's actual motor capacities influences what falls under the scope of peripersonal perception. In brief, the space that is processed as being peripersonal space is larger when one can act farther away, because of tool use, for instance (Farnè and Làdavas, 2000). Increased motor ability leads to a modification of perceptual processing of the objects that are next to the tool. Before tool use, they are processed as being in far space; during tool use they are processed as being peripersonal. This was first described by a seminal study by Iriki et al. (1996) who trained monkeys to use a rake to reach food placed too far to be reached without the tool. They found that visuo-tactile neurons, which displayed no visual response to food at this far location before training, began to display visual responses after training. A few minutes after tool use was interrupted, the visual receptive fields shrank back to their original size. Consider now cases in which motor abilities are reduced. It has been shown that after 10 hours of right arm immobilization there is a contraction of peripersonal space such that the distance at which an auditory stimulus was able to affect the processing of a tactile stimulus was closer to the body than before (Bassolino et al. 2015).[3]

We can further refine the claim about the function of peripersonal perception and propose that it originally evolved for specific types of behaviour—namely, protective ones. In short, to have a dedicated mechanism specifically tuned to the immediate surroundings of the body and in direct relation with the motor system is a relatively good solution for the purpose of the detection of close threats and for self-defence. It was the Swiss biologist Heini Hediger (1950), director of the Zurich zoo, who first noted that animals do not process space uniformly (see Møller, this volume). In particular, there is a specific zone immediately surrounding their body, described as the 'flight distance', that predators cannot approach without eliciting specific defensive responses in their prey (flight, freeze, or fight, depending on how close the predator is). Arguably, the flight distance goes beyond peripersonal space. Still the margin of safety constituted by peripersonal space can be conceived as a last resort. It is encoded in a specific way to elicit protective behaviours as quickly as possible if necessary (Graziano, 2009, 2018). In monkeys, the direct stimulation of neurons involved in peripersonal perception automatically triggers a range of protective responses, such as avoidance behaviour. The injection of bicuculline in peripersonal neurons, which disinhibits them, leads the monkeys to vividly react even to non-threatening stimuli (when seeing a finger gently moving toward the face, for instance). By contrast, the injection of muscimol in peripersonal neurons, which temporarily inhibits their neuronal activity, has the opposite effect: monkeys no longer blink or flinch when their body is under threat. We can now explain the experimental results on visual search presented earlier in this section in light of this evolutionary hypothesis. Perception must be especially thorough in peripersonal space because partial overlook may be dangerous when the potential threat is next to one's body. One can afford not to look at all the objects for a faraway visual scene but one needs to watch closely where one puts one's hand or one's foot. In other words, better safe than sorry when it is close, no matter the cost.

---

[3] There are other factors that can influence the extent of peripersonal space, such as the presence of a transparent screen between the seen object and the subject (Sambo et al., 2012) or anxiety (Lourenco et al., 2011; Sambo and Iannetti, 2013).

To summarize, peripersonal perception is always prepared to make one move. From a neural perspective, this readiness is made possible by the implementation of peripersonal perception within brain structures involved in action guidance. From a computational perspective, peripersonal perception is always prepared to make one move because it is informed by what one can and cannot do at each instant. From an evolutionary perspective, peripersonal perception is always prepared to make one move because it originally evolved for immediate threat detection and self-defence. However, although peripersonal perception is different from far perception, one may question whether it *feels* different.

10.2.3 A phenomenological signature?

We saw that there are clear peculiarities of peripersonal processing that show up in behaviour, but it is not clear whether they make a phenomenological difference. As noted by Guterstam et al. (2016, p. 44): 'The space close to our hands does not 'feel' different than the space outside our reach.' Our perceptual system may be expecting the impact of whatever is close by but we are not aware of it. Because of its tight link with action, one may be tempted to describe peripersonal perception exclusively in terms of unconscious sensorimotor processing subserved by the dorsal visual stream. It should be noted, however, that the dorsal stream is not exclusively dedicated to sensorimotor processing. Recent evidence indeed indicates that it can also be involved in perceptual identification (such as shape recognition [Freud et al., 2017]). The involvement of the dorsal stream thus does not suffice to show that peripersonal perception is exclusively a matter of sensorimotor transformation. Furthermore, if peripersonal processing were sensorimotor, visual discrimination should not be improved in close space but we have just seen that it is (for more details, see Blini et al., this volume). The fact is that the standard experimental paradigm to investigate peripersonal processing is a multisensory *perceptual* task: subjects are asked to give a verbal response about the detection of a tactile stimulus, and one measures to what extent they are influenced by a visual (or auditory) stimulus presented more or less close to them. Hence, peripersonal processing can have an effect on conscious visual (or auditory) experiences.

One can further note that, in some situations at least, proximity is associated with a specific feeling. Imagine that you are seated alone on a long bench in a waiting room and a perfect stranger comes to sit next to you and gets closer and closer. There is a distance at which you will start feeling uncomfortable. This feeling has been used as the key measure of what is called 'personal space' in social psychology, and more recently in social neuroscience (e.g. Hall, 1966; Iachini et al., 2014). Hence, the perception of proximity—or at least of social proximity—can induce a distinctive affective feeling. However, this unpleasant feeling does not seem to generalize outside the social domain. We are constantly surrounded by objects that are very close to us and yet their proximity does not make us feel uncomfortable. The question then is whether there are non-social examples of the phenomenological impact of proximity.[4] From now on, I shall argue that peripersonal perception gives rise to a sense of here-ness.

---

[4] One case comes from what is called the magnetic field illusion (Guterstam et al., 2016). Brushstrokes are applied to the participants' hidden hand and *in mid-air at some distance above* a fully visible rubber hand. Participants then report feeling a 'repelling magnetic force, a 'force field', or 'invisible rays of touch' (Guterstam et al., 2016, p. 45). However, the magnetic field feeling can simply be the result of the mislocalization of their tactile feeling and it is not clear that it really concerns peripersonal perception.

10.3 Here-ness

Consider the following examples. My book is here, in my bag. The subway station is there, at the next corner. It is raining here, in Paris. I have left my bag there, in my office. In all these cases, I describe my spatial relation to events or objects and use two different indexicals to refer to what is close and what is far. The distinction between 'here' and 'there' can be found in almost all languages[5] and constitutes the most universal example of spatial deixis (Levinson, 2004). In a nutshell, 'here' is a marker of proximity, which normally refers to a region that includes the speaker (where I am).[6] On the other hand, 'there' refers to what is farther away, a distal region more remote from the speaker (where I am not). What characterizes indexicals is that they do not have fixed referents. Many different particular locations can, on occasion, be picked out as 'here' or 'there', unlike the words 'Paris' or 'New York'. Depending where I am, the token 'here' will refer at different locations. This is a general rule of indexicals, such as here, there, I, you, or now. Still, there are differences among them. The first person depends on who the speaker is but it normally does not depend on the speaker's communicative intention: I am always the person I refer to when I say 'I'. Unlike the first person, which automatically refers to the speaker, 'here' and 'there' are what Perry (2001) calls intentional indexicals. The extent of the region that they refer to depends on the speaker's intentions and the interlocutor needs additional information to understand what they refer to. By 'here' I can mean my office, Paris, or France. By 'there', I can mean the back of the room (when I stand in front), Marseille (when I am in Paris), or the United States (when I am in France). Still 'here' respects some constraints given by the context: it always depends on where the speaker is located. By contrast, 'there' can refer to any region as long as the speaker is not located there.[7]

What is interesting for us is that this distinction is not only at the linguistic level but also at the cognitive level. Consider the following example adapted from Perry (1990) and compare the following three scenarios:

a) You see that it is raining but you do not know where you are, you only know that it is raining here.
b) You believe that it is raining in Grand Island but you do not know that it is the city you are in.
c) You believe that it is raining there.

In the scenario (a), your belief is sufficient to motivate you to take an umbrella. The spatial indexical 'here' has a distinctive cognitive significance, similar to the significance of the first person. It is because you know that the fact that it is raining is relevant *for you* that you act in the appropriate way. Perry thus suggests that we have a specific mental notion that encodes information about what is conceived as here, which is tightly connected to action: 'Let's call a notion a "here-notion" if it is associated with a self-notion and the idea of being-in' (Perry, 1990).

By contrast, in the scenarios (b) and (c), your belief does not motivate you to take an umbrella, precisely because you do not know whether it is raining where you are. Still there is

---

[5] Some languages, such as Japanese, include a third term and distinguish between the space around the speaker (*koko*), the space around the interlocutor (*soko*) and the space away from both (*asoko*).

[6] Still, one should note that technology (such as an answering machine) can distort this rule. See Perry (2001) for discussion.

[7] According to Kaplan (1989), only 'here' qualifies as an indexical, whereas 'there' is comparable to demonstratives such as 'this' or 'that' because it could almost refer to anywhere (with the notable exception of where the speaker is).

a difference between (b) and (c). There is no reference to the subject in scenario (b): your belief is simply about Grand Island. By contrast, when you believe that it is *there*, you explicitly acknowledge that it is not where you are in, which is different from simply not knowing whether this is the case or not. There is thus first-personal content, although negative. In short, there is a 'not-here' mental notion. One way to characterize it is to claim that there is a there-notion if it is associated with a self-notion and the idea of *not* being-in.

Now one may ask about the relation that the spatial notions 'here' and 'there' hold with perception and, more specifically, with peripersonal perception. At first sight, it may seem that what one sees in peripersonal space is here but the here-notion can cover a much larger area. In some sense, there is no spatial limit to the referent of 'here', which can even be the solar system. Hence, it goes far beyond what one perceives in peripersonal space, at least if one conceives of here-ness at a relatively high cognitive level. There may be, however, a more primitive here-notion, which is exploited by action. Imagine that you are hiking in the mountains and you can see the peak that you plan to reach. At the cognitive level, you can judge that you are here, meaning that you are in the Alps. 'Here' refers to the region you are in and includes the end of the trail. But at the agentive level, the peak that you want to reach and that you can already see appears to you as being *there*, where you are not. It is precisely because it is only there that you do not experience that you have achieved what you wanted to do and that you keep walking. From the point of view of the motor system, 'there' refers to the location of the goal while 'here' refers to the location of your body.

This notion of here-ness, which appears as only weakly cognitively demanding, does not require a self-notion. In particular, it is conceivable that animals that have what Peacocke (2014) calls a 'Degree 0' of self-representation can have at least a primitive version of the here-notion:

> This creature remains at Degree 0, however, because it never represents anything as standing in certain relations to itself. None of its perceptual states have *de se* contents of such forms as *that thing is that direction from me*. Rather, they have *here*-contents, such as *that thing is that direction from here* (....) Its map has, so to say, not a de se pointer I am here, but rather one saying this place on the map is here. (p. 30, emphasis in the original)
>
> Reproduced from Peacocke C. (2014) The Mirror of the World: Subjects, Consciousness, and Self-Consciousness. Oxford University Press. Oxford. 30.

Instead of self-location, it is sufficient for this primitive here-ness to be based on body location. Here-ness then refers to the place in which the body is. However, even this primitive definition of here-ness may still be too sophisticated. It indeed implies that the creature is able to represent the various segments of the body as belonging to a unified whole. One may then propose a fragmented notion of here-ness, which refers to *the places in which the various body parts are*. What is here, then, is any object or event whose felt location is encoded in its relation to these body parts. This can be a pain that you feel to be located in your left foot. This can be your hat that you feel on your head. This can also be the mosquito buzzing next to your right hand. In all cases, the frame of reference that is used is centred on the corresponding body part. The reference of the primitive notion of here-ness thus goes slightly beyond the exact location of body parts to include peripersonal space. As described earlier, the perceptual system constantly anticipates peripersonal space to become bodily space. The relation to the environment keeps changing either because the surrounding objects move or because one moves. For now, the mosquito is 5 cm from your hand but in two seconds it may well be *on* your hand. This is why objects and events perceived in peripersonal space are localized relatively to the various parts of one's body. Put it another way, peripersonal space is always 'here'; it is never 'there'. Peripersonal space is thus the space in which the body could be.

This is well illustrated by the spatial constraint that operates on the rubber hand illusion. This illusion, in which one experiences a rubber hand as being part of one's own body, works only as long as the rubber hand is presented within the limit of peripersonal space. Participants look at a left rubber hand presented in front of them, while their own left hand is hidden behind a screen. The experimenter then simultaneously strokes both the participants' hand and the rubber hand. After synchronous stimulations, participants describe that they feel as if the rubber hand was their hand (Botvinick and Cohen, 1998). Of special interest for us is that there is not a slow linear decrease of the illusion with the separation increasing between the real hand and the rubber hand. Instead, there is a drop in the illusion when the rubber hand is positioned beyond the boundaries of the peri-hand space (Lloyd, 2007).

One may be tempted to reply that there are many objects outside peripersonal space that one can perceive in relation to one's body—objects that one represents in egocentric coordinates. For instance, I can see the Eiffel Tower on the other side of Paris as being on my left. Yet the Eiffel Tower does not fall under the scope of the embodied notion of here-ness that I am developing. It is important to clearly distinguish the somatotopic frame of reference from the egocentric one.[8] Consider the following example. There is a ladybug on my right next to my right hand. The egocentric perspective of my visual experience is given relative to the posture of my head and of my torso and, if I cross my hands, my visual experience stays the same: it still presents the ladybug on my right. By contrast, the somatotopic coordinates of the ladybug have changed: it was in my peri-right hand and it is now in my peri-left hand. Now imagine that the ladybug follows my right hand. Then the egocentric coordinates of my visual experience change (on my left), but not its somatotopic coordinates (in my peri-right hand). The hypothesis is that peripersonal perception spatially singles out specific spatial areas within the large egocentric space, the areas in which one represents one's body parts to be located. On the primitive embodied definition, they correspond to the region of here-ness. There is no need for self-referential capacities, and thus many animals can enjoy this embodied here-notion thanks to their peripersonal processing.[9]

10.4 A sense of presence

The proposal is two-fold. First, there is a sense of here-ness that expresses that what one sees or hears is here—namely, where one's body parts are located. Second, the sense of here-ness is grounded in peripersonal processing and restricted to visual experiences of one's immediate surrounding. From now on, I shall refer to this notion as being peripersonally here (or 'here$_{pp}$'). My claim is that it can enter perceptual phenomenology. For some, however, the

---

[8] It should be noted that there is more than one definition of the egocentric frame. I shall use here a relatively neutral one. Within an egocentric frame of reference, a perceived object is located at a specific point relative to some axes (vertical, horizontal, longitudinal) centred on some key parts of the body (such as the head or the torso).

[9] Peripersonal processing appears to be shared by a large range of animals, and not only by primates. For instance, in rats and mice, whiskers seem to obey exactly the same predictive function as peripersonal vision: they are sensitive to remote air displacement within peripersonal space (Shang et al., 2018). Even insects and zebrafish can display stereotyped flight responses triggered by the detection of looming visual stimuli and influenced by the size of the stimulus (Rind and Santer 2004; Tammero and Dickinson 2002). I would like to thank Suliann Ben Hamed for her useful insight on this question.

sense of peripersonal here-ness may still appear as an elusive entity and they may even doubt its existence. However, as frequently the case, some of those doubts may evaporate when considering crucially differentiating cases in which the sense of here-ness$_{pp}$ is either illusory, as in VR, or disrupted, as in the syndrome of depersonalization.

10.4.1 Being there or being here?

One of the main challenges of VR and telepresence is to induce in the subject the illusion of being present in the virtual environment or at the remote location, instead of her actual physical surrounding. Presence is conceived of as a behavioural (objective) and phenomenological (subjective) response to the degree of immersion of the technology (Slater and Wilbur, 1997). In particular, it is measured by questionnaires that assess how much one experiences a sense of 'being there' (Sanchez-Vives and Slater, 2005). However, one should rather ask to what extent one experiences a sense of being *here*, meaning here in the virtual environment.[10] This is well described by Howard Rheingold, an American writer who was especially influential in the development of virtual communities. While wearing head-mounted displays connected to a remotely controlled robotic head, he saw his own physical body and he reported: 'He [Rheingold's physical body] looked like me and abstractly I could understand he was me, but I know who me is and *me is here* [in virtual environment]. He, on the other hand, was *there*' (Rheingold, 1991, my italics).

One way to characterize the sense of presence in VR is in spatial terms. It corresponds to what Slater calls the 'place illusion', which he defines as the feeling of 'being in the place depicted by the VE [virtual environment]' (Slater et al., 2010, p. 92). This dimension of presence has to be distinguished from what he calls the 'plausibility illusion', also found in VR, which he defines as the feeling of 'what is apparently happening is really happening' (Slater et al., 2010, p. 92). The two illusions can come apart: one can experience the place illusion (one experiences oneself as being present in this VE), without experiencing the plausibility illusion (one does not experience the VE as being real). Furthermore, both illusions are cognitively impenetrable. Participants in VR are fully aware that they are not located in the same place as the virtual cliff. And yet they experience the edge of the cliff as being just right here and they step back from it (Sanchez and Slater, 2005).

This dual characterization of the sense of presence can also be applied outside discussions on virtual reality to contrast visual experiences of real-life scenes with visual experiences of paintings (Noë, 2005; Matthen, 2005, 2010; Dokic, 2010). You see a tree. You see the painting of a tree. Only in the former case do you experience the tree as being present. But what do we mean by that? Most literature on the sense of presence outside VE takes the sense of presence as a unified phenomenon (for exception, see Dokic and Martin, 2017). Nonetheless, there are, at least conceptually, two different interpretations. The sense of presence can simply assert that one is spatially connected to the tree (sense of *spatial* presence). As Matthen (2010, p. 115) describes it: 'The Feeling of Presence which defines normal scene vision is, among other things, a visual feeling of spatial connection.'

So far there is no ontological commitment. The sense of presence can also assert that the tree actually exists (sense of *real* presence). Then only does it express that what one perceives exists materially or independently of one's perception of it. The centre of our interest here is

---

[10] Questionnaires are always given *after* the experiment. There-ness should be interpreted temporally and not spatially.

only the sense of spatial presence (involved in the place illusion), which is less committing than the sense of real presence (involved in the plausibility illusion).

Now consider how Matthen (2005) describes the visual experiences that we have when seeing a painting: 'Your space stops just where the space of the depicted object begins. The picture is *there*, right in front of you, but the men it depicts are not (...) The space in the picture lacks, if you will, a *here*' (Matthen, 2005, p. 316). Interestingly, in the way Matthen describes the situation, he does not seem to make a distinction between 'here' and 'there'. This can be understood in so far as the sense of spatial presence only involves the awareness of being spatially connected and one can be spatially connected even with distant objects. More precisely, Matthen (2010) argues that the sense of presence is anchored in near-space (i.e. 10 feet away) but it extends to far away objects that are connected to objects in near-space by visible paths. However, he also claimed in his earlier book: 'Our grasp of spatial relations in a distant mountain range is likely very little different from that of objects in a picture' (Matthen, 2005, p. 323).

The question thus arises: how far do we experience the sense of spatial presence? I do not want to claim that the sense of spatial presence is restricted to what is immediately surrounding me. Clearly I experience the person on the other side of the room as being present. Instead, I propose that, when this person is next to me, I experience her not only as being present but also as being *here*$_{pp}$. There is a specific type of spatial presence, a sense of peripersonal here-ness, whose definition is narrower than the sense of spatial connection. It consists in the awareness of the object as being in the place at which the various parts of the body are felt to be located.[11] It can be conceived as a tactile sense of presence (de Vignemont, forthcoming).

What is the nature of the sense of peripersonal here-ness? To answer this question, we first need to ask whether it makes a phenomenological difference to see objects and events in peripersonal space. Despite the evidence that I reviewed earlier, one may indeed still claim that there are no phenomenological consequences of the sensory specificities of peripersonal perception. If so, there is no specific phenomenology associated with the sense of peripersonal here-ness. One merely entertains the *thought* that objects and events are here$_{pp}$ but one does not experience them as being here$_{pp}$. However, this intellectual account may seem at odds with the low level of cognitive sophistication that is required by the embodied notion of peripersonal here-ness. Alternatively, one may argue that one *experiences* what one sees as being here$_{pp}$. This view is in line with the general experiential interpretation given to presence: there is a *feeling* of presence. It may come from the mode of presentation of the visual content (Matthen, 2005) or it may constitute a distinct quasi-affective experience associated with the visual experience (Dokic and Martin, 2015) but, in all cases, one experiences the world as being present. One may similarly suggest that one experiences the immediate surroundings as being here$_{pp}$. Put it another way, it makes a phenomenological difference to see objects and events in peripersonal space. This then raises a second question: what phenomenal property accounts for this difference? One possibility is that there is a distinctive phenomenal property of peripersonal here-ness that is part of our peripersonal phenomenology, and on the basis of which one can judge that what one perceives in one's immediate surroundings is here. Another possibility, which is more deflationary, is that there

---

[11] Interestingly, the place illusion in virtual reality works better when participants have a virtual body that visually substitutes their real body—as seen from a first-person perspective (i.e. while looking down, they see a virtual body). This can be taken in favour of the sense of embodied here-ness. It has been further shown that the immediate surrounding of the virtual avatar can be processed as being peripersonal (Noel et al., 2015).

is phenomenal property that grounds the sense of peripersonal here-ness, this property not being peripersonal here-ness as such, but a different one, which follows from the sensory specificities of peripersonal perception. In particular, one can propose that the readiness for impact that characterizes peripersonal perception gives rise to the experience that what one perceives is here$_{pp}$. On this latter view, there is no additional quality of peripersonal here-ness. We thus have the following three options:

(i)   No phenomenology of peripersonal here-ness.
(ii) A phenomenology of peripersonal here-ness involving a distinctive phenomenal property of peripersonal here-ness.
(iii) A phenomenology of peripersonal here-ness involving other phenomenal properties, which result from the specificities of peripersonal perception. I shall not settle this debate here.

I shall only argue that the absence of peripersonal here-ness, as displayed in the psychiatric syndrome of depersonalization, has a phenomenological impact.

10.4.2 Neither here, nor there

Patients with depersonalization report that they feel detached from the world, from their body, and from their own mental states, as if they were external observers. Of special interest to us is what is known as the 'derealization symptom'[12]: patients complain that they experience a sense of alienation from their surroundings:

It felt as if I was carried extremely far away from this world, and really far. (Dugas and Moutier, 1911, p. 22, my translation)
    I felt as if I was almost entirely separated from the world and as if there was some barrier between me and it. (Dugas and Moutier, 1911, p. 24)
    I feel detached and isolated from the world and the people in it. I feel like I am in a box of very thick glass which stops me from feeling any atmosphere. At times it is like looking at a picture. It has no real depth. (Sierra, 2009, p. 51)
    I didn't feel dreamy but as if I was physically not quite there. It was as if I was looking through a pane of glass or out of a television screen. (Sierra, 2009, p. 52)

The phenomenology of derealization includes at least three dimensions: (i) a feeling of unreality; (ii) a feeling of unfamiliarity; and (iii) a feeling of disconnection. The first aspect has attracted most attention but it is the last one that is directly relevant for our discussion: patients feel cut off from their surrounding (Mayer-Gross, 1935). One can relate their experience with what individuals suffering from migraine can also report: 'Shortly after that my vision becomes "distant". It is somewhat like seeing the world projected on a screen.' (Podoll, 2005, quoted in Sierra, 2009, p. 92). They do not feel that they share the same space with what they see and hear. Instead, they feel that the objects that they see are located where they are not. This is so although they are fully aware that they are next to them and they can act on them. As one patient well described, 'it was a permanent struggle between the involuntary

---

[12] Some have argued that derealization is a separate syndrome from depersonalisation but I shall not engage into this debate, which has little relevance for what I will discuss.

impressions and my judgement' (Dugas and Moutier, 1911, p. 5, my translation). In my terminology, these patients fail to experience what surrounds them as being here$_{pp}$ although they know that it is here$_{pp}$.

How to account for such a disruption? Billon (2017) explains all the symptoms of depersonalization by a general loss of the sense of 'myness'—that is, an inability for the patients to relate what happens to their body and to the external world to them. A patient, for instance, reported: 'it is not me who feels. I am not interested in what I appear to be feeling, it is somebody else who feels mechanically' (Janet, 1908, p. 515). According to Billon, depersonalized patients fail to feel their visual, auditory, and bodily experiences as being their own. Consequently, whatever is represented by these experiences has no direct relevance for them: a visual experience of a tree can inform a subject of the presence of the tree in front of her only if this is *her* visual experience.

The problem with this view is that the primitive notion of peripersonal here-ness does not require the use of the first person. A deficit of subjectivity should then have no effect on it. A better explanation seems to be in terms of bodily awareness. I argued earlier that peripersonal here-ness is anchored on body location and patients with depersonalization frequently describe how they feel that their body has disappeared: 'I do not feel I have a body. When I look down I see my legs and my body but it feels as if they were not there (…) it feels as if I have no body; I am not there' (Sierra, 2009, p. 28).

Such a bodily disorder may thus explain the disruption of peripersonal here-ness. In brief, to know what place here$_{pp}$ corresponds to, one needs to know where one's body is, which presupposes being aware that one has a body. At this point, one may object that the sense of disconnection does not affect only what is close to the patients, but their entire environment. Objects and events are not experienced as being here$_{pp}$, but they are neither experienced as being there$_{pp}$. They are experienced as being in a different world, a world from which one is cut off. There are, however, two possible replies. First, one can assume two distinct deficits, a disruption of bodily awareness that explains the local loss of peripersonal here-ness and a disruption of subjectivity that explains the general loss of the sense of spatial presence. Alternatively, one can propose that the sense of spatial presence also requires to be anchored in the body and if one feels disembodied—as depersonalized patients do— one can no longer relate anything to one's body, wherever the things are. If this is the correct interpretation, then it appears that the ability to experience peripersonal here-ness plays a necessary role for anchoring the sense of spatial presence in general. Bodily awareness and peripersonal awareness are indeed two sides of the same coin: one cannot lose one without losing the other. The hypothesis (to be tested) then is that depersonalized patients fail to process their environment as being peripersonal.[13] This leads them to fail to experience their immediate surrounding as being here$_{pp}$, and more generally to feel disconnected from the world.

---

[13] No sensory or motor deficits have been discovered so far in depersonalization but, since peripersonal space has never been directly tested in depersonalization, it remains an open empirical question. Furthermore, Sierra (2009), one of the leading experts on depersonalization, provides an evolutionary account that is compatible with a specific deficit of peripersonal processing. He proposes that the emotional and sensory numbing found in depersonalization results from an unwarranted activation of a hard-wired response to extreme anxiety in case of global danger, for which one is powerless. For instance, in case of earthquake, there is no point in localizing danger because it is everywhere and there is little you can do about it. This is completely different when there is a snake crawling towards you. You need to localize it so that you can retrieve your foot. This is precisely the reason for which peripersonal perception evolved: to be better prepared to react to localized threats. What Sierra describes as emotional and sensory numbing may then be nothing more than the "unplugging" of peripersonal processing. Roughly speaking, it is pointless to be ready to react because danger is overwhelming.

10.5 Conclusion

In this chapter, I argued that seeing objects and events in the immediate surrounding of one's body involves being aware that these objects and events are here. Being here then simply means being at the place at which the world and the various parts of the body can collide. The sense of here-ness is thus relatively primitive, devoid of first-personal content. Still, it is fundamental for one to feel spatially connected with the world, and thus for the world to feel present.

Acknowledgements

This work has been presented first in the winter workshop in philosophy of perception, in San Diego (USA). I am deeply grateful to all the participants for their invaluable comments. I would also like to thank Adrian Alsmith, Roberto Casati, Jérôme Dokic, Clare Mac Cumhaill, Mohan Matthen, Chris Peacocke, and Mel Slater for their insightful comments and questions.

References

Abrams, R. A., Davoli, C. C., Du, F., Knapp, W. H., & Paull, D. (2008). Altered vision near the hands. *Cognition*, *107*(3), 1035–1047.

Bassolino, M., Finisguerra, A., Canzoneri, E., Serino, A., & Pozzo, T. (2015). Dissociating effect of upper limb non-use and overuse on space and body representations. *Neuropsychologia*, *70*, 385–92.

Billon, A. (2017). Mineness First. In F. de Vignemont, & A. Alsmith (eds), *The Subject's Matter: Self-consciousness and the Body*. Cambridge Mass.: the MIT Press.

Blini, E., Desoche, C., Salemme, R., Kabil, A., Hadj-Bouziane, F., & Farnè, A. (2018). Mind the depth: visual perception of shapes is better in peripersonal space. *Psychological Science*, *29*(11), 1868–1877.

Botvinick, M., & Cohen, J. (1998). Rubber hands 'feel' touch that eyes see. *Nature*, 391, 756.

Brozzoli, C., Makin, T. R., Cardinali, L., Holmes, N. P., & Farnè, A. (2012). Peripersonal Space: A Multisensory Interface for Body–Object Interactions. In M. M. Murray, & M. T. Wallace (eds). *The Neural Bases of Multisensory Processes*. Boca Raton, FL: CRC Press.

Bufacchi, R. J., & Iannetti, G. D. (2018). An action field theory of peripersonal space. *Trends in Cognitive Sciences*, *22*(12), 1076–1090.

Canzoneri E1, Magosso E, & Serino A. (2012). Dynamic sounds capture the boundaries of peripersonal space representation in humans. PLoS ONE, 7(9):e44306.

Cléry, J., Guipponi, O., Odouard, S., Wardak, C., & Hamed, S. B. (2015). Impact prediction by looming visual stimuli enhances tactile detection. *Journal of Neuroscience*, *35*(10), 4179–4189.

Colby, C. L., Duhamel, J. R., & Goldberg, M. E. (1993). Ventral intraparietal area of the macaque: anatomic location and visual response properties. *Journal of Neurophysiology*, *69*, 902–914.

de Vignemont, F. (2018). *Mind the Body*. Oxford: Oxford University Press.

de Vignemont, F. (forthcoming a). Peripersonal perception in action. *Synthese*. doi:10.1007/s11229-018-01962-4

de Vignemont, F. (forthcoming b). A minimal sense of here-ness. *Journal of Philosophy*.

Dokic, J. 2010. Perceptual Recognition and the Feeling of Presence. In B. Nanay (ed.), *Perceiving the World*, 33–51. Oxford: Oxford University Press.

Dokic, J. and Martin, J.-R. (2015). 'Looks the Same but Feels Different': A Metacognitive Approach to Cognitive Penetrability. In A. Raftopoulos, & Zeimbekis, J. (eds), *Cognitive Effects on Perception: New Philosophical Perspectives*. Oxford: Oxford University Press.

Dokic, J., & Martin, J. R. (2017). Felt reality and the opacity of perception. *Topoi, 36*(2), 299–309.

Dufour, A., & Touzalin, P. (2008). Improved visual sensitivity in the perihand space. *Experimental Brain Research, 190*(1), 91–98.

Dugas, L., & Moutier, F. (1911). *La Dépersonnalisation*. Paris, F. Alcan.

Engel, A. K., Fries, P., & Singer, W. (2001). Dynamic predictions: oscillations and synchrony in top-down processing. *Nature Reviews Neuroscience, 2*(10): 704–716.

Farnè, A., & Làdavas, E. (2000). Dynamic size-change of hand peripersonal space following tool use. *NeuroReport, 11*(8), 1645–1649.

Freud, E., Culham, J. C., Plaut, D. C., & Behrmann, M. (2017). The large-scale organization of shape processing in the ventral and dorsal pathways. *eLife, 6*, e27576.

Gozli, D. G., West, G. L., & Pratt, J. (2012). Hand position alters vision by biasing processing through different visual pathways. *Cognition, 124*(2), 244–250.

Graziano, M. (2009). *The Intelligent Movement Machine: An Ethological Perspective on the Primate Motor System*. Oxford: Oxford University Press.

Graziano, M. S., & Gross, C. G. (1993). A bimodal map of space: somatosensory receptive fields in the macaque putamen with corresponding visual receptive fields. *Experimental Brain Research, 97*: 96–109.

Graziano, M. S. (2018). *The Spaces Between Us. A Story of Neuroscience, Evolution, and Human Nature*. Oxford : Oxford University Press.

Grush, R. (2007). Skill theory v2. 0: dispositions, emulation, and spatial perception. *Synthese, 159*(3), 389–416.

Guterstam, A., Zeberg, H., Özçiftci, V. M., & Ehrsson, H. H. (2016). The magnetic touch illusion: a perceptual correlate of visuo-tactile integration in peripersonal space. *Cognition, 155*, 44–56.

Hall, E.T. (1966). *The Hidden Dimension*. New York: Doubleday & Co.

Hediger, H. (1950). *Wild Animals in Captivity*. London: Butterworths Scientific Publications.

Hyvärinen, J., & Poranen, A. (1974). Function of the parietal associative area 7 as revealed from cellular discharges in alert monkeys. *Brain, 97*(4): 673–692.

Iachini, T., Coello, Y., Frassinetti, F., & Ruggiero, G. (2014). Body space in social interactions: a comparison of reaching and comfort distance in immersive virtual reality. *PloS ONE, 9*(11), e111511.

Iriki, A., Tanaka, M., & Iwamura, Y. (1996). Coding of modified body schema during tool use by macaque postcentral neurones. *NeuroReport, 7*(14), 2325–2330.

Janet, P. (1908). Le sentiment de dépersonnalisation. *Journal de Psychologie Normale et Pathologique*, 514–516.

Kaplan, D. (1989). Demonstratives. In J. Almog, J. Perry, &H. Wettstein (eds), *Themes from Kaplan*. Oxford: Oxford University Press, pp. 481–563.

Kelly, S. P., & Brockmole, J. R. (2014). Hand proximity differentially affects visual working memory for color and orientation in a binding task. *Frontiers in Psychology, 5*, 318.

Levinson, S. C. (2004). Deixis. In L. R. Horn, & G. Ward (eds), *The Handbook of Pragmatics* (pp. 97–121). Oxford: Blackwell.

Lloyd, D. M. (2007). Spatial limits on referred touch to an alien limb may reflect boundaries of visuo-tactile peripersonal space surrounding the hand. *Brain and Cognition, 64*(1): 104–109.

Lourenco, S. F., Longo, M. R., Pathman, T. (2011). Near space and its relation to claustrophobic fear. *Cognition*, 119, 448–453.

Matthen, M. (2005). *Seeing, Doing and Knowing: A Philosophical Theory of Sense Perception*. Oxford: Oxford University Press.

Matthen, M. (2010). Two Visual Systems and the Feeling of Presence. In N. Gangopadhyay, M. Madary, & F. Spicer (eds), *Perception, Action, and Consciousness: Sensorimotor Dynamics and Two Visual Systems*, 107–124.

Mayer-Gross, W. (1935). On depersonalisation. *British Journal of Medical Psychology*, 15, 103–122.

Noë, A. (2005). Real presence. *Philosophical Topics*, 33, 235–64.

Noel, J. P., Pfeiffer, C., Blanke, O., & Serino, A. (2015). Peripersonal space as the space of the bodily self. *Cognition*, 144, 49–57.

Peacocke, Christopher (2014). *The Mirror of the World*. Oxford: Oxford University Press.

Perry, J. (1990). Self-notions. *Logos*, 11: 17–31.

Perry, J. (2001). *Reference and Reflexivity*. Stanford: CSLI Publications.

Reed, C. L., Grubb, J. D., & Steele, C. (2006). Hands up: attentional prioritization of space near the hand. *Journal of Experimental Psychology: Human Perception and Performance*, 32(1), 166.

Rheingold, H. (1991). *Virtual Reality*. New York: Summit Books.

Rind, F. C., & Santer, R. D. (2004). Collision avoidance and a looming sensitive neuron: size matters but biggest is not necessarily best. *Proceedings of the Royal Society of London B: Biological Sciences*, 271(Suppl. 3), S27–S29.

Rizzolatti, G., Scandolara, C., Matelli, M., & Gentilucci, M. (1981). Afferent properties of periarcuate neurons in macaque monkeys. II. Visual responses. *Behavioural Brain Research*, 2(2), 147–i63.

Rizzolatti, G., Fadiga, L., Fogassi, L., & Gallese, V. (1997). The space around us. *Science*, 277(5323), 190–191.

Sambo, C. F., & Iannetti, G. D. (2013). Better safe than sorry? The safety margin surrounding the body is increased by anxiety. *Journal of Neuroscience*, 33(35), 14225–14230.

Sambo, C. F., Forster, B., Williams, S. C., & Iannetti, G. D. (2012). To blink or not to blink: fine cognitive tuning of the defensive peripersonal space. *Journal of Neuroscience*, 32, 12921–12927.

Sanchez-Vives, M. V., & Slater, M. (2005). From presence to consciousness through virtual reality. *Nature Reviews Neuroscience*, 6(4), 332.

Shang, C., Chen, Z., Liu, A., Li, Y., Zhang, J., Qu, B., Yan, F., Zhang, Y., Liu, W., Liu, Z., & Guo, X. (2018). Divergent midbrain circuits orchestrate escape and freezing responses to looming stimuli in mice. *Nature Communications*, 9(1), 1232.

Sierra, M. (2009). *Depersonalization: A New Look at a Neglected Syndrome*. Cambridge: Cambridge University Press.

Slater, M., & Wilbur, S. (1997). A framework for immersive virtual environments (FIVE): speculations on the role of presence in virtual environments. *Presence: Teleoperators & Virtual Environments*, 6(6), 603–616.

Slater, M., Spanlang, B., & Corominas, D. (2010). Simulating virtual environments within virtual environments as the basis for a psychophysics of presence. *ACM Transactions on Graphics (TOG)*, 29(4), 92.

Spence, C., Pavani, F., & Driver, J. (2004). Spatial constraints on visual-tactile cross-modal distractor congruency effects. *Cognitive, Affective, and Behavioral Neuroscience*, 4(2), 148–69.

Tammero, L. F., & Dickinson, M. H. (2002). Collision-avoidance and landing responses are mediated by separate pathways in the fruit fly, Drosophila melanogaster. *Journal of Experimental Biology*, 205(18), 2785–2798.

11

The dual structure of touch

The body versus peripersonal space

Mohan Matthen

11.1 Introduction

Peripersonal space is space defined by the reach of our limbs and the tools that we hold in them. It is the territory of direct intervention—the area where we probe and shove, the zone of the head-butt and the cross-check. By the same token, it is where others intrude with *their* bodies, the locus of greatest vulnerability. Michael Graziano (2018, p. 1) memorably calls it the 'margin of safety, bad breath zone, duck-and-flinch buffer'. Peripersonal space is, in short, where touch meets the Other, the region where self and non-self meet and act on each other.[1]

The skin forms the inner limit of peripersonal space. Our touch receptors reside there, and what we know by touch about peripersonal space is projected from the stimulation of these receptors. There is, however, a radical difference between feeling on the skin—*tactile sensation* (T), as I will call it—and *haptic perception* of things outside (H) (see Fulkerson 2015 for discussion). One important difference is that tactile sensation is felt to be private and subjective—it is awareness of things none but the subject can feel—while haptic perception delivers objective awareness of things outside—things like the pencil in your hand and the breeze playing on your face. Another difference is that (as I shall argue) haptic perception represents peripersonal space, while tactile sensation is not genuinely spatial. I will argue that, as a consequence of these differences, haptic perception is not (as some have said) merely a reformatting of tactile sensation. Rather, it provides different information and is genuinely emergent.

11.2 The self and the other

Here are two sensory events:

(T) In the darkness of early morning, I reach out to turn my phone-alarm off, and I bump against the corner of the table. I feel a sharp *thwack* on the back of my hand. This feeling is a tactile sensation.

(H) I reach past the edge of the table and grab my phone. I feel it vibrate in my hand. It is the phone that I feel vibrating, not my hand. This is haptic perception.

---

[1] De Vignemont and Iannetti (2015) suggest that distinct maps of peripersonal space are used for the distinct functions of intervention and defence. This would fit within the quite coarse-grained treatment that I offer in this chapter, which focuses on the difference between representations of the subject's own body and of external objects in contact with it.

Mohan Matthen, *The dual structure of touch* In: *The World at Our Fingertips*. Edited by Frédérique de Vignemont, Andrea Serino, Hong Yu Wong, and Alessandro Farnè, Oxford University Press (2021). © Oxford University Press. DOI: 10.1093/oso/9780198851738.003.0011.

These are complex sensory events. In both, there is awareness of the body as well as of something outside. In T, however, I am aware of something that happens to me in a way that I know nobody else can feel. In H, by contrast, it is an external object I feel—something I know you could feel as well. It is these kinds of sensory events I will be discussing—one is internally and the other externally directed. My aim is to show that there are two different sensory events here, existing side by side—the difference is not merely aspectual; it is not just a shift of attention between two ways of regarding the same data.[2]

As mentioned earlier, there are two crucial differences between T and H. The first difference is that T is *phenomenal* and *immanent*. The felt characteristics of the thwack on the back of my hand—sharpness, painfulness, transience, etc.—are qualities of my experience.[3] By contrast, H reveals *objective* and *transcendent* qualities of *the phone*. The vibration presents itself as an event that exists outside my body independently of anything I feel. It is possible that in fact the phone is *not* vibrating—that this is some kind of illusion. But the sensation produced by the *thwack* cannot be other than how it feels. However gentle and lingering the contact might actually have been, the experience itself was painful and transient—I cannot be wrong about this.

The second difference has to do with *position*. The thwack is anchored to a body part: when I move my hand away from the point of impact, the lingering sensation stays exactly where it was—on the back of my hand. I do not feel it moving. Thus, I do not feel it as occurring side-by-side with things in the material world: the sensation is not in extra-bodily space. The phone, by contrast, is (at least normally) experienced as located in external peripersonal space. Its location is sensed relative to my hand, of course, but additionally it is felt to be spatially related to other things that I see and feel near the point of contact. It is felt to be on the table, next to the bed, etc.

These points are together suggestive of Kant's dictum that space is the form of outer appearance:

> *Transcendence of Peripersonal Space*: We (tactually) experience something as existing outside the body and as the locus of objective (tactual) qualities when, and only when, we (tactually) perceive it as located in space.

In short, touch represents something as distinct from the body by locating it in peripersonal space. H-perceptions are sensory presentations of things that exist independently of ourselves, in external space. T-sensations are felt to be private episodes that others cannot literally share, and they are not felt to occur in external space.

I'll proceed as follows:

> In section 11.3, I'll explain how bodily awareness by touch, T, is qualitatively and phenomenally different from touch-experience of things outside, H.
> In section 11.4, I will argue that these states are distinct: T ≠ H.
> Finally, in section 11.5, I will argue that T is non-spatial, while H is spatial. T reveals position in a body-map; H refers an object to a location in peripersonal space. I will explain how these are different.

---

[2] Fulkerson (2015), chapter 1, is an authoritative account of the difference between tactile and haptic.
[3] I don't want to say that *my experience* is sharp—rather that it merits this description in virtue of a quality it (i.e. the experience) has.

11.3 T versus H: qualitative differences

11.3.1 Two touch experiences

Here are two examples that illustrate in more detail the different uses to which we put T and H:

1. *The art of the spin bowler*: You are a bowler in a game of cricket (or a pitcher in base-ball). You want to deliver a ball to the batsman with a finely controlled trajectory. You must grip it in a very precise manner so that, at the moment of delivery, you can use the stitching on the ball to impart spin. In order to control this, you focus on the feelings of pressure that the seam of the ball exerts on your fingers. You make sure that there is a steady feeling of resistance on the first knuckles of your index and ring fingers—'not too light, not too tight,' as bowlers say. This is something touch tells you—what is happening at and near the surface of specific body parts. This is an employment of T: *tactile* sensation or feeling.
2. *The guitarist's technique*: Now think of plucking a string on a guitar. A guitarist's aim is to create a certain effect on the string. Her effort depends on feedback she re-ceives from the string—where it is, how tight, the angle of her finger relative to the string, the motion of the string relative to the body of the guitar, and so on. In short, she has to be aware of the string. Touch provides this awareness. Of course, the gui-tarist may also be conscious of feelings on her skin: if she had a cut or an abrasion, for example, it may sting. But to pluck the string with exactly the right amount of attack, she must be aware of how she is affecting *the string*. Here, the primary ob-ject of awareness is an external object; the focus is not on the body. This is H: *haptic* perception.

T monitors the skin (and sometimes areas deeper down in the body) *as a part of the body* (Mattens 2017, Mandrigin 2019). De Vignemont and Massin (2015) say that it is awareness of *pressure* due to the resistance that the body offers to, or exerts on, an external object. I am inclined to consider T more broadly, including distinct sensations of weight, torque, stretch, and temperature, which is a distinct system[4]—and of pain as it relates to all of these. The important point here is that T is, in the sense explained in the Introduction, *phenomenal* or *immanent*. The qualities just mentioned—pressure, twist, stretch, temperature, etc.—are given phenomenally (i.e. as qualities of the experience itself).

H has different content: it informs us about the qualities and locations of external ob-jects that are in direct or indirect contact with the body. It is not, as such, about the body—the term 'perception' marks awareness of something distinct from the subject herself or her body. The qualities that H presents are attributed to things outside the body, in peripersonal space, and not to the experience itself.

It is important to note that (a) both T and H are mediated by the same cutaneous receptors but, as we shall see, (b) H requires something more.

---

[4] Weight, torque, and stretch are pressure-related, but they feel different; the feeling of warmth arises from dif-ferent receptors.

11.3.2 Proximal versus distal in touch

In perception generally, external stimuli bring about two distinct experiences, one *proximal* that reflects the state of the sensory receptors when the stimulus registers, and a distinct *distal* experience that reveals properties of the external object of perception. Colour perception is a good example: proximally, it corresponds to the total effect that illumination and object-properties have on the retina; distally, it corresponds to the reflectance or brightness of an environmental object.[5] White skin looks white even when it is viewed in the green light that filters through tree cover. But if you attend closely enough, you can notice a greenish tint in the highlights. Visual size perception is another example: proximally, it corresponds to the angle that an object subtends at the retina; distally, it corresponds to the size of the object itself. Often, the *proximal* experience is in the background and requires an act of deliberate attention, while the distal experience is foregrounded in consciousness. For example, the screen on my table looks a lot smaller than the buildings outside my window, but with a deliberate act of attention I am able to see that they take up similarly sized regions of my visual field.[6]

The proximal/distal distinction is also operative in touch. Here, T is proximal: it is an integrated construct of the response levels of cutaneous receptors. H, by contrast, is distal: it reflects the properties of external objects despite the very different effects they might have on the skin in different situations. For example, a hard surface appears hard despite the fact that one feels it through a layer of foam—the tactile feel on the skin is different through foam. Nevertheless, the object haptically appears to be hard, just as it does when it is touched without the foam. Most observers are primarily aware of the distal features of things they touch: it takes a deliberate switch of attention to focus on how the feel of the hard surface is a combination of its own hardness modified by the squishy admixture of the foam wrapping.

The unique feature of this duality in touch is that the proximal and the distal are differentiated by *location*. Tactile sensation is 'about' the skin (or a body part)—more or less where the receptors are—while haptic perception is of things that are felt to be outside, although in contact with, the body (or a tool that the subject uses to probe something at a distance). In vision, by contrast, proximal and distal colour are co-located: both appear to belong to an external object of perception. Even when we attend to the colour of the light that an object is reflecting, as opposed to the colour of the object itself, there is no shift of location. An experienced photographer might notice the greenish tint of white skin under leaf cover—but it is still the skin that looks greenish when she switches her attention in this way. We *do not* ever sense proximal colour as located in ourselves—on the retina, for example.

This difference between touch and vision yields a significant preliminary conclusion. Think again of that pale-skinned person under leaf cover. At first glance, her skin looks the

---

[5] In audition, olfaction, and gustation, the proximal stimulus is an aggregate that, in the distal stimulus, is apportioned to different sources. For example, the proximal auditory stimulus is a frequency-power distribution that pertains to the entire scene. You can hear some aspects of this—for example, the harmonies played by distinct instruments in an orchestra—but the salient experience is generally of separate voices or sources, each with their own power distribution. Here, both proximal and distal experience is of things outside the subject—the whole ambient scene or individual voices within it.

[6] Jonathan Cohen (2015) emphasizes this duality: perceptual constancy registers the sameness of distal properties while perceptual contrast registers differences in perceptual circumstances such as illumination, distance, etc. But he points out that the proximal and distal are often co-present—for example, when we see an object that is partly in shadow, we perceive both that the object is uniformly coloured and that some parts are brighter than others.

same as it does indoors in tungsten lighting of comparable brightness—that is, it looks white with a slightly pinkish cast. But, looking more carefully, the photographer notices the tint produced by the leafy filter—this is apt to affect the shot she wants to take. In cases like this, her scrutiny does not shift to a different visual state. She is still looking at her subject's skin. Moreover, you could argue, the greenish tint was visible in the original viewing, although it was the more stable object–property that originally captured her attention. There is reason to think that there is only one colour-visual state here, not two, and that the photographer's view is simply a shift of scrutiny.

With touch, matters are not so obvious. The proximal/distal shift has much greater import here. T gives information about one object—the skin—and immanent qualities like resistance, warmth, etc., while haptic perception gives information about external objects and their transcendent qualities. There is reason here to suspect something more than a merely attentional shift between different aspects of the same experience. Prima facie, at least, these are two different experiences because they present states of different objects.

But there is another side of the argument. Corresponding T and H states are simultaneously *available*, but typically you can attend to only one of them at a time. That is, if you attend to the feeling on your skin, you are less able to attend to the external object that produces that feeling, and vice versa. This leads some to say that the difference between T and H is *merely* attentional—i.e. that it is merely a matter of what you attend to in one and the same underlying perceptual state. I disagree with this, and in the next section I will argue for the idea that T- and H-states are distinct. In other words, I'll argue that:

T ≠ H.

11.4 Not one, but two

11.4.1 H and trace-integration

Now consider two touch-events:

1. *Spilled Liquid* (T*H): You are on a flight, absorbed in your book with a drink on the tray-table in front of you. Suddenly, the plane hits a bump, and you are startled by liquid splashing on your hand. At the very first instant, your experience is purely tactile: you feel an impact on your hand (T). But then the liquid trickles down the back of your hand and wrist, creating a coordinated train of sensations. Now, you become aware of cool liquid trickling down the back of your hand and beneath your sleeve (H).

   This sequence of events suggests that H emerges by integration of the temporal succession of T—over the time period in question, T has traced a continuous path on the skin. Vision can play a role here: if you happen to see the liquid splash, you may immediately feel it as an external object (H) by integration of the visual and tactual inputs.

And here's the second event:

2. *Pins and Needles* (T-alone): You are in a cramped airplane seat and you get 'pins and needles' in your foot. These prickly sensations seem to dance about, but there is no feeling of externality. The sensation is internal all along.

What is the difference between these felt experiences? It is that in the first more common kind of case, the touch-system finds a continuity in the train of T-sensations that indicates the motion of an external object against the skin. In pins and needles, the train of sensations dances randomly over your foot in a manner that cannot be ascribed to an external object. The sense of touch integrates the first: thus T*H. But it is unable to integrate, or 'make sense of,' the second: thus, T-alone.

The idea of T-alone has been overlooked. De Vignemont and Massin (2015) claim that one seems to feel something outside when there is resistance to pressure that one exerts. They write: 'in order to feel the resistance of the external world, we have to be aware that it exerts some force counteracting the force we are intentionally exerting on it' (ibid., p. 301). I agree with this as far as it goes, but it does not say how the activation of cutaneous touch receptors can yield awareness of a counteracting force from the outside. And because it does not address the emergence of H, it overlooks passive H—sensations of external pressure in the absence of force intentionally exerted in return. The awareness of liquid trickling down my hand is an example—I don't exert force on the liquid; all the same, I feel it exerting force on me. You feel the shirt on your back, but you don't exert any force on it. Thomas Reid gives another interesting example: a 'scirrhous tumour' under the skin that creates (according to him) the impression of an external object without reciprocal effort (*Inquiry*, chapter 5, section VI, see p. 25).

It is important to appreciate that H-awareness of an external object is something more than T-awareness of pressure over a contiguous and segregated area of skin.[7] That is: H ≠ {T} (where {T} denotes some collection of T-sensations). Let me make an additional point about this by considering how the closely related sub-modalities of warmth and pain are different in this respect. These sub-modalities deliver T-sensations to the skin. When I sit before a fire, contiguous parts of the skin on my face feel warm; when I feel the itch of a rash on my forearm, I am aware, similarly, of a segregated area of qualitatively homogeneous T-itch-sensations. But there is no awareness of an external object in either case (Cataldo et al. 2016). Add pressure-touch sensations, and H-perception emerges—but only if there is trace integration of T. When I hold a warm cup in my hand, or when my arm is abraded by brushing up against a rough wall, I feel similar contiguous skin-located T-sensations of heat/pain. But in these cases, I am also H-aware of an external object. The difference is pressure-touch: when it is trace-integrated, it affords awareness of something outside. This shows how H is more than contiguous {T}.[8]

To summarize: H emerges from the trace-integration of pressure T.

11.4.2 Immanence (T) versus transcendence (H)

The idea of T alone is, to some extent, an idealization. Almost always, the touch-system is able to assign an external cause to tactile stimulation and thus to produce T*H. And (as we shall see) this has led some to say, introspecting their experience in such 'normal' cases, that the difference between T and H is merely an attentional switch. When you grip the ball, you can attend to the feel of the stitching on your skin, or you can attend to the properties of the ball. Similarly, when you pluck the guitar string. Is there just one touch-state here, within

---

[7] See Matthen (2019) for an extended discussion of the same point regarding vision: visual awareness of a material object is different from visual segregation of a contiguous volume of space.

[8] I am grateful to Patrick Haggard for alerting me to this difference between pressure-touch and warmth/pain.

which you switch your attention from one aspect to another, or are there two? To help answer this question, let us examine the states themselves more closely.

Thinking, then, of T-states, one striking characteristic is what I have called *immanence*—the quality of which we become aware in a T-state as a quality of the experience itself. Consider two different T-feelings: an itch and a gentle tickle. Each of these experiences has a characteristic 'feel', or phenomenal quality, that identifies its kind. An itch is an itch because it feels like an itch; it is different from a tickle because it feels different. There is nothing more to be known about this; an itch is not what it is because (for example) it represents the secretion of a histamine; it is the kind of state it is because of how it feels. Similarly, it is not different from a tickle because its physical causes and internal processing are different, but because it feels different.[9]

By contrast, think of H-perceptual states. Let's say you have two objects in your pocket: a pen and a key. You can identify them, without looking, by feeling their shapes. There is no one tactile sensation that is associated with each shape: you may feel them from different directions, with different parts of your hand, through wool or through silk or directly. It takes a certain amount of tactual exploration to determine their shapes and thus to identify them, but each exploratory episode might be radically different from another. There is no one phenomenal quality associated with the perceptions of the same quality, whether it be shape or texture or temperature. The content of these haptic states is not defined by their phenomenal feel, but rather by the shape that is revealed. H-content is *transcendent*.

11.4.3 T-location and H-location

There is nothing more to the *kind* of a T-state than its phenomenal character. But you can be subject to many T-states of any given kind. After walking in the Canadian woods on a summer's day, you may have been bitten by mosquitos all over your hands and neck, and you may thus be plagued by many itchy experiences of the same phenomenal quality. What distinguishes these is that they are 'in' or 'on' different parts of your body.

Now, to say that an itch is 'on' a part of your body, say your forearm, needs some explication. Itches, being experiences, do not literally reside on parts of the body: an experience is not felt to be located on or in a body part in the same sense that I am *on* a bus *in* Toronto. The body part is, in some sense, a locator, but not in the sense of a spatial region that contains a material object—a sensation is not the kind of thing that resides in a spatial region. Experiences are rather 'referred to' these bodily locations: they draw sensory attention to them and make them the targets of action. An itch asks for a certain part of the body to be scratched: this is what it means to say—that it is 'in' that part.

An individual T-sensation is one that has a certain phenomenal quality referred to a certain bodily part. (Here I mean quality to include intensity—a mild itch is qualitatively

---

[9] My thesis about T-states may remind readers of a thesis about colour that Mark Johnston states this way:

> *Revelation.* The intrinsic nature of canary yellow is fully revealed by a standard visual experience as of a canary yellow thing. (1992)

And I am indeed claiming that the intrinsic nature of T-states is fully revealed by what it is like to experience them. But there is a difference. Johnston's *Revelation* purports to show how we come to know the essence of a quality, colour, that inheres in external objects. My thesis is that T-states do *not* reveal externalities: they are simply experiences of a certain kind. That said, my approach has some overlap with Johnson's. He suggests that what we know in the case of colour is an experience that actualizes a disposition; my account of T-states is that they *are* experiences.

different from an unbearable itch). A bodily sensation (such as an itch) is numerically different from another bodily sensation of the same phenomenal quality when, and only when, it is referred to a different body part. We can sum this up by defining a T-sensation as an ordered pair, <Q, p> where Q is an immanent phenomenal quality Q and p is a part of the subject's body.[10]

What is the relationship between the body part p and a location in external space? I will be discussing this question in more detail in section 11.5 of this chapter, but for the moment let us note that the answer is far from obvious. I am sitting on a bus and I feel the seat behind me. At the same spot where the seat touches me, I also feel an itch where the wool of my sweater rubs against my back. Yet, I don't feel the itch to be *in the same place* where the seat of the bus is—they are not co-located. The bus seat is *in contact with* my back but distinct from it. Further, my awareness of where the bus seat is integrated with my awareness of my own body—I can distinguish its motion from mine. Here is the conclusion we can draw from this: T (the itch) in body part p and H (awareness of something external) at body part p do not together imply that T and H are at the same location. This shows that body parts play different localizing roles in T and in H. In awareness of the bus seat, the spot on my back is (as I will explain more fully in the following sub-section) a *pointer*, not a direct locator.[11]

How does information about the bodily location of the pressure yield information about external space? Trace-integration is only part of the story. Let's call this the 'space-mapping problem'.[12]

11.4.4 Haptic content

Now, consider the individuation of H-states. As noted earlier, they are presentations of the transcendent qualities of external objects. The cup in your hand is smooth, hard, and warm. These qualities are not exhausted by experiences that you undergo: they are qualities of the cup that persist through changes in your experiences that occur as you shift your grip on the cup.

Now, just as in the case of T-states, it is possible to have two H-states that are about the same quality. You may hold a cup in each hand, compare their touch-qualities, and find that they are the same. And as with T-states, the things you haptically perceive are differentiated by their location relative to your body—one cup is in your right hand, the other in your left. However, the transcendence of the qualities makes for some important qualifications here.

---

[10] A sensation is indexed to the subject and would, in this manner, be *de se*. So, a fuller specification would be <Q, p, i>, where i is the subject. This is important for its role as a pointer to a location in external space, but I will ignore it for simplicity's sake. (See the following sub-section.) As well, Q is intensive—that is, it admits of a more and a less: some separate intensity from the Q, which would indicate a four-place specification, <Q, I, p, i>, where I measures the intensity of Q.

[11] Frédérique de Vignemont, this volume, defines peripersonal space as that region around the body in which space is multiply mapped relative to many movable bodily parts. I am sensitive, for example, to how far from my left index finger the 'm' key is, and I am also aware of how far from my right index finger it is. This is essential for communication with the motor cortex, for, if I wish to touch the 'm' key with the left finger, I have to make a different motion than if I were to do so with my right finger. Far away objects pose no such problem of differentiation. They are represented, visually, as a certain distance away from *me*.

[12] See Azañon et al. (2010) for further discussion of space-mapping. Mandrigin (2019) seems to overlook the difference between the T-map and the H-map and the allocentricity of the latter.

First, a quality like hardness is felt to be *external*. We noted earlier that tactile sensation isn't literally located in a part of the body: it is simply 'referred to' it—that is, it draws attention to this part of the body, or prompts behaviour directed to it. But a quality like hardness is felt literally to be located in a place identified by a T-sensation. You experience a feeling of pressure figuratively 'in' your hand; thereby, you feel something hard *literally* in your hand.

Technically, this can be expressed as follows: the body part in which you feel a sensation of pressure becomes a *pointer* to a region of space adjacent to the body part in which you perceive something hard. Define *T-spatial representation* as a coordinate system defined by such body-centred pointers. The question is: how much does T-spatial representation tell you about external space? This is analogous to the space-mapping problem mentioned at the end of the previous sub-section. How much information about external space is implicit in the pointer from a (subject-indexed) body part?

Second, transcendent qualities like hardness are ascribed to external *objects*—not just to spatial regions but to objects that reside in those regions. In your hand, you feel a gift-wrapped object that is hard *and* smooth *and* warm. In exactly the same part of your hand, you feel the soft cloth that is wrapped around the gift: it is also warm, but its warmth is a distinct thing. In such a case, you feel two things among which these qualities are apportioned; the qualities are not merely co-located in the space around your hand, but ascribed to different objects.[13] This kind of co-ascription doesn't arise in tactile sensation. A feeling of pressure on your hand is different from a feeling of warmth in your hand: neither is a quality of your hand—they are positionally co-referred to your hand, but distinct. This is a crucial difference between T- and H-states: in haptic perception, qualities are felt to belong to *objects* that are differentiated both from the subject's own body and from each other.

Let's sum this up in this way:

Feature Binding in H: H supports feature binding to objects; T supports only feature location relative to parts of the body.

Putting this together, we may conclude that haptic perception has content that is more complex than tactile sensation. With pins and needles, you simply feel $<Q, p>$, a tingly sensation in your foot. With the spilt liquid, you feel warmth (Q_1) and motion (Q_2) ascribed to an object O (the liquid) in a pointed location p (the region of external space just beyond the back of your hand). Thus, H-perceptual states take the form $<\{Q_i\}, O, p\}$, where one or more qualities Q_i are ascribed to an object O, in a spatial location identified by a pointer from a bodily part.

I will end this section by noting a significant detail that figures in the not-one-but-two argument of section 11.4.6. We feel things by touch not only by touching them with our body, but also by touching them with tools or other manipulable objects. For example, I can perceive the texture of a sheet of paper both by stroking it with my finger and by writing on it with a pen. I can test the give of a soft object directly and also indirectly, using a stylus or pincers. When objects are probed at a distance, they produce tactile sensations as well as haptic perceptions. With respect to the latter, they are felt to be located not at the skin but at a distance, at the end of the tool. The body part p referred to by a sensation need not be spatially next to the location p that it points to, although it often is.

---

[13] This contradicts visual field feature-placing theories such as those of Austen Clark (2000). See Matthen (2004) for discussion.

11.4.5 The non-supervenience of the haptic

We have been considering trains of tactile sensations that result in a perception of an external object when the touch-system can 'make sense' of them. It is notable that this involves a form of perceptual constancy. The spilt liquid courses down the back of your hand and up past the cuff of your shirt. Through the array of tactile sensations that result, there is a single perception of a rivulet of sticky liquid. Many sensations, one perception: different T, same H.

Equally, the same tactile sensations can lead to different (or no) haptic perceptions. Go back to the case of the spilled liquid. Consider the first moment after the splash. You have no perception of something outside you—just the feeling of something happening to you. But if that very sensation had been embedded in a temporal sequence, it might have produced a perception, and the content of that perception would depend on the character of the other components of the sequence. The 'cutaneous rabbit' (Geldard and Sherrick 1972) illustrates this. Deliver five brief mechanical pulses (2 ms each, separated by 40–80 ms) to the wrist, then 10 cm up on mid-forearm, and finally another 10 cm up near the elbow. The result is 'a smooth progression of jumps up the arm, as if a tiny rabbit were hopping from wrist to elbow.' Slow the sequence down or make it less regular and the illusion is destroyed. Same T, different H. Or, again, consider tactual form-agnosia. E.C. was a patient who had suffered an infarction in the left parietal lobe. The result in her case was a significant impairment of object recognition by touch, although other haptic capacities were spared (Reed and Caselli 1994).

Tactual form-agnosia is particularly significant because it establishes a connection with visual form-agnosia, in which patients lose the ability to recognize, or even see, shapes despite retaining visual acuity. In vision-studies, it is standard procedure to construe agnosias of all sorts to be failures of *perceptual* (as opposed to post-perceptual or cognitive) mechanisms. The parallel with visual form-agnosia strongly suggests that, although tactual form-recognition is temporally extended, it is nonetheless perceptual in nature. The same is true of the above-mentioned cutaneous rabbit: it is strongly analogous with phenomena surrounding the visual perception of motion, for example, the ϕ-phenomenon. (See Reed, Caselli, and Farah 1996 and James, Kim, and Fisher 2007 for further discussion of parallels between visual and haptic perceptual processing.)

These observations demonstrate a double dissociation between the tactile and the haptic:

LEMMA 1: You can have *different* proximal stimulations/tactile experiences (T) with the same haptic perception (H) arising. (Cutaneous rabbit illusion vs a real rabbit hopping up your arm.)

LEMMA 2: You can have the *same* proximal stimulations/tactile experiences with different haptic perceptions arising. (Tactile-form agnosia vs unimpaired subject.)

Conclusion: haptic perceptual states are *distinct from* (collections of) tactile states.

11.4.6 T versus H: not one but two

Developing a line of thought originated by Brian O'Shaughnessy (1989), M.G.F. Martin (1992) talks about touching the rim of a glass and being aware both of pressure on the skin and of the glass—these are not two but one, according to him. Martin writes:

We should think of this case **not** as one in which we have **two** distinct states of mind, a bodily sensation and a tactual perception, both of which can be attended to; **but** instead simply **one** state of mind, *which can be attended to in different ways.* One can attend to it as a bodily sensation—in which case its spatial character reveals the location of sensation—or attend to it as tactual perception of something lying beyond the body but in contact with it, so that the spatial character is that of the location of whatever it is which connects with and impedes the movement of one's body. (Martin 1992, p. 204. Emphasis added to highlight the not-two-but-one thought.)

Reproduced from Martin, M.G.F. (1992). "Sight and Touch." In Tim Crane (ed.), The Contents of Experience. New York: Cambridge University Press: 196-215.

According to Martin, the bodily sensations caused by tracing the rim of a wine glass are one and the same as the perception of the wine-glass, an external object. There are not two states here, but only one attended differently. In other words, T = H. And Martin makes the further claim that the awareness of a bodily part implicit in tactile sensation is simply the same as awareness of a spatial location. But this simply overlooks the space-mapping problem: to talk about the felt spatial location of sensation simply overlooks the fact that sensations are not felt to be spatially co-located with the external things we feel—the feeling I have on my fingertip is not felt to be on the rim of the wine glass I happen to be touching.

Let me make a brief comment about this that I'll take up in more detail in section 11.5. O'Shaughnessy and Martin are talking about a situation in which the subject is *acting*—tracing the rim of a glass in the above example. In such situations, kinesthesia provides sensory input that supplements tactile sensation, and awareness of external space ensues from the enriched sensory basis. So, it is not quite correct to say (as Martin implies above) that this example of awareness of 'something lying beyond the body' is a bodily sensation (i.e. T) in the sense that concerns me.[14]

11.4.7 Summary

The preceding discussion provides strong—I would say conclusive—reasons for rejecting the not-two-but-one idea.[15] Drawing them together:

1. *The Emergence of the Haptic*: The contrast between the spilt-liquid and the pins-and-needles cases shows that something more than T is needed for H. I argued that this 'something more' is trace integration to diagnose an external cause for the train of tactile pressure-sensations.
2. *Non-Supervenience*: No T-set is either necessary or sufficient for any given H.
3. *Transcendence of the Haptic*: Tactile sensation is quality-wise immanent; the quality presented in T is a quality of the T-experience itself. In haptic perception, however, the quality revealed is that of something that is presented as existing independently of the subject.

---

[14] Alisa Mandrigin (2019) makes the interesting claim that we don't sense *bodily* location except when we act. In other words, her idea is that I know where my hand is relative to my elbow because I know how to reach for something by flexing my arm. Her evidence for this comes from certain dissociations between bodily awareness and action in both neurologically damaged and healthy subjects. However, these subjects are tasked with locating objects in external space relative to their own body parts. (Where is that object relative to your hand?) Her argument does not show anything about T as such—tactile sensation in the absence of action.

[15] To be clear, I *don't* deny that T and H may often be different stages of the same processing stream. My claim is just that they are different. This claim is particularly important for my treatment of the spatiality of H in section 3.

4. *Externality of Haptic Location*: In the train of the difference regarding transcendence, the kind of location presented in T is different from that presented in H: T-location a bodily part; H-location is spatial. (I will say more about this in section 11.5.)

5. *The Space-Mapping Problem*: Even when T (e.g. a feeling of pressure on the finger) arises from an object that the subject touches and H-perceives (e.g. the rim of a wine glass), T is not felt to be located on the external thing. (The feeling of pressure on my finger is not felt to be on the wine glass that I am touching.)

6. *Distal Location in Tool Use*: The spatial location indicated by H isn't always adjacent to the body part referred to by the corresponding T. We feel things at the end of tools, but of course tactile sensation is always on the skin (or apparent skin, in the case of amputees who suffer from phantom-limb syndrome). This shows that spatial coincidence is not mandatory.

7. *Feature-Binding*: H-qualities can be co-bound to the same object. The cup in my hand is warm and curved and smooth. T-qualities can be co-located but not co-bound: I may feel warmth and itchiness at the same spot of my forearm, but these are not properties of my forearm.

Thus: H ≠ T.

11.5 Spatial representation

11.5.1 Bodily position versus spatial location

Having argued for the distinctness of T and H, I will now argue that T is non-spatial while H is spatial. Recall, however, that I have been saying that T is referred to a bodily position. I will now explain why the bodily-positional content of T is not spatial, or at best only sparsely spatial.

Think of a dispatcher who is in charge of a number of taxicabs; she has information about who is driving each, and whether it is taken or empty. She knows, for example, that Anil is driving Car 1, which is currently empty, and that Barbara is driving Car 2, which is taken. This is *positional* information about the two drivers—they are in Cars 1 and 2. But it does not tell her *where* in the city Anil is or how far from Barbara—the dispatcher has no explicit *spatial* information.

Still, she can make a few simple inferences. Right off the bat, she knows that Anil is in a different spatial location from Barbara—they are in different cars, and cars don't overlap spatially. This is what we might call 'background information'. And the drivers' occasional reports tell her more. Anil reported ten minutes ago that he was dropping a passenger off at Union Station; Barbara said five minutes ago that she was at the airport. Since the dispatcher knows where these landmarks are, she can figure out the minimum and maximum distances between Anil and Barbara. We can call this 'update information'. As time passes, update information degrades: the taxis are constantly moving and update information from an hour ago tells us little about the position of the cabs now.

A spatial representation consists of a set of points, or locations, and certain metrical relations among these points—canonically, these relations will include or imply distance and direction. The taxi-dispatcher has a representation of a certain set of locations—the cars and some of their non-spatial properties. But she knows little about the distance and directions of one location to another. She can make certain inferences from other information but, since the locations move about both in space and relative to one another, background information is sparse and update information constantly degrades and must constantly be renewed.

Like the taxi dispatcher, touch is poorly informed about space. Cutaneous receptors are embedded in a body map consisting of minimal bodily locations—little bits of skin within which distinct positions cannot be discriminated—tactile receptive fields, or 'RFs', as they are called. And we can assume that the system is so constructed as to take advantage of constant background information. For example, some of the RFs are at a fixed distance from one another—those on a single non-deformable part of the body, such as the forearm or the back of the head, are same-sized. And there are certain limits on distance that follow from body-size. But information about space is generally changeable. For example, feeling a touch on your finger and another touch on your shoulder doesn't suffice to tell you how far apart the two touches are, because your finger moves in space relative to your shoulder. Here, update information is needed—information provided by the proprioceptive sense, not by touch. As in the case of the taxis, background and update information provide some clues as to spatial relations. But it is worth noting that update information is generally provided by senses other than touch—specifically proprioception, kinesthesis, and (crucially) vision.

In peripersonal space, in particular, vision and proprioception (or awareness of bodily position) work together with touch to construct awareness of spatial location. You experience a prick on your finger; you can see that the finger is in contact with a needle on the table. Consequently, you know that the prick on your finger (positionally identified by touch) is caused by an object on the table (spatially located by vision)—you are able, for instance, to use your other hand to move the needle away. Interestingly, it has been found that neurons that encode bodily position in the motor areas of the brain are sensitive to visual experience of those bodily positions (Brozzoli, Ehrsson, and Farnè 2014). In the case just mentioned, the neuron that records cutaneous stimulation of the finger is activated also by visual information that records objects near the finger. These bimodal neurons provide a link between bodily position and peripersonal space. However, it is unclear *how* the link is computed or maintained. How does the brain collate tactile information with visual?

11.6 Reid's *Experimentum Crucis*

Thomas Reid posed a classic, and as yet unsolved, problem for the idea that T is spatial. Imagining a blind man who 'by some strange distemper' had lost his conception of extension and space, Reid writes:

> Suppose ... that a body is drawn along his hands or face, while they are at rest. Can this give him any notion of space or motion? It no doubt gives a new feeling; but how it should convey a notion of space or motion, to one who had none before, I cannot conceive. (*Inquiry*, chapter 5, section VI, see p. 25)

Reid thinks that this thought experiment demonstrates that touch sensations lack spatial content.

Now, before we can properly assess this argument, we must take note of the anchoring of tactile sensation to bodily position. On any part of the body, there is a minimum distance,

d, between two points on the skin such that two simultaneous stimuli separated by d can be discriminated at two—two stimuli separated by a shorter distance feel like a single stimulus. This distance d is known as the 'two-point threshold'. (The value of the two-point threshold is different in different parts of the body—we will come back to the importance of this point in the following section.) We can imagine that the skin is divided into little cells, or RFs, each constituting one minimal part of sensitivity—the size across an RF is, in other words, equal to the two-point threshold for that part of the body. The RFs are units of bodily position.

Consider, then, Reid's blind man who has no conception of extension or space. Somebody strokes his face. As they do so, he experiences a brief stimulation of one RF and then another and then another. Reid's claim is that this succession of sensations does not 'convey a notion of space or motion.' Is this credible?

Go back to the case of the taxi dispatcher. Suppose that she has a list of positions that the taxis could occupy: call these A, B, C, etc. These are her RFs. She also knows (as background information) that, in order to be at A and then later at C (or vice versa), you have to be in place B at a time in between. Now suppose she finds that Car 1 reports being in place A, followed by place B, followed by place C. Can she abstract from this information a 'notion of space or motion?' No, because she has no information about distance or direction, or about the trajectory of the car. AB could be a lot shorter than BC; it could be at right angles; the driver might have been driving faster along BC, etc. In short, in the absence of further information about the relationship of the 'places', the dispatcher might know that a car has changed places, but not how far or in what direction it has moved. This, in essence, is Reid's point.

Now, of course, the touch-perceptual system may possess a good deal more background information than our taxi-dispatcher (Spence, Pavani, and Driver 2004). Nevertheless, it is instructive to note that its information is limited. Consider:

1. If somebody draws their finger along your nose, you experience a sequence of tactile sensations (T), but gain no information about the shape of your nose. Yet, if you were to trace your nose with your finger (H), you would feel in your finger the shape of your nose.
2. Suppose a liquid was intravenously injected along your femoral vein, and you feel the coldness (T) coursing from groin to toe. You could not tell from the progress of the disturbance whether your leg was bent or straight. Yet, if you traced your leg with your finger (H), you can easily gain this information.
3. If a small vibrator is securely attached to your finger and turned on (T), it feels stationary even when somebody (including you) moves your hand or finger.
4. If two small vibrators are attached, one to your forearm and the other to your ribcage on the opposite side, you cannot tell which one is higher from the ground.

Cases 1 and 2 are those in which T contains less spatial information than the corresponding H, while 3 and 4 are cases in which T lacks some spatial information (although the equivalent H might also). Update information could, of course, be added to these cases and this might be sufficient for spatial structure. But as we have seen, this information may come from modalities other than touch.

I conclude from the above that, at the very least, Reid is correct to say that T lacks some of the structure that we would expect in a spatial representation.

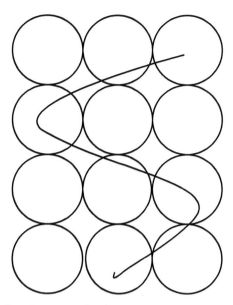

Figure 11.1 An S-shaped tracing across the skin.

11.7 Receptive field path-integration

Now, despite the examples just presented, it seems that, even though T is deficient in its representation of space, it is not entirely lacking in it. For it is hard to deny (although Reid seems to) that, when somebody draws their finger along your nose or face, you experience something like motion. Such an impression requires, at the very least, some awareness of adjacency relations and of direction. In Figure 11.1, the impression of motion demands that the touch-system is able to determine not only that certain RFs are activated in sequence, but that the second is adjacent to the first, and that the third continues on, and that there is a reversal of direction at the third RF, etc.[16]

Where does the awareness of these relations come from? Is it innate in touch? Or does it require input and calibration from other sources, particularly vision, proprioception, and kinesthesia? A full discussion of this question goes beyond present knowledge—and certainly beyond the scope of this chapter—but I'll conclude with a brief overview.

Fardo et al. (2018) investigated passive S-shaped tracings on the back of the hand. They asked participants to bisect the distance between the start and finish locations of the tracing—that is, the (unstimulated) midpoint of the straight line that joins these locations. Subjects were found to be successful in identifying this midpoint, with a slight bias towards the larger half of the S (in cases where the shape was not symmetrical from top to bottom). The task requires an integration of the back-and-forth movements of the tracing to yield the direct line from start to finish. The authors write: 'Our results are consistent with a path of tactile path integration that constructs a RF-based organization of "skin space"' (p. 241).

---

[16] Jean Nicod (1930/1924) has a rigorous discussion of the requirements of such impressions of space and motion. See also Haggard et al. (2017).

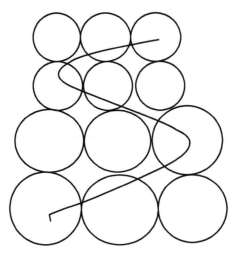

Figure 11.2 An S-shaped tracing over areas of differing spatial resolution.

Now, this is an impressive conversion of bodily position (the RFs) to spatial measurement (the bisection task). But the experiment is conducted under two special constraints. First, the tracings are all along the back of the hand and spatial discrimination is constant within this area—that is, the size of the RFs is the same throughout, and RFs can function as a proxy for distance. Also, the speed of the tracing is more or less constant. Thus, time elapsed between one RF and the next serves as a proxy for speed.

The question to ask is: what would happen if the tracing crossed over into an area where sensitivity was different—across the wrist line, for example, or over into the torso with the hand held adjacent? Then we would have a tracing more like Figure 11.2. If it assumed equal size, the touch-system would come up with a midpoint that is biased toward the more sensitive region. For a successful bisection in this scenario, therefore, the perceptual system would require information about the size of different RFs. Where would this information come from? In personal correspondence, Patrick Haggard suggested the following two possibilities, which I have embellished somewhat.

1. The system could calibrate the size of RFs by assuming constant speed for any given moving stimulus. A simple strategy of this sort would predict some wrong answers because obviously some stimuli do not move with constant speed. But a somewhat more complex strategy could calibrate RF relative to some subset of stimuli. However, the most obvious strategies of this kind would involve self-awareness: some stimuli are likely to be moving with constant speed when the subject herself is moving with constant speed, since then anything that she brushed up against would satisfy the condition. But this restriction involves self-awareness through a non-touch modality. Moreover, there is good reason to think that the touch-system makes systematic errors across areas of differing spatial discrimination.[17] So it is doubtful that it successfully uses this method.

---

[17] The question of pattern detection across different areas of the body was investigated (inter alia) by Haggard and Giavagnoli (2011), who report: 'judgements about the alignment of tactile patterns spanning two distinct body parts were surprisingly poor' (ibid. 73).

2. The system could use external measures to calibrate motion—specifically vision, proprioception, and kinesthesis. But this is, in effect, to acknowledge Reid's point that touch by itself is incapable of genuine spatial representation.

The conclusion to be drawn from this section is that spatial representation across body parts is computationally difficult for the touch-system and, although a constant speed assumption could help, it is both empirically dubious and somewhat implausible that this is the way the system actually works.

11.8 Conclusion

Does pure touch represent bodily position non-spatially? The question can be reformulated as follows: does haptic perception (which *is* spatial) take its spatial information from other modalities? I have offered two reasons for thinking that the answer to these questions is 'yes'. The first is phenomenological: it seems that we feel touch on our skins but not on objects in contact with our skin. The second is neurophysiological: cutaneous response is anchored to minimal body parts, or RFs, and the spatial extent of these RFs varies across the body. Together, these considerations support the idea that haptic perception (H) is (contrary to some influential philosophical accounts) distinct from tactile sensation (T), and that H requires input from non-touch modalities. Since tactile sensation is experientially immanent, while haptic perception is transcendent, my argument is strongly suggestive of Kant's thesis that things are presented as existing objectively when they are represented spatially.

Acknowledgements

I gave this paper to audiences in Antwerp, Paris, and Durham and received very helpful feedback, particularly from Keith Allen, Roberto Casati, Jérôme Dokic, Pierre Jacob, Kevin Landy, Clare Mac Cumhaill, Elisabeth Pacherie, Maarten Steenhagen, and Gerry Viera (as well as from a number of other questioners whose names I don't know). I am particularly grateful to Solveig Aasen, Michael Barkasi, Jonathan Cohen, Becko Copenhaver, Thomas Crowther, and Matt Fulkerson for extensive comments on earlier written versions. I owe special thanks to Ophelia Deroy, Patrick Haggard, and Frédérique de Vignemont for indispensable help with the structure of the argument.

References

Azañon, E., Longo, M.R., Soto-Faraco, S., and Haggard P. (2010). 'The posterior parietal cortex remaps touch into external space.' *Current Biology* 20/14 (July 27): 1304–1309.

Brozzoli, C., Ehrsson, H.H., and Farnè, A. (2014). 'Multisensory representation of the space near the hand: from perception to action and interindividual interactions.' *Neuroscientist* 20/2: 122–135.

Cataldo, A., Ferrè, E.R., di Pellegrino, G., and Haggard, P. (2016). 'Thermal referral: evidence for a thermoceptive uniformity illusion without touch.' *Scientific Reports* 6: 35286 (Oct 24). doi:10.1038/srep35286

Clark, Austen (2000). *A Theory of Sentience*. Oxford: Oxford University Press.

Cohen, Jonathan (2015). 'Perceptual Constancy.' In M. Matthen (ed.) *Oxford Handbook of the Philosophy of Perception*. Oxford: Oxford University Press: 621–639.

de Vignemont, Frédérique and Iannetti, G.D. (2015). 'How Many Peripersonal Spaces?' *Neuropsychologia 70*: 327–334.

de Vignemont, Frédérique and Massin, Olivier (2015). 'Touch.' In M. Matthen (ed.) *Oxford Handbook of the Philosophy of Perception*. Oxford: Oxford University Press: 294–313.

Fardo, Francesca, Beck, Briana, Cheng, Tony, and Haggard, Patrick (2018). 'A mechanism for spatial perception on human skin.' *Cognition 178*: 236–243.

Fulkerson, Matthew (2015). *The First Sense: A Philosophical Study of Human Touch*. Cambridge MA: MIT Press.

Geldard, Frank A. and Sherrick, Carl E. (1972). 'The cutaneous "rabbit:" a perceptual illusion.' *Science 178/4057* (October 13): 178–179.

Graziano, Michael (2018). *The Spaces Between Us: A Story of Neuroscience, Evolution, and Human Nature*. New York: Oxford University Press.

Haggard, Patrick and Giovagnoli, Giulia (2011). 'Spatial patterns in tactile perception: is there a tactile field?' *Acta Psychologica 137*: 65–75.

Haggard, Patrick, Cheng, Tony, Beck, Briana, and Fardo, Francesca (2017). 'Spatial Perception and the Sense of Touch.' In F. de Vignemont and A.J.T. Alsmith (eds) *The Subject's Matter: Self-Consciousness and the Body*. Cambridge MA: MIT Press: 97–114.

James, T.W., S. Kim, and J.S. Fisher (2007). The neural basis for haptic object processing. *Canadian Journal of Experimental Psychology 61/3*: 219–229.

Johnston, Mark (1992). 'How to speak of the colors.' *Philosophical Studies 68*: 221–263.

Mandrigin, Alisa (2019). 'The where of bodily awareness.' *Synthese*. doi:org/10.1007/s11229-019-02171-3

Martin, M.G.F. (1992). 'Sight and Touch.' In Tim Crane (ed.), *The Contents of Experience*. New York: Cambridge University Press: 196–215.

Mattens, Filip (2017). 'The sense of touch: from tactility to tactual probing.' *Australasian Journal of Philosophy 95/4*: 688–701.

Matthen, Mohan (2004). 'Features, objects, and places: reflections on Austen Clark's *Theory of Sentience*.' *Philosophical Psychology 17/4*: 497–518.

Matthen, Mohan (2019). 'Objects, seeing, and object-seeing.' *Synthese*. doi.org/10.1080/00048402.2019.1603246

Nicod, Jean (1930/1924). *Geometry in the Sensible World*. In his *Geometry and Induction*. London: Routledge and Kegan Paul. (Translated by R.F. Harrod from *La géométrie dans le monde sensible*, Paris: F. Alcan, 1924.)

O'Shaughnessy, Brian (1989). 'The sense of touch.' *Australasian Journal of Philosophy 67/1*: 37–58.

Reed, Catherine L. and Caselli, Richard J. (1994). 'The nature of tactile agnosia: a case study.' *Neuropsychologia 32*: 527–539.

Reed, Catherine L., Caselli, Richard J., and Farah, Martha J. (1996). 'Tactile agnosia: underlying impairment and implications for normal tactile object recognition.' *Brain 119*: 875–888.

Spence, Charles, Pavani, Francesco, and Driver, Jon (2004). 'Spatial constraints on visual–tactile cross-modal distractor congruency effects.' *Cognitive, Affective, & Behavioral Neuroscience 4/2*: 148–169.

12

Sameness of place and the senses

Alisa Mandrigin and Matthew Nudds

12.1 Introduction

The spatial ventriloquism effect occurs when there is spatial discrepancy between roughly synchronous visual and auditory stimuli.[1] The effect is produced when, say, a ventriloquist produces speech sounds without moving her lips, at the same time as manipulating the mouth of a dummy. Various aspects of the effect have been studied in the lab, with three main findings. Two involve localization responses: there is mislocalization of both visual and auditory stimuli in the direction of one another; and there is a similar mislocalization after-effect for visual or auditory stimuli presented independently of one another after a period of consistent spatial discrepancy between audio-visual stimulation (see Bertelson, 1999, for an overview).

Vision and audition ground awareness of their objects as having locations in space. Given the patterns of localization responses in studies of ventriloquism, we might think of the effect as merely a case in which where we hear a sound to come from, and where we see a material object to be, each shift as a result of a multisensory interaction between vision and audition. Visual information about the location of a material object impacts auditory processing of information about the location from which a sound has emanated. Auditory information about the location from which a sound has emanated impacts visual processing of information about the location of a material object.

But, the consequence of ventriloquism is not simply a shift in the apparent location of what we see and hear, as measured by our spatial responses. The change is such that what we hear appears to come from what we see.[2] The speech produced by the ventriloquist appears to come from the dummy's mouth. One way to understand that is spatially: what we hear appears to come from the same place as what we see, or, if we think we represent sound sources in audition, the source of what we hear appears to be at the same place as what we see.

Illusions such as the ventriloquism effect can serve to highlight features of everyday experience that would otherwise go unnoticed. It is not just in the bad (illusory) case that what we hear might seem to be at, or to come from the same place as what we see. In the ordinary case when I watch and listen to you speaking, I see the movement of your lips and I hear the sounds you produce. The sounds appear to come from the place I see your lips to be. Touch also grounds awareness of its objects as having locations in space. The things that we see, hear, and touch can appear to us to be in or coming from the same place.

What is it in general for two things to appear to be in the same place across modalities? A first thought might be that it is to experience places in such a way that we can draw the following inference, where F is an auditory feature, and G is a visual feature:

---

[1] Hereafter, for the sake of brevity, called 'the ventriloquism effect' or 'ventriloquism'.

[2] There may be still more to the illusion: it might be claimed that, additionally, ventriloquism produces an illusory cross-modal experience of a speech event, or an illusory cross-modal experience of causation.

Alisa Mandrigin and Matthew Nudds, *Sameness of place and the senses* In: *The World at Our Fingertips*. Edited by Frédérique de Vignemont, Andrea Serino, Hong Yu Wong, and Alessandro Farnè, Oxford University Press (2021). © Oxford University Press. DOI: 10.1093/oso/9780198851738.003.0012.

> That (visual) place is G.
> That (auditory) place is F.
> Therefore that (visual) place is F and G.[3]

That is, if we visually experience a place as having some visual feature, and auditorily experience the same place as having an auditory feature, we can immediately draw the conclusion that the place has both auditory and visual features. But how is this possible? To say that we can immediately draw the conclusion means that no collateral information, no further premise to the effect that the place we see is identical to the place we hear, is required to make the inference. We have two distinct applications of a demonstrative concept—two judgements to the effect that 'that place is x'—from which we can conclude that 'that place is F and G'. And the thought might be that I have two experiences that ground the application of the same demonstrative concept. Hence, I can draw the conclusion about *that place* on the basis of my visual and auditory experiences of it.[4] This is certainly one way that we can think of the connection between our experiences of places across the senses.

But consider again the ventriloquism effect. The third measure used in studies of the ventriloquism effect is that of 'perceptual fusion'. Subjects are asked to indicate when they have 'an impression of common origin' (Bertelson, 1999, p. 348). That is, they are asked when they have the impression that the visual and auditory stimuli are at the same place. In many studies, subjects report that the visual and auditory events *appear* to them to have occurred in the same location, despite the two stimuli being spatially disparate (Bertelson & Radeau, 1981, Choe et al., 1975; Jack & Thurlow, 1973; Thurlow & Jack, 1973; Witkin et al., 1952).

In ventriloquism, I *experience* the source of the sound as at the same place as the visually perceived object. Sameness of place is part of the content of experience. Experience presents the places as one and the same, such that no drawing of the conclusion is required. We don't need to make the inference because perceptual experience is such that the identity of place will be apparent. So, we need an account that does more than explain how it is that we can perceive two things or two features in audition and vision such that it is transparent to thought that the same place is being experienced. We need to explain how it is that their shared location is perceptually apparent to us: we need to explain how experience presents visible, audible, and tactual features as features of a single place.

12.2 A common reference frame

When we experience the same place in vision and touch, what is required for it to be apparent to us that it is the same place? One answer is that sameness of place will be apparent if we pick out or identify places in the same way irrespective of the sense modality of experience.

When we pick out places, we must do so relative to something. That is, we need a frame of reference within which, or relative to which, the place can be picked out. The frame of

---

[3] In the case of ventriloquism, this is, of course, not true: that visual place and that auditory place are not in fact the same place, so the inference is not valid.

[4] In such a case, it will always be possible for someone to fail to reflectively appreciate that the place picked out in one experience is identical to the place picked out in the other experience, perhaps due to inattention. Thank you to an anonymous reviewer for drawing our attention to this way in which we might fail to appreciate identity of place. Still, the suggestion is that they will nevertheless have reasonable grounds to draw the inference: if they considered the matter, it would not be rational for them to fail to draw the conclusion that the place they visually experience is identical to the place they auditorily experience.

reference provides a means for picking out a place by locating it relative to some other thing or things. If a place is picked out on two occasions in the same way relative to the same frame of reference, then it will be the same place. Applied to perceptual experience, the idea is that experiences that present places must do so relative to a frame of reference. So places are presented in experience, and have their identity, relative to that frame of reference. One way in which two features can be presented in experience as being at the same place is by being presented as being in the same relative place within the same frame of reference.

If our perceptual experiences in different sense modalities present places in a frame of reference that identifies places in the same way, then it might be tempting to assume that, when two experiences identify the same place, it will be apparent to the perceiver that it is the same place.

According to Evans' account of spatial content (1982, 1985), spatial locations are represented in perceptual experience in an egocentric frame of reference. To label a reference frame as 'egocentric' is to say that it individuates locations relative to the perceiver's body.[5] For Evans, perceptual experience has egocentric spatial content in virtue of its relations to behaviour. To perceive the cup's egocentric spatial location is to be disposed to engage in a repertoire of behaviours that are appropriate, given the spatial location of the cup relative to oneself. That is, to be represented as having a location is to be represented as having a property whose significance is cashed out in terms of relations to a certain behavioural repertoire.[6]

Now, according to Evans, because 'there is only one behavioural space' (1985, p. 390), there is therefore only one egocentric space. If both vision and audition represent things that are in the same place—just to my right, say—they each represent their respective objects as reachable in the same way—that is, as requiring the same movement in order to reach them. Since the behavioural repertoire required to reach an object will be the same irrespective of modality, and since for Evans egocentric content identifies the locations of things by means of the bodily behaviours that they afford, by being represented at the same egocentric location, the location of something that is seen and something that is heard will be represented in the same way.

If we understand egocentric spatial content in this way, then we will endorse the following. When I hear the ringing of the doorbell and see the steam rising from the kettle, my perceptual awareness of the spatial locations of each of the events will be cashed out for each event in terms of dispositions to perform bodily movements. When I hear the whistle of the kettle and see the steam rising from the very same kettle, my awareness of the spatial location of these events is to be explained in terms of my dispositions to perform the same set of bodily movements. I see that place (in front of me) is F, and hear that place (in front of me) is G. If Evans is right, then I am certainly in a position to judge that that place (in front of me) is F

---

[5] The term 'egocentric' has come to be used across a variety of disciplines to talk about spatial representations at different levels of explanation. For example, psychologists and cognitive neuroscientists suggest that spatial information is encoded in early sensory processing in a number of different egocentric reference frames, each taking a different part of the body as its point of origin. This is a claim about sub-personal representation. Philosophers have also claimed that conscious perceptual experience locates objects and events by means of an egocentric reference frame. But this is a personal level claim: the claim that conscious perceptual experience has egocentric content.

[6] Campbell (2005) suggests that a natural way to understand Evans' claim here is in terms of affordances (Gibson 1979). Egocentric spatial content is constituted by its implications for behaviour in that identifying the egocentric location of an object is identifying the current affordances of the object given its location relative to one's own body. Put simply, places are tied to responses.

and G. I am able to think 'that place in front of me is F', and 'that place in front of me is G', and I can conclude that 'that place in front of me is F and G'.

But it is tempting also to think that, if places are individuated in visual and auditory experience in the same way, it will be apparent that it is the same place that is F and G. Can we explain how sameness of place can be experientially apparent simply by appealing to the idea that experiences in different sensory modalities individuate places in the same way?

According to Peacocke (1986), when a location is represented in experience in the same way twice, then the identity of the location will be experientially apparent to the perceiver. This will be so whether both things are experienced within the same modality, or across modalities. In discussing perceptual experience of distances, he says that:

> we should require that if μ is the manner in which one distance is perceived and μ' is the manner in which a second distance is perceived by the same subject at the same time, and $\mu = \mu'$, then the distances are experienced as the same by that subject (Peacocke, 1986, p. 5).

Peacocke uses the term 'manner' to speak of the way in which an object or property is presented in perceptual experience. To give the content of a perceptual experience, he suggests, we must give an account of the manner or way in which the things that are perceived are perceived to be.

Applied to the experience of places, if x and y are at the same place and if x's location is represented in vision in the same manner or way (and at the same time) as y's location is represented in audition, then, according to Peacocke, x and y will be experienced as at the same location. Indeed, Peacocke stipulates that spatial locations must be represented in the same way across modalities on the grounds that it is this that allows us to be perceptually aware of the sameness of direction, size, and so forth, across sense modalities:

> [...], there is a second requirement on the manners. This is the requirement that they be *amodal*, in the sense that the same manner can enter the content of experiences in different sense modalities. You may hear a birdsong as coming from the same direction as that in which you see the top of a tree: we would omit part of how the experience represents the world as being were we to fail to mention this apparent identity. It also makes sense to say that something feels roughly the same size as it looks. (Peacocke, 1986, p. 6)
>
> Reproduced from Peacocke, C. (1986). Analogue Content. Proceedings of the Aristotelian Society Supplementary Volume, 60: 1-17.

We agree with Peacocke that to fail to mention that experience presents the identity of places across the sense modalities would be to leave out part of how experience presents the world as being. But, with Millikan (1991, 2000), we want to deny that sameness of reference frame will deliver a representation of sameness of place. Millikan (1991, 2000) points out a number of mistakes that can result from not adequately distinguishing between the personal and the sub-personal, and, in particular, attributing a property at one level because we have reason to attribute it at the other level. We earlier introduced the notion of an egocentric reference frame at the personal level, but talk of egocentric reference frames can also be pitched at the sub-personal level. To say that we have evidence of the existence of egocentric reference frames at the sub-personal level is to say something about the neural vehicles of representational content. To claim that egocentricity is a personal-level feature is to make a claim about the representational content of experience: it is a claim about what is represented. The mistakes that Millikan has in mind stem from confusing the content of experience—what is conveyed to the subject by her experience—with properties of the vehicles of representation, and vice versa.

One of the mistakes that Millikan draws attention to is that of confusing the sameness of the vehicles of representation with the representation of sameness (2000, p. 2). That is, we make the assumption that, because there are two instantiations of the same vehicle at the same time, the relation of sameness that holds between the vehicles of representational content will itself feature as part of the content of perceptual experience. In general, it is the mistake of assuming that all properties of the vehicles of representation will equally be part of the content of experience. Since vehicles are real, concrete things, then, if they are the same, they will bear the sameness relation to one another. The mistake arises if we then project, without argument, the sameness of the vehicles into the content of perceptual experience. That is, if we don't keep in mind that vehicles of representation are distinct from their representational content, then we might make the mistake of assuming, without offering any argument, that sameness will be part of the content of experience. To use Millikan's terminology, we're guilty of 'externalizing' the sameness of the vehicles—the sameness of the sub-personal neural vehicles of representation—into the content of experience. The claim, then, is that there is a difference between perceptually representing a location twice in the same way, and perceptually representing the identity of the location experienced in two sense modalities.

This can be backed up by considering how relations between places are represented. Suppose that your hand is hidden beneath a table, and we project a spot of light onto the surface of the table. Would it be possible for you to tell where the spot is in relation to your hand? In other words, would you be able to tell that the spot of light is to the left or right of your hand? One way you could tell would be to judge where the spot is in relation to some part of your body—your nose, perhaps—judge where your hand is in relation to that same body part, and then work out where the spot is in relation to your hand. For example, you might judge that the spot is $30°$ to the right of your nose and that your hand is $20°$ to the right of your nose, and work out that the spot is $10°$ to the right of your hand. In this case, we are supposing that both locations are given in the same way, but you still have to work out the non-egocentric relation between the two things: between the spot and your hand. Millikan's point becomes clear when we notice that it would make no difference to this example if the spot and your hand are in the same place. You would still need to make a judgement about the egocentric location of your hand and the spot of light, and then work out that they are in the same place. The working out might seem to be a trivial step, but it is still a step.

So, we want to argue, being represented in the same way is not in general sufficient for us to experience sameness: some further conditions must be met. If the location of x is represented in vision, and the location of y is represented in audition, and x and y have the same location, it does not follow merely from the fact that their location is represented in the same way that it will appear that x and y are in the same location. Sameness of representational content does not automatically produce an experience of sameness.

Even if we can show that we are aware of locations in such a way that it is transparent when two things have the same location—that is, they are represented in such a way that we are in a position to draw the inference that they have the same location—that still falls short of what we want to explain. In ventriloquism cases, what we hear appears to come from the same place as what we see. We don't have to reflect on our experience in order to judge that this is the case, just as we don't have to reflect on our experience when we see two things as in the same place. Sameness of location is part of the content of our experience. We need to explain how that is so. Appealing to the *way* locations are represented does not do that.

So, the conclusion we can draw from this is not that we do not experience sameness of location when locations are represented in the same way. Sameness of location is part of the content of our experience and we need to explain how that is so. What we should conclude, though, is that simply appealing to the *way* locations are represented does not help us to provide that explanation.

12.3 Cross-modal experiences

If all experience gave us was a place presented in the same manner, or reference frame, we would only be able to become aware that the source of a sound is in the same place as something that we see when we engage in an act of reflection. But there is an alternative and stronger sense in which I might be aware of sameness of place across modalities: I don't have two experiences—visual and auditory—of the same place, but rather a cross-modal experience of a place. Rather than making two distinct demonstrative judgements from which I can draw the conclusion that *that place is F and G*, I have a single cross-modal experience that grounds the judgement 'that place is F and G', where F and G are features perceived with different senses.

What does the idea that we have cross-modal experiences of the same place (that is, of a single cross-modal experience that grounds the judgement 'that place is F and G', where F and G are features perceived with different senses) imply about hearing and seeing something at the same place? There are two ways we might develop this idea.

The first is that we think of places as individual things, so hearing and seeing something as the same place is a matter of hearing and seeing the same individual place. It turns on the question of whether we can experience the same individual as such with two or more senses. Of course, we can experience the same thing in more than one sense: I can see a cube that I hold in my hand, for example. But the question is whether it is somehow perceptually apparent to me that it is the same thing that I both see and touch. If we treat places as individuals, then we might think that the issues that arise for this question will be the same as those that arise with respect to our experience of sameness of place. We'll set this question aside here (see Bayne 2014; Nudds 2014; O'Callaghan 2016 for discussion of cross-modal experiences of objects).

The second is that hearing something as coming from the same place as something that you can see involves being aware of the spatial relation that exists between them. For two things to be in the same place is for them to be spatially related to each other in a certain way (e.g. next to each other, on top of each other), or in a shared spatial relation of a certain kind to a third thing (e.g. on the table, in the box). So to perceive two things as in the same place might instead involve perceiving the spatial relation between those things. Can we experience the spatial relations between things perceived across the senses?

If two objects are located using an egocentric frame of reference, it is possible to compute where they are relative to one another. So, if object A is represented as being 20° to my right, and object B is represented as being 30° to my right, then I can work out that object B is 10° to the right of object A. If asked the colour of the pen to the left of the coffee cup, I can project my own left and right onto the cup and use that frame of reference to identify the relevant pen. For us to work this out, we engage in some reasoning. However, it might be that in many cases reasoning about the locations of objects relative to one another isn't necessary, because experience represents the spatial relations between things by representing both egocentric and allocentric spatial locations.

The idea here is that there is a contrast we can draw between our working out the spatial relation between two things we experience, and experience representing the spatial relation between those two things. This matches the contrast we have drawn above between working out sameness of place on the basis of two applications of the same demonstrative concept, and sameness of place being perceptually apparent to us. The suggestion is that there are processes or mechanisms that encode the spatial relations between the objects that we experience in a way that makes the information available in some distinctively experiential way.

There is empirical evidence that the visual system encodes not only the egocentric location of objects, but also their allocentric location (that is, the location that objects stand in to one another). Using a series of reaction time experiments Gordon D. Logan (1995) has shown that we are able to direct visual attention to an object by computing the spatial relation between one object and another. In Logan's experiments, subjects were presented with a visual array and a cue that directed their attention to an item in the array based on its spatial location relative to the cue.[7] Subjects were required to respond by indicating which item in the array had been cued. There was no way of performing the task successfully other than by using the cue to select the object that satisfied the relation between the cue and the target, or by guessing (Logan, 1995, p. 119). Logan found that reference frames 'can be rotated and translated across space according to the intentions of the observer and they can be aligned with the intrinsic axes of attended objects' (1995, p. 169). The speed and accuracy of subjects' responses indicate that subjects can direct their attention without engaging in reasoning. Instead, the computations required to specify which location is to be attended to are carried out by the visual system. This suggests that *within vision* allocentric spatial relations can be represented.

There are, then, two ways in which we can be aware of the spatial relations between two things that are given egocentrically in experience. It is possible to derive information about the spatial relations between objects from information about their egocentric location. We can derive this information non-perceptually by calculation or reasoning, but it is also possible for this information to be derived perceptually via the operation of sub-personal computational processes.

A perceptual system may or may not carry out the computational procedure. For example, it may be that, while both vision and audition represent egocentric information about location, only vision computes intrinsic (i.e. allocentric) locations, so only vision represents the spatial relations between the things it represents. Whether computations are performed by a system is an empirical matter.

We can draw the same contrast between working out the spatial relations between two things and experiencing the spatial relations between those things for two entities experienced in different sense modalities. On this approach, experiencing sameness of place across the senses will just be a version of experiencing the spatial relations between two things experienced in different sense modalities.

And just as we can ask whether the visual system or the auditory system computes allocentric locations, we can ask whether the spatial relations between objects experienced in different sense modalities are computed. That is, do the perceptual systems generate multisensory spatial representations? The notion of a multisensory spatial representation takes us beyond the idea of sameness of reference frame across the senses and of places being represented in the same way. It doesn't follow simply from the fact that locations are given egocentrically in vision and in audition that we are perceptually aware of the relations

[7] Logan's paradigm was based on experiments by Eriksen & Hoffman (1972).

between the entities we experience in the two senses. What is required is that the perceptual systems compute the allocentric locations of things experienced in the different sense modalities.

What would multisensory spatial representation get us that representing space in the same way in the different sense modalities does not? Our answer is that it would ground a certain kind of experience of unity across the senses that goes beyond judgements of sameness of place. Multisensory spatial representations would play a distinctive experiential role that isn't provided by any of the senses operating on their own. But are there representations of this kind?

12.4 Peripersonal space representations

While researchers working on the science of perception have historically tended to study each sensory system in isolation from the others, there is a large and growing body of evidence indicating that the sensory systems interact with one another. Various multisensory effects, including the spatial ventriloquism effect, indicate that, for example, processing in regions traditionally associated with vision can impact processing in regions traditionally associated with touch, and so forth. What is more, some areas of the brain involved in the processing of sensory stimulation seem to be multisensory: some neurons within these areas respond to stimulation in more than one sense. A number of regions have been identified as being multisensory in this way, including the superior colliculi (see, e.g. Meredith & Stein 1986), and parts of the parietal and frontal cortex involved in what have come to be known as 'peripersonal space representations'.

The term 'peripersonal space' (PPS) refers to the area immediately around the perceiver, usually extending not more than 30 cm from the body (Rizzolatti et al. 1981a, 1981b). Evidence of multiple multisensory representations of PPS comes from a number of sources: single cell recordings in monkeys; neuropsychological studies in humans with disorders of spatial attention resulting from lesions in the parietal and frontal cortex; and neuroimaging of neurologically healthy humans.

For example, Rizzolatti and colleagues (1981a, 1981b) recorded the activity of single neurons in ventral premotor areas of the macaque brain. They found that some neurons responded to both tactile stimuli on specific parts of the body and to visual stimulation. Importantly, these 'bimodal' neurons were responsive to visual stimuli presented within the PPS of the particular body part that elicited a tactile response, but not to visual stimuli presented outwith PPS. The tactile and visual receptive fields of these neurons are aligned with one another, and remain in register even when, for example, the limb of the monkey was passively displaced (Graziano et al. 1994). Bimodal neurons in these areas are not only responsive to visual and tactile stimuli: Graziano and colleagues (1999) found that some neurons in premotor area F4 respond to both tactile stimulation on a body part and auditory stimuli presented within PPS.

Evidence of the existence of multisensory representations of PPS is not limited to neurophysiological studies involving animals. For example, neurological patients with lesions to the frontal and parietal cortex often display contralesional extinction. That is, when two stimuli are presented at once, one in each hemifield, the stimulus presented on the contralesional side is not detected. This is so even though a single stimulus presented on either side can be detected. Contralesional extinction is found within a sense modality (i.e. when the two stimuli are both visual), but also across the senses. This is taken as evidence of multisensory PPS representation because cross-modal extinction is stronger when the visual stimulus is presented within PPS compared to when it is presented in far space (di Pellegrino et al. 1997; Làdavas et al. 1998a, 1998b).Di Pellegrino and Làdavas conclude that:

> Collectively, these studies reveal that the primate brain constructs multiple, rapidly modifiable representations of space, centered on different body parts (i.e., hand-centered, head-centered, and trunk-centered), which arise through extensive multisensory interactions within a set of interconnected areas in the parietal and frontal cortex. (di Pellegrino & Làdavas 2015, p. 127)

Should we think of these PPS representations as multisensory representations of space of the kind that would allow for sameness of place across the senses to be perceptually apparent to us?

It is worth noting that the term 'representation of space' will typically be used to mean different things in perception science and in the philosophy of perception. All the perception scientist might be committed to in positing a representation of space is the existence of a neuron or set of neurons that respond differentially to the spatial location of a stimulus. So, a neuron that responds differentially when a visual stimulus is within 10 cm of the perceiver's hand and when it is outwith 10 cm of the perceiver's hand will count as representing space (or spatial location). So, on the perception scientist's understanding of the term, PPS representations will certainly count as representations of space.

But what we are interested in is something more. We want a structure that is capable of delivering information about the spatial relations that a number of different objects stand in to one another. In particular, we want a structure that delivers information about the spatial relations that objects perceived through different sense modalities stand in to one another. What reasons might we have to extrapolate from the empirical work to the claim that PPS representations are representations of space?

Peripersonal space representations are typically described as being map-like, or as mapping the space around the body (see, for example, Grivaz et al. 2017, p. 603 and Graziano & Gross 1993, p. 107). A map represents a number of places, as well as the distance and direction of each of those places from each of the other places that is represented, and perhaps also such that for any two places represented, every place in between them is represented. A nucleus is typically described as being map-like if the neurons that make up the nucleus are arranged topographically, so that there is an isomorphism between the spatial relations between particular neurons and the spatial relations between the receptive fields of those neurons.

But having this kind of structural organization doesn't seem to be sufficient for a representation to count as a representation of space, unless there is some way for the perceptual system to be able to extract information about the spatial relations between places from the spatial relations between neurons in the structure in question. In thinking of representations of space as map-like, what matters is the *kind* of information that is represented—spatial relations between places being encoded, for example—rather than the structure of the representation itself. What makes something a spatial representation is that it has the *functional* properties that maps have.

Gareth Evans (1982, 1985) raises this issue in his discussion of egocentric spatial content. He questions how it is that sensory inputs, which convey spatial information only in so far as they stand in some kind of systematic relation to other possible sensory inputs, come to have spatial significance for the perceiving subject. According to Evans:

[...] an egocentric space can exist only for an animal in which a complex network of connections exists between perceptual input and behavioural output. A perceptual input—even if, in some loose sense, it encapsulates spatial information (because it belongs to a range of inputs which vary systematically with some spatial facts)—cannot have a spatial significance for an organism except in so far as it has a place in such a complex network of input-output connections. (1982, p. 154)

Reproduced from Evans, G. (1982). The Varieties of Reference.
Oxford: Clarendon Press.

Evans' proposal is that what connects a visual or auditory representation with physical space or environmental locations is, in part, dispositions to move in certain ways with respect to locations in physical space. In the absence of such connections, we lose our grip on the idea that visual or auditory experience has spatial content at all. So, one way in which we might try to establish that peripersonal representations should count as multisensory representations of space is by establishing that they bear the right kind of connections with motor behaviour.

On first glance, things might seem promising. Many of the visual-tactile neurons in PPS areas respond not only to visual or tactile stimulation, but also during motor action (Brozzoli et al. 2012, p. 451; de Vignemont 2018). And it has been suggested that the connections between PPS representations and the motor system indicate that one of the main functions of PPS representations is to facilitate interactions between the perceiver and objects in her environment (Rizzolatti et al. 1981b; Brozzoli et al. 2012; Brozzoli et al. 2014).

But while there is evidence that areas responsive to sensory stimulation in PPS are closely connected with the motor system and that peripersonal representations have a role to play in the guidance of motor action, it is not clear that they will also underpin conscious perceptual experience in the way that would be required for peripersonal representations to provide for an awareness of sameness of place across the senses.

First, some of the areas involved in PPS representation that are not part of the premotor cortex are part of the dorsal stream of visual processing (Brozzoli et al. 2012). Recent empirical work has been taken to support our making a functional distinction between vision for perception (subserved by ventral processing) and vision for action (subserved by processing in the dorsal stream). Evidence of the impact of lesions in either the dorsal or ventral streams in macaques (Ungerleider and Mishkin 1982), from neurological patients with visual form agnosia or blindsight on the one hand and optic ataxia on the other hand (Goodale & Milner 1992; Milner and Goodale 1995), and differences in the verbal and motor responses made by healthy human subjects to illusory stimulation have been taken to indicate that visual information is processed differentially depending on the task to which it is put. Taken together, these cases appear to offer us a double dissociation between these two functions, indicating that the processing on which each supervenes is localized in different parts of the brain. The dorsal visual stream is (now) typically conceived of as being 'for action'—devoted to the programming and control of motor acts—while the ventral stream in vision is thought of as being 'for perception', bringing to awareness information about objects required to identify and remember them (Goodale & Milner 1992).

The dual visual systems hypothesis is controversial. Yet, the claim that we have an unconscious, online system dedicated to the processing of visual information for the guidance of motor actions, and the fact that it is within this system that PPS representations are to be found, should alert us to the possibility that peripersonal representations do not provide for our perceptual awareness of the spatial relations between objects experienced in different sense modalities.

Second, the connections to motor action that have been uncovered look to give us, at most, perception of egocentric location only. What we seem to have is a representation of the spatial location of a visible object relative to the perceiver's body. But we have been arguing that representation of spatial location relative to the perceiver is not enough to give us a representation of sameness of place across the senses. We need something further.

What makes my Ordnance Survey map a useful tool for understanding the spatial relation between two geographical features is not that there exists a piece of paper with two marks on it, each of which I can use to guide motor actions towards the respective geographical feature that the mark represents. What allows me to grasp the spatial relation that the two geographical features stand in to one another is that I am able to see where the marks on the paper are in relation to one another. Similarly, what will make a topographically organized layer of neurons a representation of space of the kind we are interested in is the existence of a mechanism that extracts the spatial relation between cells that are active in that layer so that the spatial relation that the two relevant stimuli stand in to one another can itself be represented. In fact, the topographically organized layer of neurons in isolation would not count as a representation of space: it would be the layer of neurons *together with* the mechanism that gathers and compares relative spatial positions of active neurons that would be considered a representation of space of the sort we are after. Evidence of the existence of PPS representations, even if they have a topographic arrangement, does not give us a structure that can support our conscious awareness of the unity of space unless we have evidence of an associated mechanism that extracts relevant information about the spatial relations between neurons.

Finally, it is not clear that peripersonal representations exhibit the spatial specificity for them to underpin our conscious awareness of the spatial relations between entities experienced through different sense modalities. For example, the tactile receptive fields of bimodal neurons in premotor area F4 are considered to be large, encompassing in some cases whole body parts such as the hands, and many have overlapping receptive fields. For this reason, PPS neurons are thought to form only 'a crude map of the body surface' (di Pellegrino & Làdavas 2015, p. 127).

We therefore have a structure that may underpin our awareness of only the rough or approximate spatial relations that the objects we perceive stand in to one another. For example, investigation has revealed a distinction between PPS cells that respond to visual stimuli that are within 10 cm of the macaque's hand, and those that respond to visual stimuli that are between 10 cm and 30 cm of the macaque's hand. If we assume the existence of a mechanism that compares responses in these two types of cells, this would allow for presentation as of a visible object being anywhere within a region of space that is between 10 cm and 30 cm from an object near to the subject's hand.

In response to this, it might be pointed out that many PPS neurons exhibit superadditive effects. If the macaque is presented with both a visual and tactile stimulus within the receptive fields of a particular PPS neuron, then activity in the neuron will supersede the sum of the activity produced by the same neuron in response to a visual stimulus alone and a tactile stimulus alone. Could this at least underpin an awareness that a seen object and a felt object are in the same place? Superadditivity, the thought might be, could encode for sameness of place across vision and touch.

Again, the problem is that of specificity. Many PPS neurons respond to both tactile stimulation on a body part, and to visual stimulation within a region of space around the body part in question. So the same neuron will respond when a visible object is directly adjacent to an object that touches the relevant body part of the perceiver, and when a visible object is located 9 cm from an object that touches the body part. Depending on the fineness of grain with which we are interested, we might consider both of these two cases to be ones in which something that is seen is in the same place as something that is felt. However, it seems to be at odds with our perceptual awareness, which admits of finer grained distinctions. It is possible, for example, for typical perceivers to visually distinguish a case in which two pencils that are within arms' reach are lying in parallel next to one another from a case in which two pencils that are within arms' reach are lying in parallel but 9 cm apart.

PPS neurons therefore don't seem to have the specificity required to account for the determinacy of our perceptual awareness. What is more, many of the neurons investigated respond to somatosensory stimulation on either of two unrelated parts of the body. A single neuron might respond to either touch on the right hand or to touch on the right cheek. Rizzolatti and colleagues therefore conclude that:

> […] it is difficult to imagine that neurons that respond equally well to a stimulation of different, spatially segregated parts of the body have as their main function that of informing the organism where the stimulus is located. This is even more the case if they also respond to visual stimuli. (Rizzolatti et al. 1981b, p. 160)

While PPS representations may, in some sense, map the space immediately around a perceiver, we agree with Rizzolatti and colleagues that it doesn't look as though they do so in a way that will underpin the perceiver's perceptual awareness of the spatial relations that the objects she perceives in the different senses stand in to one another.

12.5 Conclusion

In ordinary, everyday situations, we don't take ourselves to be encountering distinct and discrete spaces when we perceive the world through more than one sense modality (Eilan 1993). I might hear a sound whose source I cannot see because it is too far away. Still, I don't take the place at which that sound source is located to be isolated from the visible objects I see in my immediate environment. We are able to answer questions about the spatial relations between the things that we see, hear, and touch. We can, for example, say that the place the high-pitched beeping is coming from is to the left of the flashing light.

From an ecological perspective, all that we need is the capacity to make use of sensory information in order to perform successful motor actions within a suitable time frame. Given that we have multiple sensory systems that process information about the location of objects and features in our environment, there is a problem to be solved: how is it that the spatial relations between things perceived through different modalities come to be resolved so that we can act successfully on the world using sensory information from more than one sense at a time? The problem might be solved in a number of ways. A computation that maps location L (as encoded in one sensory system) onto location L′ (as encoded in another sensory system) might be built into the architecture of the perceptual systems. Or there might be a common way of encoding spatial information across the sensory systems, so that no translation is required. Howsoever the problem is solved, it doesn't matter for our survival that we perceive the unity of space. Yet, in some cases, the objects we encounter in different senses appear to us to be located within a single space, or, even, in the same place. The speech I hear seems to come from the same place as the movement of the lips that I see. This is highlighted for us in the case of spatial ventriloquism, when, for example, the voices of the actors perceptually appear to come from the place that we see lips moving on the cinema screen. That is, it perceptually seems to us that the actors' voices come from the same place at which we see the lips move.

Empirical evidence suggests that there exist peripersonal representations that encode multisensory information about the region of space that immediately surrounds the body, that contribute to goal directed actions, and that play a role in mechanisms that protect the body. The existence of peripersonal representations generates a puzzle for accounts of perception—namely, what is the relation between the peripersonal representations that figure in the empirical discussions and our everyday perceptual experience of ourselves and the world?

Here we have considered whether PPS representations might underpin our conscious appreciation of the unity of the space we encounter in the different sense modalities. That is, we have examined whether PPS representations might play a role in our conscious awareness of cross-modal spatial relations, including sameness of location. Many peripersonal neurons are multisensory, their tactile receptive fields have rough somatotopic organization, and there is spatial alignment between their tactile receptive fields and their visual or auditory receptive fields. They are often said to represent the space immediately surrounding the perceiver. However, neurophysiological studies on macaques reveal that PPS representations have a number of properties that suggest they are ill-suited to underlie conscious awareness of the unity of space across the senses: the receptive fields for neurons are large and overlapping, a single neuron will often be responsive to somatosensory stimulation on different, spatially separated parts of the body, and the same neuron will respond to both somatosensory stimulation on a body part, and visual stimulation in the space around that body part. Moreover, these bimodal neurons are mostly found in the premotor cortex or in parts of the visual dorsal system, suggesting they play a role in controlling the way we interact with our environment, without impacting on conscious experience. It is still possible that PPS neurons might play a role in our consciousness awareness of the spatial relations between objects perceived through more than one sense modality. There might, for example, be a mechanism that extracts information about the spatial relations between receptive fields of PPS neurons in order to generate a representation of the spatial relations between stimuli. Until this has been investigated, though, we think that we should withhold from judgement about whether PPS representations underpin a cross-modal representation of space.

References

Bayne, T. (2014). The Multisensory Nature of Perceptual Consciousness. In D. Bennett, & C. Hill (eds) *Sensory Integration and the Unity of Consciousness,* Cambridge, MA.: Massachusetts Institute of Technology Press, pp. 15–36.

Bertelson, P. (1999). Ventriloquism: A Case of Crossmodal Perceptual Grouping. In G. Aschersleben, T. Bachmann, & J. Müsseler (eds.) *Cognitive Contributions to the Perception of Spatial and Temporal Events.* Amsterdam: Elsevier Science B.V., pp. 347–362.

Bertelson, P., & Radeau, M. (1981). Cross-modal bias and perceptual fusion with auditory-visual spatial discordance. *Perception & Psychophysics, 29*: 578–587.

Brozzoli, C., Ehrsson, H.H., & Farné, A. (2014). Multisensory representation of the space near the hand: From perception to action and interindividual interactions. *Neuroscientist,* 20(2): 122–135.

Brozzoli, C., Makin, T.R., Cardinali, L., Holmes, N.P., & Farnè, A. (2012). Peripersonal space: A multisensory interface for body-object interactions. In M.M. Murray & M.T. Wallace (Eds.) *The neural bases of multisensory processes.* Boca Raton, FL.: CRC Press.

Choe, C.S., Welch, R.B., Gilford, R.M., & Juola, J.F. (1975). The "ventriloquist effect": Visual dominance or response bias? *Perception & Psychophysics*, 18, 55–60.

Campbell, J. (2005). Information-processing, phenomenal consciousness and Molyneux's question. In J.L. Bermúdez (ed.) Thought, Reference, and Experience: Themes From the Philosophy of Gareth Evans. Oxford: Clarendon Press, pp. 195–219.

di Pellegrino, G. & Làdavas, E. (2015). Peripersonal space in the brain. *Neuropsychologia*, 66, 126–133.

di Pellegrino, G, Basso, G. Frassinetti, F. (1997). Spatial extinction on double asynchronous stimulation. *Neuropsychologica*, 35, 1215–1223.

Eilan, N. (1993). Molyneux's question and the idea of an external world. In N. Eilan, R. McCarthy and B. Brewer (eds.) Problems in the philosophy and psychology of spatial representation. Oxford: Blackwell, pp. 236–256.

Eriksen, C.W., & Hoffman, J.E. (1972). Temporal and spatial characteristics of selective encoding from visual displays. *Perception and Psychophysics*, 12: 201–204.

Evans, G. (1982). The Varieties of Reference. Oxford: Clarendon Press.

Evans, G. (1985). Molyneux's Question. In his Collected Papers. Oxford: Clarendon Press, p. 364–399.

Gibson, J.J. (1979). *The Ecological Approach to Visual Perception*. Boston, MA: Houghton Mifflin.

Goodale, M.A. and Milner, A.D. (1992). Separate Visual Pathways for Perception and Action. *Trends in Neurosciences*, 15(1), pp. 20–25.

Graziano, M.S.A., Yap, G.S., & Gross, C.G. (1994). Coding of visual space by premotor neurons. *Science*, 266, 1054–1057.

Graziano, M.S., Reiss, L.A., & Gross, C.G. (1999). A neuronal representation of the location of nearby sounds. *Nature*, 397, 428–430.

Graziano, M.S. and Gross, C.G. (1993). A bimodal map of space: somatosensory receptive fields in the macaque putamen with corresponding visual receptive fields. *Experimental Brain Research*, 97, 96–109.

Grivaz, P., Blanke, O. & Serino, A. (2017). Common and distinct brain regions processing multisensory bodily signals for peripersonal space and body ownership. *NeuroImage*, 147: 602–618.

Jack, C.E. & Thurlow, W.R. (1973). Effects of degree of association and angle of displacement on the "ventriloquism" effect. *Perceptual and Motor Skills*, 37, 967–979.

Làdavas, E., di Pellegrino, G., Farnè, A., Zeloni, G. (1998a). Neuropsychological evidence of an integrated visuotactile representation of peripersonal space in humans. *Journal of Cognitive Neuroscience*, 10, 581–589.

Làdavas, E., Zeloni, G., Farnè, A. (1998b). Visual peripersonal space centred on the face in humans. *Brain*, 121, 2317–2326.

Logan, G.D. (1995). Linguistic and Conceptual Control of Visual Spatial Attention. *Cognitive Psychology*, 28: 103–174.

Meredith, M.A. & Stein, B.E. (1986). Visual, auditory, and somatosensory convergence on cells in superior colliculus results in multisensory integration. *Journal of Neurophysiology*, 56(3), 640–662.

Millikan, R. (2000). *On clear and confused ideas*. Cambridge: Cambridge University Press.

Millikan, R.G. (1991). Perceptual content and Fregean myth. *Mind*, 100(4), 439–459.

Milner, A.D. & Goodale, M.A. (1995). *The Visual Brain in Action*. Oxford: Oxford University Press

Nudds, M. (2014). 'Is Audio-Visual Perception Cross-Modal or Amodal?', in D. Stokes, M. Matthen and S. Biggs (eds) *Perception and Its Modalities*. Oxford: Oxford University Press, pp. 166–188.

O'Callaghan, C. (2016). Objects for multisensory perception. *Philosophical Studies*, 173: 1269–1289.

Peacocke, C. (1986). Analogue Content. *Proceedings of the Aristotelian Society Supplementary Volume*, 60: 1–17.

Rizzolatti, G., Scandolara, C., Matelli, M. & Gentilucci, M. (1981a). Afferent properties of periarcuate neurons in macaque monkeys. I. Somatosensory responses. *Behavioural Brain Research*, 2, 125–146.

Rizzolatti, G., Scandolara, C., Matelli, M. & Gentilucci, M. (1981b). Afferent properties of periarcuate neurons in macaque monkeys. II. Visual responses. *Behavioural Brain Research*, 2, 147–163.

Thurlow, W. R. & Jack, C. E. (1973). A study of certain determinants of the "ventriloquism effect". *Perceptual and Motor Skills*, 36: 1171–1184.

Ungerleider, L.G. and Mishkin, M. (1982). Two Cortical Visual Systems, in D.J. Ingle, M.A. Goodale, and R.J.W. Mansfield (eds) *Analysis of Visual Behavior*, Cambridge, MA: Massachusetts Institute of Technology Press.

de Vignemont, F. (2018). Peripersonal perception in action. *Synthese*. doi.org/10.1007/s11229-018-01962-4

Witkin, H. A., Wapner, S., & Leventhal, T. (1952). Sound localization with conflicting visual and auditory cues. *Journal of Experimental Psychology*, 43: 58–67.

13

The structure of egocentric space

Adrian Alsmith

13.1 Introduction

In *The Varieties of Reference*, Gareth Evans rightfully notes that any talk of 'egocentric space' picks out not a special *kind* of space, but rather a special *way* of representing space (Evans, 1982, p. 157). Shortly afterwards, he claims that 'a fundamental point of similarity' between auditory and haptic (and presumably visual) experience is that their spatial content can be specified in egocentric terms.[1] And that it is 'a consequence of this that perceptions from both systems will be used to build up a unitary picture of the world. There is only one ego-centric space, because there is only one behavioural space' (Evans, 1982, p. 160). Although I agree with Evans' conclusion, I share the suspicions of any reader who thinks that the rea-soning here might have moved a little quickly.

For one thing, they might not accept what one might call the 'Evansian' conception of egocentric space. On that conception, perceptual experience is egocentrically structured in virtue of its connection to the subject's capacity to act appropriately in relation to what she perceives.[2] Thus, Dominic Gregory writes: 'On the face of it [...] it is bizarre to think that auditory and visual directions derive from apparent action-involving relationships between sights, sounds, and our conscious selves' (Gregory, 2013, p. 38). For another, they might not see exactly the relevance of this idea to the matter of building up 'a unitary picture of the world' from the various egocentrically structured presentations of that world available to the perceiver.

The purpose of this chapter is to offer an indirect defence of the Evansian conception of egocentric space. It will do so by showing how it resolves a puzzle concerning the notion of 'a unitary picture of the world', as 'one egocentric space'. It shows how one can experi-ence the world from multiple egocentric perspectives, unified relative to a single egocentric perspective.

I will begin by outlining, in section 13.2, several natural assumptions about egocentric perspectival structures. I will then show, in section 13.3, that a subject's experience, both within and across her sensory modalities, may involve multiple structures of this kind. This

---

[1] Evans' language actually suggests the stronger view that the spatial content of perceptual experience *must* be specified in egocentric terms. It can surely also be specified, for example, in terms of relations between objects themselves, or in terms of their relation to a more fundamental frame not essentially involving some privileged entity. But whether it can be specified in these ways, without presupposing some egocentric specification, is an issue that we cannot address here.

[2] See Evans (1982, pp. 154–162; 1985, pp. 383–389). This broad idea is, of course, not original to Evans. He evi-dently takes inspiration from Pitcher (1971), as well as Taylor's (1978) discussion of Merleau-Ponty (1962/2002), who in turn was heavily influenced by Husserl (see especially his 1952/1989, 1973/1997).

Adrian Alsmith, *The structure of egocentric space* In: *The World at Our Fingertips*. Edited by Frédérique de Vignemont, Andrea Serino, Hong Yu Wong, and Alessandro Farnè, Oxford University Press (2021). © Oxford University Press.
DOI: 10.1093/oso/9780198851738.003.0013.

raises the question of how perspectival unity is achieved, such that these perspectival structures form a complex whole, rather than a merely disunified set of individually, distinctively structured experiences.

In section 13.4, I consider a variety of accounts: switch accounts, according to which perspectival structures are themselves individually unified, but do not form a complex whole; sensory accounts, according to which egocentric perspectival structures are unified relative to a dominant sensory modality's perspective; transformation accounts, according to which egocentric perspectival structures are unified in virtue of coordinate transformations between spatial representations encoding information in distinct egocentric frames of reference; and ultimate accounts, according to which egocentric perspectival structures are unified in virtue of mutual anchoring relations to an ultimate anchor.

In section 13.5, I will return to the Evansian conception to show that it provides us with a further kind of account—an agentive account—according to which egocentrically structured experiences present the world in relation to a part of a single thing, the body as a dynamic unity.

13.2 Perspectival structures

13.2.1 Egocentric structures

A perspective is an essential feature of any experience that exhibits a perspectival structure, for a perspectival structure is an organization of content determined by a perspective. *Egocentric* perspectival structures are such because their elements are systematically organized within an egocentric frame of reference.

Minimally, an egocentric frame of reference is one in which locations on an axis are individuated relative to a privileged point on that axis. Typically, egocentric frames of reference have multiple orthogonal axes. Points on each axis can be denoted in a coordinate system. For example, a Cartesian coordinate system assigns the number zero to the point at which the axes intersect. A metric would provide intervals of a certain scale, such that changes in value would correspond systematically to changes in position on each axis. The extent to which a value on each axis deviates from the zero-point would thus serve to determine locations in the frame of reference.

There are also a range of familiar egocentric locative expressions. In English, we have static terms such as *nearby, far away, straight ahead, to the left*, or *to the right*; we also have dynamic terms such as *coming closer* or *moving leftward*. These can serve to describe spatial relations in an egocentric frame of reference, where the axes of the reference frame are labelled with canonical terms for egocentric directions such as *leftward*, *upward*, or *forward*.

13.2.2 Limitation structures

Issues concerning the unity of perspective are centrally issues concerning egocentric structure. But another, important kind of perspectival structure that characterizes perceptual experience concerns the ways in which the appearance of objects is determined by the characteristic limitations of one's senses (Martin, 1992; Richardson, 2010). This kind of perspectival structure—call it *limitation structure*—has been the subject matter of much disagreement since at least the seventeenth century, particularly as it figures in the Argument from Perspectival Variation.[3] A key premise of that argument is that, strictly speaking, three-dimensional objects do not appear as such visually. Rather, their shapes and sizes are distorted, as if lines were projected from the objects themselves through an image plane to a single point, as in some form of graphical perspective. Thus, what distinguishes limitation structures from egocentric structures is that the former (but not the latter) involve a particular determination of the apparent sizes and shapes of objects relative to a given perspective, although the extent and nature of this determination has long been controversial.

Thankfully, what is of most relevance to our discussion here is the less controversial (if closely associated) idea of a sensory field. This is roughly the idea, the totality of what can be sensed by a subject in a given modality at a time and can be described by picking out a region of space.[4] For instance, if the immediate objects of sight were indeed planar, then the visual field would have *only* horizontal and vertical dimensions.[5] But what is key to our discussion here is not how many dimensions sensory fields possess, but the ways in which they are characteristically bounded on those dimensions. Indeed, it is a familiar idea that the boundaries of the visual field are a consequence of eye position, head morphology, the position and structure of the eyes, etc. In this way, the eyes constitute a perspective determining a limitation structure of a kind, the *human* visual field, one that is rather different from that of, for example, a hammerhead shark. For the same reason, the human visual field is different from the human auditory field in that the former is bounded on the vertical and horizontal planes in ways that the latter is not.[6]

13.2.3 More on egocentric structures

The typical purpose of describing locations in egocentric terms is to capture how they stand not merely in relation to the privileged point that structures the frame of reference. Rather, the purpose is to capture how they stand in relation to a particular object on which the frame of reference is *centred*. We will need to say more about this shortly, but it will suffice for the present to say that a frame of reference is *centred* upon an object when its zero-point falls within an object's boundaries and its axes are aligned with salient structures within the object. For instance, the parts of many creatures (including the human animal) can be regularly divided along horizontal, vertical, and sagittal planes. Running labelled axes along these planes will enable a systematic description of locations in egocentric terms relative to the object in question.

I will assume that being egocentrically structured is minimally sufficient for any content to be perspectival. I will also assume that experience has content, and thus that, if its content

---

[3] See, e.g. arguments for sense data presented by Russell (1912/1959, pp. 10–11) and Broad (1927, p. 240), surely inspired by remarks from Locke (e.g. 1690/1997, pp. 143–144) Berkeley (e.g. 1732/2008, p. 21) and Hume (e.g. 1739–1740/2007, p. 56),

[4] Here we can embrace a point made by Soteriou (2013, pp. 115–119) that, from the fact that one's awareness is constrained to such a region, it does not follow that one is aware of the region itself as some object over and above its contents.

[5] For discussion of the difficulties of excluding visible depth from one's characterization of visual experience, see Schwitzgebel (2006); Smith (2000). J.J. Gibson's work on occlusion is perhaps the most compelling demonstration of the need for a three-dimensional conception of the visual field (Gibson et al.,1969). For more general discussions of a three-dimensional conception of the visual field, see Clark (1996); Martin (1992).

[6] A further disanalogy is in the possibility that there may be a coherent notion of a temporal auditory field analogous to the spatial field of vision (Soteriou, 2013). I note also that it is not clear whether there is a notion of a tactile field with anything like the kinds of phenomenological connotation with which philosophers have imbued that notion in the visual case, but see Haggard and Giovagnoli (2011); Serino et al. (2008).

is structured in one of the ways described earlier, then that experience's content is perspectival. The distinction between experience and content is important but, for ease of expression, I will collapse the distinction and speak merely of experiences being perspectival or perspectivally structured.

One can express the structural difference that egocentric perspective makes to perceptual experience in the following terms: a subject's perceptual experience is perspectivally structured in so far as the world she experiences is presented *from* an egocentric perspective. It should be noted, though, that, as described, an egocentric perspectival structure does not involve any connection between a perspective and the body of the subject (cf. Gregory, 2013, p. 38).

What is right about this, I think, is that being perspectivally structured is a property not only possessed by perceptual experiences. It is a property possessed by images (both mental and physical) and by various ways of representing space (both mentally and physically). Indeed, in a broader sense of the notion of a perspective, it characterizes *any* representation that is structured in relation to some privileged entity.

But consideration of these broader connotations ought not to distort our characterization of perceptual experience as perspectival. For when considering perceptual experience, in the central case, what it is for the world to be presented to a subject from a perspective is for it to be presented in relation to the subject's body (see, e.g. Husserl, 1952/1989, p. 166).[7] Indeed, as will become evident shortly, the problem of the unity of egocentric perspective emerges when considering the specific constraints characterizing the relations between bodies and perspectival structures.

13.2.4 Anchoring and embedding

The notion of *centring* provides only a correspondence between a body and a perspective. This may just as well be coincidental. It fails to capture how the comportment of the subject's body can affect the perspectival structure of her experience. For this we need to introduce the further notions of *anchoring* and *embedding*. To illustrate *anchoring*, take Christopher Peacocke's Buckingham Palace example:

> Looking straight ahead at Buckingham Palace is one experience. It is another to look at the palace with one's face still toward it but with one's body turned toward a point on the right. In this second case the palace is experienced as being off to one side from the direction of straight ahead, even if the view remains exactly the same as in the first case. (Peacocke, 1992, p. 62)

In this imagined case, the left–right axis is systematically causally related to one's torso, such that changing its spatial properties (e.g. its orientation) affects one's perspectival experience. In short, that part of the structure is *anchored* to the torso.

The conceptual distinction between *centring* and *anchoring* can be grasped by considering a more fanciful case imagined by P.F. Strawson, in which a multi-bodied subject's experience may be determined by the *location* of one body—call this body *Loco*—and the *orientation* of

---

[7] I should also note that I lack the space to even briefly address broader questions concerning whether perceptual experience being perspectivally structured is in itself sufficient for it to have *de se* content (Schwenkler, 2014). For discussion of these issues, see Alsmith (2017).

a second body—call this *Oriento*. Thus, while it would be possible for this multi-bodied subject to see whatever would be visible from *Loco's* location, she cannot see in every direction at once; it is the orientation of *Oriento* that determines her line of sight, and thus the view she experiences at any given time (Strawson, 1959/2003, p. 90). In short, her perspective is centred on *Loco* while it is anchored to *Oriento*.

Clearly this departs drastically from our conception of the perspectival structure of ordinary perceptual experience. For that is typically expressed as a combination of *centring* and *anchoring*, such that one perceives the world *from* the very body that determines the structure of one's experience. To capture what is missing, we need a notion that refers to cases in which a perspective is located within an object that determines the structure of an experience at that location. Call this *embedding*: a perspective is *embedded* within an object when it is located within an object by virtue of its structure (or some part of its structure) being *anchored* to that object. *Embedding* is thus a composite notion that requires both *centring* and *anchoring*.

13.3 The problem

13.3.1 The complex structure of visual perspective

With these basic ideas on the table, we can begin to express the problem that will occupy us in the pages to follow. In the Buckingham Palace example, describing the perspective in question as *anchored* to the torso captures the fact that the structure of the (imagined) experience changes as a consequence of the rotation of the torso. But it leaves open the question of whether the perspective is *embedded* within the torso. In simpler terms, would one perceive the palace *from* the *torso*?

One core element to the problem here is that the body is not a structureless lump. It is an articulated object with parts that are each to some extent independently mobile, and thus capable of varying degrees of alignment, complicating the question of their contribution to the egocentric structure of our experience. Thus, what might seem odd about the idea of *seeing from the torso* is that there is another candidate body-part from which one visually perceives—namely, one's *head*—for the head causally affects the structure of one's experience in broadly the same respects. To paraphrase Peacocke: it is one experience to look at an object in front of one's face; it is rather another to look at that object when it is off to the side of one's face.

This brings out another core element of the problem—namely, the intuition that we possess a *single* perceptual perspective upon the world, such that our perspectivally structured experience is unified in relation to that perspective. This intuition seems especially problematic when also considering the perspectival structure of sensory modalities other than vision, as will shortly become clear. But the problem has some bite when considering vision alone. Indeed, one can imagine a subject with one's head and torso misaligned, with a visible object placed at 15° relative to her torso and −15° relative to her head. Her visual experience of the object might equally well be characterized as being 'to the right' and 'to the left'.[8] Yet, on

---

[8] Indeed, tests of intuition concerning this kind of scenario in a perspective-taking task have repeatedly demonstrated that individuals' spatial judgements are influenced by both the avatar's torso position and head position, with clear individual differences in the weightings assigned to each body-part (Alsmith, Ferrè, & Longo, 2017; Longo, Alsmith, & Ferrè, 2020).

the assumption that her experience of the object is unified in relation to a single perspective, it cannot seem to her to be in both directions at once.

In more general terms, given the complexity of the egocentric structure of perceptual experience, in virtue of what is that structure unified? This is the problem of the unity of perspectival experience, and it will form the focus of the remaining discussion.

13.3.2 The complex structure of auditory and haptic perspective

Discussions of perspective in both psychology and philosophy have generally focused on the specific case of visual perspective. This reflects a broader trend of disproportionate focus on vision in the study of perception, only partially justified by the dominance of the visual sense upon others (Stokes & Biggs, 2014). But we should not be held captive by the visual when thinking about perspectival phenomena in general. And although it is possible to express the problem of the unity of perspectival experience when considering sight alone, it is more acute when one considers the non-visual senses. For both auditory and tactual experience present their objects in an egocentrically structured manner.[9] Moreover, each provides clear examples of intramodal complexity in their egocentric structure.

It is in haptic perception that we can most easily discern tactual experience as egocentrically structured.[10] Take a simple case, such as feeling an object in one hand that extends beyond one's grip. The feel of the object would be structured relative to a perspective embedded within the hand. Parts of the object would be perceived from the hand, in the sense that the perspective of the egocentric frame of reference would be centred within the hand (e.g. within the palm, according to one's grip). And at least part of the overall egocentric structure would be causally dependent upon salient structures within the hand. These need not be labelled in canonical egocentric terms (e.g. it might suffice to label a direction as *thumbwards*). As one manipulates the object, pushing part of it *thumbwards*, bringing another part of it towards the centre of one's grip, one's overall experience of the object changes, in a manner organized according to its egocentric structure.

Haptic perception can involve a variety of egocentric structures (Oldfield & Phillips, 1983) depending on the purpose of a given task (Millar, 2008, chapters 2 and 5), and even how long it lasts (Zuidhoek et al., 2003). This might seem in tension with the fact that philosophers sometimes describe the perspectival structure of tactual experience as if it were simple (e.g. Evans, 1982; Gregory, 2013; Nichols & Horgan, 2015). But this is largely, I suspect, to facilitate ease of expression: it is useful sometimes to describe the structure of a part of a subject's tactual experience in isolation. This should not mask the complexity of a subject's typical tactual experience when taken as a whole. For instance, pushing a piece of furniture into the corner of a room might involve perspectives embedded within one's hands and feet, and perhaps anchored to one's hips and torso.

---

[9] Not all would be inclined to agree with such claims in full generality. For instance, Strawson writes that 'such expressions as "to the left of", "spatially above", "nearer", "farther" have no intrinsically auditory significance' (Strawson, 1959/2003, p. 65). But, as O'Callaghan (2010) points out, this does not sit easily with empirical research on spatial hearing. And even if we admit that Strawson's sound world would be a 'no-space world', it should not push us to the conclusion that, in our world, auditory experience does not have egocentric structure (see, e.g. Mershon & King, 1975). With respect to touch, some might reject the description of this form of perceptual experience as perspectivally structured, because they believe that '[nothing] in the tactual sphere corresponds to the fact of optical perspective' (Blumenfeld, 1937, p. 56). But here we should be careful not to deny that touch is egocentrically structured because it does not have the kind of *limitation structure* exhibited by another modality.

[10] See, though, Parsons and Shimojo (1987) for work on egocentric structures in passive touch.

As introspectively obvious as the complexity of haptic perceptual experience might seem, so it might seem equally obvious that the structure of auditory experience is simple. However, work on spatial hearing suggests otherwise, in so far as a subject may hear sounds in relation to distinct auditory perspectives as a function of the spatial position of each sound's source.

Neelon, Brungart, & Simpson (2004) studied the egocentric structure of hearing by adapting methods originally designed to study the location of the estimated 'egocentre' in binocular vision. These methods, first developed by W.C. Wells and E. Hering in the eighteenth and nineteenth centuries, respectively, essentially involve adjusting a line (along a rotating rod, or an imagined axis between two or more points) until it is judged by the participant to be pointing directly at herself. By conducting the task at a series of radial directions and extending the line in each judgement to pass through the observer, an examiner could determine the participant's estimated egocentre as the point of their intersection (Howard & Templeton, 1966; for a historical and methodological review, see Ono, 1981). Using an auditory form of this task, Neelon et al. (2004) found a systematic direction-dependent shift for the estimated position of the auditory egocentre. For sound sources ±30° from midline, the pattern of responses clearly indicated an egocentre located towards the front of the head. Responses to more lateral sources (beyond a region roughly ±60° about the midline) indicated an egocentre located further back, roughly at the centre of the interaural axis—that is, just between the ears.

13.3.3 Split perspectives

It will be useful to have a device for expressing the problem in its various guises simultaneously. For that purpose, I offer the case of *Split*, a subject whose total perceptual experience is of a virtual world. This virtual world is a conjunction of perfect visual, auditory, and tactual worlds, each of which is a facsimile of a world genuinely perceptible through these sensory modalities. *Split's* perceptual experience is structured by perspectives embedded within and anchored to three humanoid virtual bodies. There is no topological connection between the virtual parts of each of *Split's* bodies. And each is independently mobile. *Audio*, from which she hears the audible properties of an auditory world, is thus a distinct body from *Visuo*, from which she sees the visible properties of a visual world, and *Tactuo* is a third body, from which she feels the tangible properties of a tactual world.

Assume further that *Split's* auditory experience is entirely structured by the shape and position of *Audio's* head, within which *Split's* interaural and interocular perspectives are embedded. *Split's* visual experience is causally connected to a different body—*Visuo*—and has a similarly complex relation to that body. For it is structured both by the shape and position of *Visuo's* head, within which *Split's* visual perspective is embedded, but also *Visuo's* torso, to which the left-right axis of her visual experience is anchored, as in Peacocke's Buckingham Palace example above. *Split's* tactual experience presents potentially the most complexity, with potentially as many anchoring and embedding relations as there are parts of *Tactuo's* body engaged in haptically perceiving the tangible properties of the world. But as noted above, we can simplify the situation somewhat by considering only the part of *Split's* tactual experience structured according to a perspective embedded within the hand.

The purpose of the example here is, of course, not to show that this strange subject's experience is anything like ours. Rather, the purpose is to provide a case in which parts of our experience, which are usually intimately connected, are separate, so that we can better examine how they come to be unified. *Split* is thus precisely the model of a subject whose experience is not unified according to a single, multimodal perspective. Note that this disunity is not a consequence of *Split's* bodies occupying different positions. Indeed, we could assume a mapping between the visual, audible, and tactual worlds. And as her bodies are virtual, *Split* is able to situate them such that they simultaneously fill exactly the same volume of virtual space. For ease of reference, let's call this version of *Split* '*Split-juxtaposed*'. Even though, broadly speaking, *Split-juxtaposed* visually, auditorily, and tactually experiences from a single place, she does so in very different ways from three different bodies.

What exactly, then, is it about *Split's* multiple embodiment in virtue of which her experience is disunified? Put differently, what kind of relation would be required between her three sensory bodies, such that her experience of the world would be unified according to a single egocentric perspective?

Before moving on to consider a variety of responses to the problem expressed here, it is worth saying a little to clarify its relation to issues concerning the unity of consciousness between the senses. Indeed, raising the bare question of how 'a unitary picture of the world' in the form of 'one egocentric space' is built up from egocentrically structured experience in different modalities—as per Evans' remarks at the start—might suggest that the problem is really one of how modality-specific perspectival structures are united.

If this were right, then what *Split's* experience would lack is what Ayers calls an 'integrated sensory field', in which there are not 'several sets of apparent directions' corresponding to each sensory modality and associated or identified with one another, but rather a single multimodal field 'of which we are aware in different but essentially integrated ways' (Ayers, 1993, p. 164). But, as I hope to have shown in sections 13.3.1–13.3.2, the problem plausibly runs deeper: for each sensory modality might itself present multiple 'apparent directions', raising again the question of how the complex structure of each modality is unified relative to a single perspective.[11]

13.4 Accounts of perspectival unity

13.4.1 Switching accounts

Despite the problem at hand being distinct from questions concerning unity relations between the senses, one might take inspiration from a certain approach to the latter in giving an account of the former. It is commonly assumed that a subject's perceptual experience at a time can be multisensory, in that it may consist in multiple modality-specific experiences (or modality-specific contents). But while it is true that perceptual processing is highly multisensory, as Spence & Bayne (2014) note, it is difficult to shake the sceptical view that 'all such cross-modal interactions take place "below" (or perhaps "prior to") the level of awareness' (p. 107). Moreover, it is possible that *switches* between modality experiences might occur at a temporal scale beneath our capacity for introspective discrimination—all of which makes it hard to find dispositive evidence that perceptual experience is multisensory in the relevant respect, and to rule out the alternative unisensory view that 'a subject's awareness of

---

[11] This could be more precisely expressed by creating further separations within *Split's* embodiment, such that each of her bodies was qualified not only with reference to a sensory modality but also to a certain egocentric organization of the contents of her experience from that body in that modality. But this would, I think, increase the difficulty in imagining the case in ways that would not really justify the precision we might gain.

the world involves frequent and rapid alternations (or switches) between different modalities (ibid., p. 102).

The unisensory view suggests a similar approach to the Gordian knot of perspectival unity. For it is clear that much of the force of the problem is in the assumption that the egocentric structure of perceptual experience is complex. But, if this is taken to imply that this complexity is a feature of the egocentric structure of one's perceptual experience at a time (or, better, within a specious present), the empirical justification for this assumption might seem rather weak. It hardly rules out the alternative possibility that the structure of one's perceptual experience is never complex in this way. Building upon this, one might pursue an account according to which our experience at any given time is only ever simple. Any appearance of complexity—and thus any basis for any substantive problem of perspectival unity—would then be explained as a consequence of rapid switching between perspectives.

There is, of course, a clear respect in which this kind of account is consistent with the idea that there is just 'one egocentric space'—or, more elaborately, the idea that a subject's experience is synchronically unified relative to a single perspective. For there is just a single perspective from which the subject experiences the world at a time.

I do not at this point have any knock-down argument against a switching account. But given the similarity between this account and the dispute between the unisensory and multisensory views of perceptual experience, the grounds on which the account should be evaluated ought to be correspondingly similar. Thus, the extent to which one will find a switching account compelling depends on whether there are convincing cases of experiences that could not occur when considering a single perspectival structure alone.[12] In our discussion thus far, the key cases would then be the visual experience of an object placed equidistant from the midlines of a misaligned head and torso (see the elaboration on Peacocke's Buckingham Palace example in section 13.3.1) and the auditory experience of an object placed around ±60° from the midline, where head and torso are aligned (see the discussion of Neelon et al., 2004, in section 13.3.2).

13.4.2 Sensory accounts

One explanation of Neelon et al.'s (2004) results is especially worth examining because it suggests the general shape of what one might call a sensory account of the unity of perspectival experience. Using similar methods to Neelon et al. (2004), Sukemiya, Nakamizo, & Ono (2008) tested sighted, late-blind and congenitally blind subjects. They found that the estimated auditory egocentres for sighted and late-blind subjects were near the midpoint of the interocular axis, similar to Neelon et al.'s finding for frontally located auditory stimuli, whereas the egocentre for the congenitally blind was near the midpoint of the interaural axis, similar to Neelon et al.'s laterally located stimuli. Such a 'concordance of the locations of the visual and auditory egocentres', as Sukemiya et al. (2008) write, would thus eliminate any potential 'cross-modal mismatch between the visual and auditory directions of an audiovisual object' (p. 1549).

There are two sets of facts here that might suggest an approach to the problem at hand. An auditory egocentre is located at the site of a visual egocentre, and this only obtains auditory stimuli that fall within the boundaries of the visual field in sighted subjects, or where those boundaries are for late-blind subjects. These are just what one would expect if the egocentric structure of sight were dominant, such that it served to unify the structure of one's perceptual experience in other modalities. That is, to the extent that objects and events perceived in non-visual modalities fall within the scope of the subject's (previous) visual field, the egocentric structure of her auditory experience will be dominated by her visual perspective. Generalizing this approach, one might say that the perspectival structure of a subject's experience is unified, in so far as the spatial contents of experience in each modality adopt the structure of her visual perspective.

They shortcoming of this approach demonstrates the difficulties inherent to any solution to the problem that appeals to the perspectival structure of a sensory modality as a means of unifying others. As noted in section 13.2.2, each sensory modality involves distinctively bounded sensory fields, partly due to the characteristics of the sense organs themselves and partly their bodily position. For the unity of perspectival structure to be merely a matter of one sense dominating the others, that unity will be constrained to the possible overlap between their respective fields. Accordingly, for any such approach, the 'in so far as' qualifier will be essential in describing the experience of any creature whose sensory fields across their modalities do not wholly overlap. The case of audiovisual perspectival unity in virtue of visual perspective makes this especially clear, with at most a third of the auditory field's 360° horizon being thus unified.

13.4.3 Transformation accounts

It may have entered the reader's thoughts (especially given the present volume) that, if there is a problem of perspectival unity at all, it is only made more acute when considering peripersonal spatial representation—for one of the most carefully and repeatedly documented facts concerning peripersonal spatial representation is that it involves integrating sensory information in frames of reference structured around various body-parts. Early important work in this area used single neuron recording in macaque monkeys to reveal both cortical and subcortical neural populations with multisensory receptive fields. For instance, cells that responded to tactile stimulation on a part of the hand (Rizzolatti et al., 1981a) would also respond to visible objects close to the hand and only poorly for visual stimuli at a greater distance (Rizzolatti et al., 1981b). Cells have been documented with similar visuotactile receptive fields for the surface and surrounding regions of various body-parts, such as the head, neck, and trunk (Fogassi et al., 1996; Graziano & Gross, 1993), with some cells exhibiting responses to not only visual and tactile, but also auditory stimuli (Graziano, Reiss, & Gross, 1999).

The embedding of these multisensory receptive fields is demonstrated in various ways. If a body-part for which a cell has a tactile receptive field is moved, the field within which the cell will be responsive to visible objects will shift accordingly. For instance, a cell might have a tactile receptive field for the dorsal surface of the forearm and a visual receptive field for the adjacent region above. When the arm moves, the cell's response again increases or decreases as a function of the visible object's distance from the arm; when the arm is moved out of sight, the cell ceases responding to visual stimuli altogether, even though the retinal location of the stimuli are unchanged. In short, these cells respond to multisensory stimuli in egocentrically structured frames of reference embedded in particular body-parts.

Similar results have been found in the study of human spatial attention, in both neuro-logically normal subjects (Spence, Pavani, & Driver, 1998, 2004) and those suffering from visuotactile and audiotactile extinction (di Pellegrino, Làdavas, & Farnè, 1997; Farnè & Làdavas, 2002). Taken together, this work suggests that the brain systematically integrates tactile information, concerning the superficial surface and/or orientation of a particular body-part, with visual and auditory information concerning the space surrounding that part, in egocentric frames of reference the perspectives of which are embedded within that part.

Robert Briscoe's comments on the relevance of this research for egocentric structure of perceptual experience are worth quoting at length:

> This subpersonal representational arrangement seems to be reflected at the personal level […] I may perceive, e.g., that the object is closer to my right hand than to my left hand, or above my waist, or below my chin, etc. Such body-relative spatial information […] is part of the content of a visual experience of an object and is reflected in its phenomenology. (Briscoe, 2009, pp. 425–426)

We are capable of such a broad range of egocentrically structured perceptual experiences, he claims, in virtue of our proprioceptive awareness of our body-parts. As he puts it: 'my visual experience of an object may convey information about its location relative to any part of my body (seen or unseen) of which I am proprioceptively aware' (ibid., p. 425).

It is arguable that proprioception can play this role, because it is critical in solving a problem faced by the brain, so to speak, which is very similar (and closely related) to that which we have called the problem of perspectival unity: how is it that neural structures are able to integrate sensory information, if that information is encoded in such a great variety of egocentric frames of reference? In essence this is a coordinate transformation problem–a problem of mapping coordinates from one frame of reference to another. Achieving such a mapping requires some means of identifying locations specified in one frame with loca-tions specified in another (Clark, 2010; Driver & Spence, 1998). Part of the solution to this problem may be that these various sources of information are integrated in virtue of being combined with proprioceptive information concerning the body-parts within which the egocentric structures are embedded. If proprioception provides information about positions of body-parts relative to one another, this information can be used to achieve coordinate transformations between egocentric frames of reference (Briscoe, 2019, section 6).

Let us assume (as is eminently plausible) that coordinate transformation can serve to inte-grate information from distinct egocentric frames of reference. Would this then be sufficient to solve the problem of perspectival unity? It is not at all obvious that it would be sufficient on its own, precisely because it might be that coordinate transformation is achieved 'on the fly', using a flexible mechanism for integrating information in any frame available to the system (Pouget, Deneve, & Duhamel, 2002). This would then be compatible with a form of partial unity, such that various egocentric structures may be unified with one another (pairwise, for instance) without each structure being unified with every other (cf. Bayne, 2010, pp. 36–45).

13.4.4 Ultimate accounts

Ultimate accounts supplement the bare notion of coordinate transformation with the notion of an ultimate frame of reference anchored to a particular body-part. Thus, on such accounts, all the egocentric frames of reference comprising the complex structure of a subject's egocentric perspectival experience are mapped onto a single, ultimate frame.

There are independent motivations for such an ultimate frame being anchored to the head or to the torso. A great number of spatially informative sensory organs are found in the head: the eyes, ears, and the vestibular labyrinth. As Sherrington noted, the latter is a particularly significant source of spatial information, in that the vestibular system 'maintains not merely a limb in flexion or extension, but a posture of the whole animal in regard to gravitation (Sherrington, 1907, p. 480). But, morphologically speaking, the head is not the most major body-part and it is one of the most mobile. By contrast, the torso is the most major body-part and the least mobile, making it perhaps the most stable anchor for the construction of a consistent egocentric representation (Blanke, 2012; Grush, 2000).

It is not clear on what grounds we should evaluate these two possibilities. We should question the need to make such a choice, for it is not at all obvious that opting for one over the other would be optimal. Each presents a robust means by which cognitive systems might perform coordinate transformations between reference frames available to the system. Depending on the current needs of the system, a head-embedded frame might serve best, for instance, where a task ultimately requires comparison of visual and vestibular information (Ionta et al., 2011; Pfeiffer et al., 2013); in other circumstances, a torso-embedded frame might be the optimal choice (Serino et al., 2015).

Indeed, it is evident that cognitive systems exhibit a great deal of flexibility in their use of spatial information (Millar, 2008). They are able to use hybrid frames involving combinations of body-part anchored frames (e.g. Carrozzo & Lacquaniti, 1994) and idiosyncratic frames for transformation between body-part anchored frames (e.g. Chang & Snyder, 2010; Gazzaniga, Ledoux, & Wilson, 1977). Such flexibility seems a virtue. It only seems a vice when placed in an inappropriate context of trying to find a single, spatially unified representation for all purposes.

13.5 Agency and perspectival unity

13.5.1 Agency and egocentricity

What I hope to have shown in the course of this chapter is that it is not a trivial matter how a subject can 'build up a unitary picture' from the multiple ways in which her perceptual experience of the world is egocentrically structured, such that she experiences 'one egocentric space' (Evans, 1982, p. 160). In this final section, I now hope to fulfil the other aim stated at the start, to show how this matter is addressed by the Evansian conception of egocentricity, according to which being in possession of '[egocentrically structured] perceptual information at least partly consists in being disposed to do various things' (Evans, 1985, p. 383).

We can develop an agentive account of perspectival unity from this basic idea and its elaboration in the writings of, for example, José Luis Bermúdez (1998, 2005), Bill Brewer (1992, 1993), Robert Briscoe (2009, 2019), and Christopher Peacocke (1986, 1992), for it follows from this that each of a subject's egocentrically structured experiences will have as a common constraint the agentive capacity of the subject. Moreover, each will represent objects and events with respect to a perspective embedded in a part of the very same whole, within the body as a dynamic unity. We do not suffer the disunity of *Split's* experience, not merely because of the integration of the information encoded within and between our sensory systems, but rather as a consequence of their integration with our capacities for bodily action.

13.5.2 The agentive unity of egocentric perspective

Many theorists who are broadly sympathetic to the Evansian conception show some degree of sensitivity, at least, to the idea that there might be some issue. For instance, Peacocke writes: 'Actually, in the specification of the representational content of some human experiences, one would need to consider several such systems of origins and axes, and to specify the spatial relations of these systems to one another' (1992, p. 63). These remarks—and similar ones made by proponents of an intimate connection between egocentrically structured perception and action (see, e.g. Bermúdez, 1998, pp. 140–142; Briscoe, 2009, pp. 424–426; 2019, section 6) are somewhat brief. But what they suggest is roughly that, to the extent that there is a problem here, it can be resolved by establishing spatial relations between egocentric structures.

In the last section, we considered three kinds of account that did just that. Sensory accounts establish spatial relations by means of overlap in sensory fields, transformation accounts by means of coordinate transformation, and ultimate accounts by means of an ultimate frame of reference serving as a universal for such transformations. I have argued that each of these kinds of accounts is insufficient to resolve the problem at hand, but there is a sense in which each picks out one of the many means by which egocentric structures can be related. Indeed, *pace* the denial of synchronic perspectival unity, there may also be a lesson to be learned from switching accounts. For whether or not we do experience the world from multiple perspectives at a time, it is certainly plausible that, even as a general rule, our attentional focus may be such that a single egocentric structure may be more prominent than the others.

An agentive account serves to supplement the bare idea of spatially relating egocentric structures by providing a means for all such structures to be unified in their relation to a single agent. What the other accounts provide is a further specification of the particular forms that the complex egocentric structure of a subject's experience might take and some of the mechanisms that might be involved. Thus, a subject's experience may be unified according to a single perspective, such that one particular egocentric structure within a particular modality shapes the focus of her engagement with the world. Nevertheless, this structure might be synchronically unified with several others in virtue of coordinate transformations supported by a common anchor.

13.5.3 Evidence for action-orientated peripersonal representations

In closing, I want to illustrate a little further how the agentive account manages the contrast between *Split* and ordinary subjects, by appealing to the well-established action-orientated function of peripersonal representations.

As di Pellegrino and Làdavas write, peripersonal spatial representations are 'probably best described as multisensory-motor interfaces, which serve to encode the location of nearby sensory stimuli to generate suitable motor acts' (2015, p. 131). Evidence for the action-orientated function of peripersonal spatial representation abounds. Visuotactile integration has been shown to be sensitive to the functional range of action effectors located in the relevant regions (Làdavas, 2002; Làdavas & Serino, 2008). Indeed, this is of key behavioural significance, precisely because that range determines the system's physical contact with objects of interest (de Vignemont, this volume). This has been demonstrated by extending the functional range of effectors through the active use of a tool such as a rake or hockey stick. Using single neuron recordings, Iriki, Tanaka, & Iwamura (1996) found that cells with visuotactile receptive fields surrounding the hand were sensitive to visual stimuli presented around the distal end of an actively used tool, held in that hand. Similarly, cross-modal extinction effects have been found at the tip of an actively used tool (Farnè, Bonifazi, & Làdavas, 2005; Farnè & Làdavas, 2000). And visual distracters placed at the tip of an actively used tool had a cross-modal congruency effect similar to visual distracters from LEDs on subjects' hands (Holmes, Calvert, & Spence, 2007).

These 'multi-sensory motor interfaces' are also affected by a range of basic features of visual or auditory stimuli, such as direction and velocity, but also more complex features such as positive or negative valence (Bufacchi & Iannetti, 2018). The latter suggests a broadly useful way of classifying the action-orientated functions of peripersonal spatial representations into appetitive and defensive functions (de Vignemont & Iannetti, 2015), although I agree with the caveat noted by Klein (this volume) that it is only in limit cases that we can easily individuate actions on these bases, because so many of our actions involve acting carefully to fulfil our needs.

13.5.4 Split agentive perspectives

How might possessing this further property of peripersonal spatial representation (*viz.* being action-orientated) affect *Split's* case? Assume that it would be possible to integrate the sensory systems of *Visuo*, *Audio*, and *Tactuo*, such that they formed action-orientated multisensory representations of the space surrounding the corresponding parts of each body. The integration on the sensory side of the interface would necessitate that the motor outputs would affect all *Split's* bodies simultaneously. This is perhaps almost inconceivable, because it would be a remarkably dysfunctional mechanism for *Split*. And the situation is no better for *Split-juxtaposed*. What, for instance, would be the appropriate action in a case in which *Visuo* is presented with a positively valued visual stimulus and *Audio* is presented with a negatively valued auditory stimulus in the same apparent direction?

The trouble here is that *Split's* bodies, even when juxtaposed, do not form a part of the same physical whole in the facsimile-physics of her virtual worlds. They occupy the same location, but this is not in virtue of their unity: it is merely in virtue of a mapping between her visual, auditory, and tactual worlds. Consequently, the connections between *Split's* complexly structured perspectival experience and her actions would involve individually establishing how things are with respect to the body with which she acts. The only way in which *Split's* or *Split-juxtaposed's* situation would be workable would be to sever the direct connection between the multisensory representation and action inherent in peripersonal spatial representation. In this way, she would be able to flexibly distinguish between, for example, what her visual world afforded to *Visuo* and what her auditory world afforded to *Audio*.

But this is precisely the difference that makes the difference in our case for, in our case, there is no distinction between the body with which we act on the basis of what we see, hear, or feel. Each of the ways in which we experience the world perspectively is unified in virtue of its structural connection to a single thing, the one and only body with which we directly act. This, I think, is the best way of making sense of Evans' dictum: 'There is only one egocentric space, because there is only one behavioural space' (Evans, 1982, p. 160).

Acknowledgements

This work was supported by the grant ERC-2017-STG (757698) awarded to Joshua Shepherd. The author also wishes to thank two anonymous referees for insightful comments that resulted in significant improvements.

References

Alsmith, A. J. T. (2017). Perspectival structure and agentive self-location. In F. de Vignemont & A. Alsmith (eds), *The Subject's Matter: Self-consciousness and the Body* (pp. 263–288). Cambridge, MA: MIT Press.

Alsmith, A. J. T., Ferrè, E. R., & Longo, M. R. (2017). Dissociating contributions of head and torso to spatial reference frames: The misalignment paradigm. *Consciousness and Cognition, 53,* 105–114. doi:org/10.1016/j.concog.2017.06.005

Ayers, M. (1993). *Locke: Epistemology and Ontology.* London: Routledge.

Bayne, T. (2010). *The Unity of Consciousness.* Oxford: Oxford University Press.

Berkeley, G. (1732/2008). An essay towards a new theory of vision. In D. M. Clarke (ed.), *Berkeley: Philosophical Writings* (pp. 1–66). Cambridge: Cambridge University Press.

Bermúdez, J. L. (1998). *The Paradox of Self-Consciousness.* Cambridge, MA: MIT Press.

Bermúdez, J. L. (2005). The phenomenology of bodily awareness. In A. L. Thomasson & D. W. Smith (eds), *Phenomenology and Philosophy of Mind* (pp. 295–316). Oxford: Oxford University Press.

Blanke, O. (2012). Multisensory brain mechanisms of bodily self-consciousness. *Nature Reviews Neuroscience, 13*(8), 556–571.

Blumenfeld, W. (1937). The relationship between the optical and haptic construction of space. *Acta Psychologica, 2,* 125–174. doi:org/10.1016/S0001-6918(37)90011-8

Brewer, B. (1992). Self-location and agency. *Mind, 101,* 17–34.

Brewer, B. (1993). The integration of spatial vision and action. In N. Eilan, R. A. McCarthy, & B. Brewer (eds), *Spatial Representation: Problems in Philosophy and Psychology* (pp. 294–315). Oxford: Oxford University Press.

Briscoe, R. E. (2009). Egocentric spatial representation in action and perception. *Philosophy and Phenomenological Research, 79*(2), 423–460.

Briscoe, R. E. (2016). Multisensory processing and perceptual consciousness: Part I. *Philosophy Compass, 11*(2), 121–133. doi:10.1111/phc3.12227

Briscoe, R. E. (2019). Bodily awareness and novel multisensory features. *Synthese.* doi:10.1007/s11229-019-02156-2

Broad, C. D. (1927). *Scientific Thought.* London: Harcourt, Brace and Company, Inc.

Bufacchi, R. J., & Iannetti, G. D. (2018). An action field theory of peripersonal space. *Trends in Cognitive Sciences, 22*(12), 1076–1090. doi:10.1016/j.tics.2018.09.004

Carrozzo, M., & Lacquaniti, F. (1994). A hybrid frame of reference for visuo-manual coordination. *NeuroReport, 5*(4), 453–456.

Chang, S. W., & Snyder, L. H. (2010). Idiosyncratic and systematic aspects of spatial representations in the macaque parietal cortex. *Proceedings of the National Academy of Sciences, 107*(17), 7951–7956.

Clark, A. (1996). Three varieties of visual field. *Philosophical Psychology, 9*(4), 477.

Clark, A. (2010). Crossmodal cuing and selective attention. In F. Macpherson (ed.), *The Senses: Classical and Contemporary Perspectives* (pp. 375–396). New York: Oxford University Press.

de Vignemont, F., & Iannetti, G. (2015). How many peripersonal spaces? *Neuropsychologia, 70*, 327–334.

di Pellegrino, G., & Làdavas, E. (2015). Peripersonal space in the brain. *Neuropsychologia, 66*, 126–133.

di Pellegrino, G., Làdavas, E., & Farnè, A. (1997). Seeing where your hands are. *Nature, 380*, 730.

Driver, J., & Spence, C. (1998). Cross–modal links in spatial attention. *Philosophical Transactions of the Royal Society of London. Series B: Biological Sciences, 353*(1373), 1319–1331. doi:10.1098/rstb.1998.0286

Evans, G. (1982). *The Varieties of Reference.* Oxford: Oxford University Press.

Evans, G. (1985). *Collected Papers.* Oxford: Oxford University Press.

Farnè, A., Bonifazi, S., & Làdavas, E. (2005). The role played by tool-use and tool-length on the plastic elongation of peri-hand space: A single case study. *Cognitive Neuropsychology, 22*, 408–418.

Farnè, A., & Làdavas, E. (2000). Dynamic size-change of hand peripersonal space following tool use. *NeuroReport, 11*, 1645.

Farnè, A., & Làdavas, E. (2002). Auditory peripersonal space in humans. *Journal of Cognitive Neuroscience, 14*(7), 1030–1043. doi:10.1162/089892902320474481

Fogassi, L., Gallese, V., Fadiga, L., Luppino, G., Matelli, M., & Rizzolati, G. (1996). Coding of peripersonal space in inferior premotor cortex (Area F4). *Journal of Neurophysiology, 76*, 141–157.

Gazzaniga, M. S., Ledoux, J. E., & Wilson, D. H. (1977). Language, praxis, and the right hemisphere. *Neurology, 27*(12), 1144.

Gibson, J. J., Kaplan, G. A., Reynolds, H. N., & Wheeler, K. (1969). The change from visible to invisible. *Perception & Psychophysics, 5*(2), 113–116. doi:10.3758/BF03210533

Graziano, M., & Gross, C. (1993). A bimodal map of space: Somatosensory receptive fields in the macaque putamen with corresponding visual receptive fields. *Experimental Brain Research, 97*, 96–109.

Graziano, M., Reiss, L. A., & Gross, C. G. (1999). A neuronal representation of the location of nearby sounds. *Nature, 397*(6718), 428.

Gregory, D. (2013). *Showing, Sensing, and Seeming: Distinctively Sensory Representations and Their Contents.* Oxford: Oxford University Press.

Grush, R. (2000). Self, world and space: The meaning and mechanisms of ego-and allocentric spatial representation. *Brain and Mind, 1*, 59–92.

Haggard, P., & Giovagnoli, G. (2011). Spatial patterns in tactile perception: Is there a tactile field? *Acta Psychologica, 137*(1), 65–75.

Holmes, N. P., Calvert, G. A., & Spence, C. (2007). Tool use changes multisensory interactions in seconds: Evidence from the crossmodal congruency task. *Experimental Brain Research, 183*(4), 465–476. doi:10.1007/s00221-007-1060-7

Howard, I. P., & Templeton, W. B. (1966). *Human Spatial Orientation.* London: John Wiley & Sons.

Hume, D. (1739–1740/2007). *A Treatise of Human Nature.* Oxford: Clarendon Press.

Husserl, E. (1952/1989). *Ideas Pertaining to a Pure Phenomenology and to a Phenomenological Philosophy (second book): Studies in the Phenomenology of Constitution* (R. Rojcewicz, & A. Schuwer, trans.). Dordrecht: Kluwer Academic Publishers.

Husserl, E. (1973/1997). *Thing and Space: Lectures of 1907* (R. Rojcewicz, trans.). Dordrecht: Kluwer Academic Publishers.

Ionta, S., Heydrich, L., Lenggenhager, B., Mouthon, M., Fornari, E., Chapuis, D., … Blanke, O. (2011). Multisensory mechanisms in temporo-parietal cortex support self-location and first-person perspective. *Neuron, 70*(2), 363–374.

Iriki, A., Tanaka, M., & Iwamura, Y. (1996). Coding of modified body schema during tool use by macaque postcentral neurones. *NeuroReport, 7*, 2325.

Làdavas, E. (2002). Functional and dynamic properties of visual peripersonal space. *Trends in Cognitive Sciences, 6*, 17–22.

Làdavas, E., & Serino, A. (2008). Action-dependent plasticity in peripersonal space representations. *Cognitive Neuropsychology, 25*(7–8), 1099–1113.

Locke, J. (1690/1997). *An Essay Concerning Human Understanding.* London: Penguin Classics.

Longo, M. R., Rajapakse, S. S., Alsmith, A. J., & Ferrè, E. R. (2020). Shared contributions of the head and torso to spatial reference frames across spatial judgments. *Cognition, 204*, 104349. doi:org/10.1016/j.cognition.2020.104349

Martin, M. G. F. (1992). Sight and touch. In T. Crane (ed.), *The Contents of Experience. Essays on Perception* (pp. 196–215). Cambridge: Cambridge University Press.

Merleau-Ponty, M. (1962/2002). *Phenomenology of perception* (C. Smith, trans.). London: Routledge.

Mershon, D. H., & King, L. E. (1975). Intensity and reverberation as factors in the auditory perception of egocentric distance. *Perception & Psychophysics, 18*(6), 409–415.

Millar, S. (2008). *Space and Sense.* New York: Psychology Press.

Neelon, M. F., Brungart, D. S., & Simpson, B. D. (2004). The isoazimuthal perception of sounds across distance: A preliminary investigation into the location of the audio egocenter. *Journal of Neuroscience, 24*(35), 7640–7647.

Nichols, S., & Horgan, T. (2015). The zero point and I. In S. Miguens, G. Preyer, & C. Morando (eds), *Pre-reflective Consciousness: Sartre and Contemporary Philosophy of Mind* (pp. 155–187). New York: Routledge.

O'Callaghan, C. (2010). Perceiving the locations of sounds. *Review of Philosophy and Psychology, 1*(1), 123–140.

O'Callaghan, C. (2017). *Beyond Vision: Philosophical Essays.* Oxford: Oxford University Press.

Oldfield, S. R., & Phillips, J. R. (1983). The spatial characteristics of tactile form perception. *Perception, 12*(5), 615–626.

Ono, H. (1981). On Wells's (1792) law of visual direction. *Perception & Psychophysics, 30*(4), 403–406. doi:10.3758/BF03206159

Parsons, L. M., & Shimojo, S. (1987). Perceived spatial organization of cutaneous patterns on surfaces of the human body in various positions. *Journal of Experimental Psychology: Human Perception and Performance, 13*(3), 488–504. doi:10.1037/0096-1523.13.3.488

Peacocke, C. (1986). Analogue content. *Proceedings of the Aristotelian Society: Supplementary volume, 60*(1), 1–18. doi:10.1093/aristoteliansupp/60.1.1

Peacocke, C. (1992). *A Study of Concepts.* Cambridge, MA: MIT Press.

Pfeiffer, C., Lopez, C., Schmutz, V., Duenas, J. A., Martuzzi, R., & Blanke, O. (2013). Multisensory origin of the subjective first-person perspective: Visual, tactile, and vestibular mechanisms. *PLoS ONE*, *8*(4), e61751.

Pitcher, G. (1971). *A Theory of Perception*. Princeton: Princeton University Press.

Pouget, A., Deneve, S., & Duhamel, J.-R. (2002). A computational perspective on the neural basis of multisensory spatial representations. *Nature Reviews Neuroscience*, *3*(9), 741–747. doi:10.1038/nrn914

Richardson, L. (2010). Seeing empty space. *European Journal of Philosophy*, *18*(2), 227–243. doi:10.1111/j.1468-0378.2008.00341.x

Rizzolatti, G., Scandolara, C., Matelli, M., & Gentilucci, M. (1981a). Afferent properties of periarcuate neurons in macaque monkeys. I. Somatosensory responses. *Behavioural Brain Research*, *2*(2), 125–146.

Rizzolatti, G., Scandolara, C., Matelli, M., & Gentilucci, M. (1981b). Afferent properties of periarcuate neurons in macaque monkeys. II. Visual responses. *Behavioural Brain Research*, *2*(2), 147–163.

Russell, B. (1912/1959). *The Problems of Philosophy*. Oxford: Oxford University Press.

Schwenkler, J. (2014). Vision, self-location, and the phenomenology of the 'point of view'. *Noûs*, *48*(1), 137–155. doi:10.1111/j.1468-0068.2012.00871.x

Schwitzgebel, E. (2006). Do things look flat? *Philosophy and Phenomenological Research*, *72*(3), 589–599.

Serino, A., Giovagnoli, G., de Vignemont, F., & Haggard, P. (2008). Spatial organisation in passive tactile perception: Is there a tactile field? *Acta Psychologica*, *128*(2), 355–360.

Serino, A., Noel, J.-P., Galli, G., Canzoneri, E., Marmaroli, P., Lissek, H., & Blanke, O. (2015). Body part-centered and full body-centered peripersonal space representations. *Scientific Reports*, *5*, 18603. doi:10.1038/srep18603

Sherrington, C. (1907). On the proprio-ceptive system, especially in its reflex aspect. *Brain: A Journal of Neurology*, *29*(4), 467–482. doi:10.1093/brain/29.4.467

Smith, A. D. (2000). Space and sight. *Mind*, *109*(435), 481–518.

Soteriou, M. (2013). *The Mind's Construction: The Ontology of Mind and Mental Action*. Oxford: Oxford University Press.

Spence, C., & Bayne, T. (2014). Is consciousness multisensory? In D. Stokes, M. Matthen, & S. Biggs (eds), *Perception and its Modalities* (pp. 96–124). Oxford: Oxford University Press.

Spence, C., Pavani, F., & Driver, J. (1998). What crossing the hands can reveal about visuotactile links in spatial attention. *Abstracts of the Psychonomic Society*, *3*(13).

Spence, C., Pavani, F., & Driver, J. (2004). Spatial constraint on visual-tactile cross-modal distractor congruency effects. *Cognitive, Affective and Behavioural Neuroscience*, *4*(2), 148–169.

Stokes, D., & Biggs, S. (2014). The dominance of the visual. In D. Stokes, M. Matthen, & S. Biggs (eds), *Perception and its Modalities* (pp. 350–378). Oxford: Oxford University Press.

Strawson, P. F. (1959/2003). *Individuals: An Essay in Descriptive Metaphysics*. London: Routledge.

Sukemiya, H., Nakamizo, S., & Ono, H. (2008). Location of the auditory egocentre in the blind and normally sighted. *Perception*, *37*(10), 1587–1595.

Taylor, C. (1978). The validity of transcendental arguments. *Proceedings of the Aristotelian Society*, *79*, 151–165.

Zuidhoek, S., Kappers, A. M., Van der Lubbe, R. H., & Postma, A. (2003). Delay improves performance on a haptic spatial matching task. *Experimental Brain Research*, *149*(3), 320–330.

PART III

THE SPACE OF SELF AND OTHERS

14

Peripersonal space, bodily self-awareness, and the integrated self

Matthew Fulkerson

14.1 Introduction

The self—or at least our sense of it—seems unified, coherent, and simple. We are aware of ourselves as agents who possess a single body, and often believe and act in coherent, consistent ways. This feeling is so transparent in typical experience that we can fail to realize how fragile it really is. Trauma and pathology can reveal the hidden seams in our sense of self. For instance, Clive Wearing was a successful musician and conductor living in England. At 47 years old he contracted herpesviral encephalitus, which eventually resulted in the almost complete loss of his hippocampus. Since his illness, he no longer has the ability to form new memories or to adequately control his emotions.[1] He has at best a diminished sense of self. Or consider the case of Ian Waterman, who was 19 years old when his afferent nerve fibre demyelinated, leaving him with a complete loss of proprioceptive awareness. Regaining control over his body using vision alone was an arduous process that took years.[2]

In both these cases, we see vivid evidence of self-related dissociation. Wearing lost his capacity to form and access new memories. He cannot extend his intentions and thoughts into the future. He cannot form long-term attachments. Waterman's loss also represents a kind of traumatic upheaval of self-awareness, but one based on his *bodily* awareness. Unlike Wearing, he maintained his psychological integrity, but he lost awareness and control of his body. He could no longer feel the limits of his bodily self, where *he* ended and the world began. He could not engage or act on the things around him. Just sitting up and maintaining balance was all but impossible for him. This loss of bodily awareness and control deeply undermined his sense of self-location, self-control, and bodily ownership.

Cases like these raise a number of interesting and important questions. What are the constituent elements of the self? What processes coordinate and unify their operations? What impact do these processes have on our agency, experience, and sense of purpose?

There is a sizable literature on the self. This work spans multiple fields, levels of detail, and concern. Theoretical options span the gamut from the idea that there is no self, only a minimal self, to the idea of a full Cartesian inner agent. Topics of interest concern self-knowledge, agency, moral responsibility, and the unity of the self. Evidence from clinical dissociations, self-report, and behavioural studies are used to build models of the self.

---

[1] His story has been the subject of many works, including several documentaries, books, and articles. See, for example, Sacks (2007) and Wilson and Wearing (1995). Wearing's case is similar to that of Henry Molaison, the famous amnesic patient, HM, whose story was recently told in Corkin (2013).
[2] This story has also been widely discussed. See, especially, Cole (1995).

Matthew Fulkerson, *Peripersonal space, bodily self-awareness, and the integrated self* In: *The World at Our Fingertips*. Edited by Frédérique de Vignemont, Andrea Serino, Hong Yu Wong, and Alessandro Farnè, Oxford University Press (2021). © Oxford University Press. DOI: 10.1093/oso/9780198851738.003.0014.

Consider the Cartesian inner agent. This agent serves the function of unifying the disparate elements of the self.[3] Such a view gives a nice account of agency and moral responsibility. But it faces both philosophical and empirical challenges. As Dennett has forcefully argued, the supposition of a single place—a so-called 'Cartesian Theater' where everything comes together—generates both regress worries and threatens an unnecessary dualism (Dennett and Kinsbourne 1992; Dennett 1981). In addition, it proposes a single system—perhaps even a module—that plays this functional role. We currently lack sufficient empirical evidence for such a dedicated system that spans the entirety of elements related to the self.[4]

Consider another alternative: the view that the self is unified by narrative, that in some sense we are the author of our own existence, both a writer of and an actor in the events occurring around us (Dennett 1991; Schechtman 2011). This view seems promising, but often ends up fictionalizing the mind, or undermining the very thing it is supposed to explain (Strawson 2009). Yet another view is that the self is the emergent result of the dynamic coupling of its individual components, *enacting* or bringing forth the self as the emergent result of such coupling (Morf and Mischel, 2012). All these various views have their advantages, and their weaknesses. I do not have a full account of the self along these lines. Instead, I have a suspicion. And this is that the best way to approach understanding the self is from the bottom up, by looking at how psychological construction occurs for those important constituent elements that are an essential part of the self.

In what follows, I explore how we might come to think about an alternative, decentralized account of the sense of self. This account is inspired by recent work on bodily awareness. Bodily awareness, I take it, is one of the most fundamental elements of self-awareness. And our awareness of our bodies also seems unified and coherent. But recent empirical and theoretical investigations have supported a more complex and integrated understanding of how bodily awareness arises. This is especially true in understanding the different forms of spatial awareness. This work starts with the recognition that integration problems are ubiquitous in psychology. Examples of integration can be found in perception, emotion, attention, motor control, decision making, and many other domains. Rather than requiring any central agent or integrator, unity in these domains seems to occur in virtue of active, flexible rules of integration.

In particular, peripersonal space (PPS)—the intermediate spatial frame of reference immediately around our bodies and defined by our potential for active engagement—is itself a multimodal construct that coordinates and unifies a diverse set of perceptual and bodily inputs with skilled action (Rizzolatti et al. 1997). This example is apt because there is no evidence for, nor need to postulate, a single dedicated peripersonal module or system that subserves our awareness of peripersonal space: most current theories hold that peripersonal representations are constructed through the competitive interactions and exchanges among its many constituent elements. These interactions are rule-governed, flexible, task sensitive, and ultimately cooperative in nature. The end goal of such interactions is a more unified,

---

[3] We could think of this as something similar to a global workspace model of consciousness (Baars 2005). The workspace is not identical to the conscious self, but serves as the (functional) location where everything comes together. So, too, on this proposal, the inner agent isn't identical to the self or the sense of self, but serves as the primary functional glue that grounds both.

[4] See Northoff et al. (2011) for a discussion of the methodological issues around isolating specific brain regions with such an inner agent. As we will see, my own view suggests that the integration occurs in a distributed manner through the cooperative interactions of the constituent elements themselves, without any additional component or system doing the work for them.

cohesive, and functionally advantageous form of spatial awareness that guides our successful, active engagement with the world. These forms of integration give us a key insight into how the self as a whole might be constructed.

The idea is ultimately a fairly traditional one. It is the view that perhaps the sense of self is just the result of the very same kinds of integrative processes that operate in these other domains. The problem of unifying the self is nothing more than the totality of the other decentralized forms of integration. It is a kind of *bundle theory* (the self consists in many parts, all somehow bundled together), but one where we have a more empirically respectable account of the glue that holds everything together.

Well, such is the hope. It is still early days in our understanding of integration even at the level of sensory interactions. Many elements of self-integration await further investigation and analysis. At present, we are not in a position to lay out a detailed account of the various processes that function to integrate the self. This is one reason to focus on PPS, an area of intense empirical and philosophical engagement where much progress is being made (for instance, see this volume). Importantly for my purposes, PPS awareness seems to play a crucial role in our overall sense of self. This is because there is strong evidence that PPS plays a central role in bodily self-awareness (BSA). BSA is a form of awareness that involves our sense of self-location (knowing where we are in space and the limits of our bodily selves) and felt bodily ownership (our sense of our body as ours). BSA in turn is a major constituent of our overall sense of self. Seeing how PPS representations can be integrated and unified without any central agent, without narrative, without dynamic emergence, can serve as a kind of existence proof for how we should think of the unity of self overall.

My discussion will focus on three claims: (i) peripersonal space is itself the integrated result of competitive exchange between a diverse range of multimodal inputs; (ii) peripersonal space is a strong candidate for grounding (part of) our sense of bodily self-awareness (BSA); and (iii) bodily self-awareness is a critical constituent of our overall sense of self.

Taken together, these reflections make plausible the idea that the self is constructed out of more basic elements, and that the processes of integration involved need not posit any central agent or single unifying process (like self-narration). This is not (yet) an account of self-integration, of course. Instead, it is an attempt to make plausible a certain class of models based on our current understanding of how some of the key underlying components actually fit together.

14.2 Peripersonal space and cooperative integration

I believe that the sense of self is best understood as the product of what I call 'cooperative integration'. In some contexts, this cooperation occurs literally: one system facilitates and aids another. This occurs, for instance, when visual inputs help disambiguate speech in noisy environments (Jones and Jarick 2006). The two distinct sensory channels combine forces to generate an overall best interpretation of distal events. In other contexts, the cooperation is more competitive, often bordering on combative. For instance, sometimes when two modalities disagree, one *suppresses* the other signal. Such competitive exchanges involve local winners and losers, but the exchange itself serves to increase global accuracy and reliability. In this way, many psychological forms of integration function something like a courtroom. Adversarial in practice, but in spirit aiming towards a larger cooperative goal.

While such integration effects are commonly found throughout our psychologies, in a wide range of different domains, I will focus here on sensory interactions. This is both because it is the domain I know best, but also because it most closely aligns with the sorts of integration effects we see in the construction of peripersonal space and bodily self-awareness.

Multisensory interactions are heterogeneous, and occur at all levels of sensory processing and between all modalities (Fulkerson 2014).[5] These interactions involve discrete modalities, but do not themselves seem to have anything like a separate system or inner agent supervising or integrating the contributions of each modality.[6] What we do not see, in other words, is anything that plays the role of a judge in the courtroom. Sensory interactions may be akin to a tribunal, but they lack any single functional authority keeping things in order. There is no modular system that specifically functions to assess and unify the potential disagreements between sensory modalities. Instead, we see a variety of models for sensory integration that operate via flexible rules of integration. There are many popular models of such integration, and I do not think we should expect there to be a single model that explains all such interactions. Like the interactions themselves, the principles of integration are likely to be heterogeneous and varied. Still, it can be useful to have some more specific models in mind as we work through the evidence. For this reason, in what follows, I will focus mostly on Bayesian approaches like the *maximum likelihood estimation* (MLE) model (Ernst and Banks 2002; Helbig and Ernst 2007).

According to the MLE model, each sensory system performs an estimation operation on an environmental property. Assuming the noise in each modality is independent with a uniform Bayesian prior, then the contributions of each modality is weighted to minimize variance (and, by hypothesis, thereby maximize the accuracy of the estimate): '[T]he MLE rule states that the optimal means of estimation (in the sense of producing the lowest-variance estimate) is to add the sensor estimates weighted by their normalized reciprocal variances' (Ernst and Banks 2002, p. 430).

This model suggests a flexible rule for integrating sensory inputs that functions in an efficient and highly adaptive way. According to the MLE model, for instance, vision will dominate in contexts where *pattern* is most salient (since vision displays less variance when detecting pattern). In contexts where *uniform intensity* is more salient (say, for fabric texture), haptic touch will dominate. This process is modulated by attention, is task- and context-sensitive, and occurs relatively late in perceptual processing (Klatzky and Lederman 2010, p. 222). The model has shown excellent promise in explaining a wide range of sensory interactions. As noted, it is just one class of model and far from the settled view even for those cases where it seems to work well.[7] In addition, it seems poorly suited for many paradigm forms of sensory interaction. Flavour, for instance, is clearly the result of multisensory integration, but it is not, it seems, best explained by interactions that are principally directed at maximum accuracy. These limitations are important to keep in mind—the ultimate story of how the self is integrated will likely involve a large number of distinct principles of integration.

These features of multisensory integration are important when we look at peripersonal space. This spatial frame is essential for successful engagement with the world, but it also requires coordinating and integrating information from a wide variety of sources in an efficient manner. It has long been known that we represent space in different ways (Derdikman and Moser 2010; Millar 2008). Each modality seems to provide distinct forms of spatial awareness

---

[5] See O'Callaghan (2008, 2017) for some recent work on the philosophical implications of the pervasive multimodality of perception.

[6] Although, of course, the nature of these interactions can put pressure on the idea that we have distinct, self-standing modalities in anything like the classical sense. See recent debates in Shimojo and Shams (2001); O'Callaghan (2012); Fulkerson (2011).

[7] See, for example, Chandrasekaran (2017); Seilheimer et al. (2014).

of the external environment. This involves different spatial coordinates, resolutions, and frames of reference. These different signals need to be brought into alignment with each other. Add to this the fact that we also have a complex system of bodily awareness, often with coordinates defined relative to body parts rather than to object locations in external space (Kappers 2007). When we are walking down the street, a pain felt in our finger is stationary in bodily space but moving in external space. The representations of these spaces differ: internal space is thought to involve an egocentric, body-based map (perhaps *many* of them [de Vignemont 2018]) while external space uses an allocentric coordinate system. A key issue is that acting in the world requires quickly and smoothly coordinating all these inputs across both domains. Our motor system, for instance, must use external, allocentric spatial information to locate objects of interest in the environment, but it must also guide our bodily movements, which requires knowing where the body parts are located relative to each other.

Solving these translation problems is a difficult computational and practical task (cf. Wu 2011). Part of the solution involves an intermediate spatial frame of reference that is constructed through the cohesive integration of both sensory and bodily spatial inputs. As Holmes and Spence (2004) say in the abstract of their influential paper on the relation between body schema and peripersonal space, 'The effective piloting of the body to avoid or manipulate objects in pursuit of behavioural goals requires an integrated neural representation of the body (the 'body schema') and of the space around the body ("peripersonal space")' (p. 94). They continue by quickly summarizing the evidence concerning this space:

> [R]esults from neurophysiology, neuropsychology, and psychophysics in both human and non-human primates … support the existence of an integrated representation of visual, somatosensory, and auditory peripersonal space. Such a representation involves primarily visual, somatosensory, and proprioceptive modalities, operates in body-part-centred reference frames, and demonstrates significant plasticity. Recent research shows that the use of tools, the viewing of one's body or body parts in mirrors, and in video monitors, may also modulate the visuotactile representation of peripersonal space. (p. 94)
>
> Reproduced from Holmes, N. P. and Spence, C. (2004). The body schema and multisensory representation (s) of peripersonal space. Cognitive processing, 5(2):94–105.

As they note, peripersonal space is essential for effective active engagement with the world. While our sensory information comes from a variety of distinct (although interacting) sources, we need to act coherently and consistently on that information. Solving these coordination problems involves cooperative integration.

While some of these inputs are relatively basic, others are themselves constructs derived from even more basic elements. For instance, one of the elements involved in PPS is information concerning self-motion. This information is made available both consciously and unconsciously. PPS representations make use of information about how our bodies are moving: direction, orientation, velocity, acceleration, etc. This information is critical for accurately updating PPS coordinates. These elements, however, are not simple atoms of information. Instead, they too are the result of integration. For instance, one element involved here is *self-motion awareness* (SMA). Despite its name, SMA need not be thought of as an entirely conscious form of awareness: it is rather just our capacity to keep track of whether and how we are moving. SMA is itself the result of competitive exchange. It requires integrating information across a wide range of diverse bodily and sensory inputs to derive accurate information about the movement and orientation of our bodies. As Campos and Bülthoff (2012) describe it:

During almost all natural forms of self-motion, there are several sensory systems that provide redundant information about the extent, speed, and direction of egocentric movement, the most important of which include dynamic visual information (i.e., optic flow), vestibular information (i.e., provided through the inner ear organs including the otoliths and semicircular canals), proprioceptive information provided by the muscles and joints, and the efference copy signals representing the commands of these movements. (p. 595)

> Reproduced from Campos, J. and Bulthoff, H. (2012). Multimodal integration during self-motion in virtual reality. In The neural bases of multisensory processes, pages 603–628. CRC Press.

We can better understand the relative contributions of these constituents and the principles by which they integrate by using clever experiments to pull them apart.

Just such an experiment was conducted by Frissen et al. (2011). Using a novel experimental set-up, they were able to (largely) dissociate the contributions to self-motion awareness from vision, proprioception, and the vestibular system. This was difficult because these systems (in these kinds of tasks) are always working together. They had subjects stand on a large rotating disc. On the top of the disc was a treadmill, and directly in front of the subjects was a monitor that displayed a moving scene that provided optic flow information. There was also a pointer that the subject used to perform a constant updating task. This elaborate set-up allowed for the independent manipulation of three different variables: vestibular awareness of acceleration from the rotating disc, proprioceptive information from the walking motion on the treadmill, and visual feedback about forward motion provided by optic flow on the monitor. By manipulating these variables and having subjects perform self-motion tasks, the researchers were able to determine the relative contributions of each of the constituent systems to the overall sense of forward motion. They found the following: 'Overall, the results demonstrate evidence for the integration of proprioceptive and vestibular information that is qualitatively consistent with the MLE model' (p. 172).

In such self-motion tasks, we find that proprioception, vision, and the vestibular system smoothly integrate their different inputs according to the relative accuracy of the individual inputs. Because these integrations are so smooth and automatic, only extremely clever set-ups like this allow for them to be separated experimentally.

This was not the only study to use such a set-up, of course. Later studies used virtual reality to independently manipulate the sensory inputs (Campos and Bülthoff 2012). The results of these experiments suggest that each subsystem contributes to self-motion awareness along the lines proposed by the MLE model of sensory interaction:

> Overall, this study highlights the fact that even when cues to self-motion provide redundant information about distance traveled, each contributes to the final estimate. It is also clear that body-based information provides a particularly important role in estimating traveled distance during walking and that both visual and vestibular information contribute to passive self-motion perception with a slightly higher weighting of vision. Finally, both sources of body-based cues contribute to walked distance estimation with a higher weighting attributable to vestibular inputs. (Campos et al. 2012)
>
> Reproduced from Campos, J. and Bulthoff, H. (2012). Multimodal integration during self-motion in virtual reality. In The neural bases of multisensory processes, pages 603–628. CRC Press.

Here we see that self-motion awareness crucially involves exactly the same processes of integration that occur in other sensory interactions. The result is a specialized form of awareness that is not mediated by any single modality or function subsystem. Instead, it is the result of the distributed operations of cooperative integration.

An important element of these patterns of integration is how dynamic they are. They are not rigidly applied: instead, they are updated as cue reliability changes. For instance, Fetsch et al. (2009) showed dynamic cue reweighting in animal models (using monkeys) as the reliability of the cues was manipulated. An important element of this process involved the contributions of vestibular awareness (Angelaki and Cullen, 2008). While often recessive in experience, our vestibular system contributes immensely to our self-consciousness (especially our sense of location and orientation in the world).

When translated into the domain of spatial awareness, we see that the coordinate systems generated for PPS also involve integrated inputs across a wide range of sensory modalities. Indeed, among the inputs to PPS are MLE-compatible bundles like SMA, which is precisely *not* to say that PPS itself is the result of an MLE integrative process—we still await more detailed proposals here. At present, we know that PPS is the result of integrative processes, and we have a plausible account of how some of these integrated elements themselves are brought together. This can give us some insight into how the processes might go.

There are some important lessons here. At even the most basic levels, we see mental systems with their own ways of doing things. There are multiple distinct informational channels that provide overlapping information about the world and our place in it. These channels are not distinct, however. They interact and coordinate with one another at multiple levels, applying a variety of flexible rules that balance the need for internal coherence and the survival requirements of accuracy. Such coordination requires solving many practical and computational problems. The intermediate spatial frame of reference is the result of these coordination problems and the ways our nervous systems typically solve them. It is an elegant solution, although involving the complex interplay of many dissociable elements. We need to coordinate our perceptual inputs with our motor systems smoothly. The information about where things are located needs to be converted into information about where to move in the world based on where we are currently located and moving. PPS provides the bridge connecting these domains, and provides a shared coordinate system for active engagement. From the perspective of the subject, however, it is almost invisible. We only know about it because of pathological dissociations and from extremely novel experimental set-ups like those discussed above.

All this is a useful lesson for thinking about the sense of self. It too can seem completely transparent to experience. Cases of dissociation reveal its fissures and constructed nature, but don't completely reveal the nature of the constituents or how they might be coordinated. PPS is thus something of a proxy for these larger coordination problems. Thankfully, this is not just a helpful analogy. PPS itself plays a critical role in bodily self-awareness, and so we can start to make some small progress in understanding how the sense of self is constructed.

14.3 Peripersonal space and bodily self-awareness

We have seen that PPS is constructed by coordinating inputs from many distinct sensory and bodily networks. This process involves integrating the potential disagreements from the contributing elements in order to have a coherent and useful functional output. We started this discussion by wondering how the self could be constructed, and about whether or not the processes that generated PPS could apply as well to the self as a whole. A key element here is that the information generated by PPS involves information not principally about external stimuli, but about the self (the bodily self and its active engagement with the world). Indeed, the evidence here is promising, in particular because PPS seems to be a critical element of our sense of bodily self-awareness (BSA).[8]

BSA involves our awareness of ourselves as a bodily self. While it is a form of awareness, and so typically conscious, it's also something that involves a lot of under-the-hood activity and (like vestibular awareness and PPS) is often recessive or implicit in experience. It does not stand out in the same way as a visual or haptic experience—but exerts its influence in more subtle ways. For this reason, I want to remain neutral here about the extent to which a subject is consciously aware of, or able to access, the elements of bodily awareness. What *is* important to me is that BSA seems to be the result of multisensory integration.[9] As Noel et al. (2015) describe it:

> A fundamental aspect of our sense of self as subject of conscious experience is the experience of the bodily self, that is, the feeling of being located in a particular space within a body we own and control. Empirical data demonstrate that the feeling of owning a body (self-identification), as well as the sense of being located within the boundaries of that body (self-location), are fundamentally rooted in the congruent and cohesive integration of multiple sensory modalities within the spatio-temporal dimensions of the physical body. (p. 55)
>
> Reproduced from Noel, J.-P., Pfeiffer, C., Blanke, O., and Serino, A. (2015). Peripersonal space as the space of the bodily self. Cognition, 144:49–57.

On this view, bodily self-awareness consists of two other more fundamental elements.[10] One is the sense of bodily ownership, the sense that one's body is one's own. The other is bodily self-location, the sense of where one is located and the limits of one's bodily self. There are many debates in the literature about the nature of these elements. Some, like Martin (1992), collapse the distinction, holding that the felt apparent limits of one's body capture the sense of bodily ownership (we feel as ours that which is within the apparent limits of our bodies). Others defend phenomenal accounts of ownership that can pull apart (in some cases) from our sense of bodily location (De Vignemont 2018). Others defend more cognitive accounts of the sense of ownership as a kind of judgement rather than a basic feeling.

The evidence suggests that at least one key aspect of BSA—the element of self-location—might be partially accounted for by PPS. Peripersonal space might provide the coordinates in which we feel ourselves to be located. If this is right, then perhaps PPS provides the frame for a key element of our sense of bodily self, and does so through processes of coordination and integration.

Let me briefly describe two recent studies that strongly suggest a key role for PPS in the construction of bodily self-awareness.

Noel et al. (2015) used a full body illusion (FBI) to manipulate both felt bodily location and self-identification. As they describe it:

> During the FBI subjects see a virtual body (avatar), placed 2 m in front them, being stroked, while synchronously receiving a congruent tactile stimulation on their physical body. Under such circumstances participants report to identify with the virtual body (change in self-identification), and feel displaced toward the virtual body (change in self-location). (p. 49)

---

[8] Sometimes also called 'bodily self-consciousness'.

[9] See de Vignemont (2014) for a persuasive argument for this claim.

[10] This list is not standard. Others, for instance Bermúdez (2018), offer a more fine-grained taxonomy of the primitive forms of bodily self-awareness. Like Bermúdez, I think there are more than two constituents involved in BSA, but the focus on these two is still a useful explanatory simplification.

This illusion demonstrates the malleability of these elements of bodily self-awareness. The authors test the hypothesis that PPS representation is bound not to the physical body, but to the illusory location of the body during FBI. They first induced FBI in subjects using a camera and head-mounted display, which allowed subjects to witness, from behind, a virtual image of their own bodies as they experience either synchronous or asynchronous visuo-tactile stimulation. They determined the spatial limits of the subject's PPS using reactions to incoming audio signals (generated by an array of speakers on either side of the subject). Afterwards, through a questionnaire, they assessed changes in a subject's bodily self-consciousness. They found that the synchronous condition did indeed extend the limits of PPS:

> When participants received a tactile stimulation on their physical body while viewing a synchronous stimulation administered to a virtual body seen at a distance, they reported a greater feeling of being directly touched by the stimulus touching the virtual body, of feeling touch at the location of the virtual body, and of feeling to drift forward toward the virtual body, indicating a shift in the experienced location of the self from their physical body toward a virtual replacement of it. (Noel et al. 2015, p. 54)
>> Reproduced from Noel, J.-P., Pfeiffer, C., Blanke, O., and Serino, A. (2015).
>> Peripersonal space as the space of the bodily self. Cognition, 144:49–57.

This demonstrated that the coordinated integration of tactual, visual, and auditory elements was constructing a novel spatial representation of peripersonal space. This extended space, generated by the activity of integration, then produced a strong effect on the subject's own sense of bodily awareness. As they summarized, this finding is significant:

> The focus and main new finding from the present study is that the FBI was associated with a shift in the representation of the PPS ... More importantly, we show that the center of the PPS representation is not bound to the physical body, but it is centered at the experienced location of the self. Normally self-location and body location coincide, and so does PPS. However, if body location and self-location are dissociated, for instance by means of conflicting multisensory stimulation, PPS representation shapes congruently with the change in self-experience. More generally, the present findings suggest that PPS can be considered as a representation of the self in space, which may mediate interactions between the individual and the environment. (Noel et al. 2015, p. 55)
>> Reproduced from Noel, J.-P., Pfeiffer, C., Blanke, O., and Serino, A. (2015).
>> Peripersonal space as the space of the bodily self. Cognition, 144:49–57.

This is one potential gloss of the findings. If correct, it would suggest that an important constituent of BSA, the sense of self-location, is either coextensive with, or depends upon, the centred frame of PPS. The self is located, in other words, at the center of PPS. What are not quite settled in this discussion are the dependence relations between bodily location, PPS and BSA. For instance, it is not clear that *any* extension of PPS generates variations in BSA. Using a tool, for instance, is known to extend PPS (Farnè and Làdavas 2000; Canzoneri et al. 2013), but it is not clear that holding a tool always involves a change in self-location. When I pick up a tennis racket, I may start to integrate multisensory spatial inputs at a great distance from my felt location, extending PPS, but this does not seem to alter the felt location of my bodily self in any way. What is unique about the FBI data above is that the extension of PPS occurs during a full body illusion, in which the illusion is pushing the sense of self forward.

Another recent study by Salomon et al. (2017) supplies some additional evidence for thinking that PPS plays a critical role in the generation of self-location. More interestingly, they show that such integration can occur *unconsciously*: '[I]ntegration of bodily signals within the peripersonal space (PPS) underlies the experience of the self in a body we own (self-identification) and that is experienced as occupying a specific location in space (self-location), two main components of bodily self-consciousness (BSC). (Salomon et al. 2017, p. 174).'

The authors investigated whether the signals that mediate both PPS and its role in BSA could be unconscious. They once again induced FBI in a subject (as in the Salomon et al. study above) and this time, instead of an array of audio speakers, they used the method of continuous flash suppression (CFS) to present subjects with a looming visual object that was, for the subject, not conscious. They did this by showing a moving Mondrian pattern to the subject's dominant eye, while presenting a looming virtual object to the weaker eye. The strong pattern in the dominant eye suppressed conscious awareness of the other signal (CFS is itself a wonderful example of competitive interaction). But the information from the weaker eye was still processed and taken up by the visual system. In addition, the unconscious signal influenced PPS representations. This allowed the authors to construct a series of experiments that demonstrated facilitation effects for the integration of the unconscious visual signals, and the ability to manipulate PPS boundaries with unconscious visual stimuli. As they summarize: 'The present study brings novel comprehensive evidence that multisensory integration in PPS does not require conscious awareness and, importantly, that these unconscious multisensory processes modulate the phenomenological content of BSC [bodily self-consciousness]. (Salomon et al. 2017, p. 181).'

We have now seen two promising studies purporting to show a solid link between PPS and self-location. Obviously, we need more than two studies (see also Noel et al. 2018 for a more recent summary).

The role played by peripersonal space suggests the following highly speculative hypothesis: the self (itself) might just be what we get when all the more basic elements of are brought into coordination by some of the principles of integration discussed above. This coordination would involve multiple sensory modalities, the integration of many distinct body-part centred frames of PPS, and even more far-flung functional capacities and elements (emotional, affective, homeostatic, and cognitive drives and wants, for examples). The main upshot (I think) is that the self, despite lacking a central agent, decider, or constructor, can be bundled together through processes of integration to generate something more than the sum of their parts.

14.4 Bodily self-awareness and the sense of self

We have now considered some of the evidence that PPS representations are the result of flexible, decentralized processes of integration, and that these representations seem to play an essential role in grounding our sense of bodily awareness. Now we turn to the relation between bodily self-awareness and our overall sense of self. This section is even more speculative than what has come before.

Philosophical investigations into self-awareness often focus on high-level capacities like introspection and our internal awareness of our own intentional states. Many works focus especially on these forms of self-awareness (see, for example, Carruthers 2011). Returning to our original examples, we can see that Clive Wearing lost an important element of this capacity. Without a hippocampus, he was unable to store and access new memories. But his deficits were even more severe. He was also unable to assess his other intentional states or determine his own intentions, desires, and abilities (see Sacks 2007 for the details).

The focus here on PPS and BSA suggest that an emphasis on these high-level, top-down capacities can obscure the important structure of actual self-consciousness, which is often hidden and implicit. Ian Waterman's loss, for example, involved little to no cognitive deficits. What he lost was something much more basic. Reflecting on the discussion above, we can see that the inputs he lost are typically essential for constructing PPS representations. Waterman was unable to locate his body directly, and therefore was missing the essential self-location coordinates that would be needed to generate PPS representations. We can also explain his ability to eventually recover bodily control through visual feedback. Vision after all is a major contributor to PPS representations, and so could be used to partially make up for his proprioceptive deficits.

All these reflections suggest a deep and important connection between bodily self-awareness and our overall sense of self. The feeling of bodily location and ownership especially seems to be an essential ingredient of our overall sense of self.

Bermúdez (2018) begins his recent volume of essays on the bodily self by highlighting the foundational role played by self-awareness in grounding our awareness of the environment, of our own thoughts, and the capacity for constructing rich narratives about our lives: 'The fundamental source for these (and many other) abilities and achievements is our capacity for self-consciousness, or self-awareness' (p. 1). In the later essays, he defends the idea that our overall capacity for self-awareness is built out of more fundamental forms of self-awareness, including, especially, the many elements of bodily awareness. Such primitive forms of awareness, like somatic proprioception, provide information about the embodied self, mark the boundary between self and other, and open up the body as a target of introspective awareness (Bermúdez 2018, 40–41).

These capacities plausibly provided by bodily awareness suggest a sense of self that is not only composite, but also one structured by more fundamental elements that open up increasingly sophisticated forms of self-awareness. Introspecting my fine-grained desires or constructing elaborate narratives about my exploits are certainly important aspects of my sense of self, but deploying them seemingly depends on the more fundamental forms of bodily awareness subserved by PPS.

The many increasingly fantastic deviations of self-awareness found in Dennett's classic piece, 'Where am I?', involve disruptions of bodily self-awareness (found in Dennett 1981). It seems that the body, and our awareness of it, plays a major role in the felt unity of self. This is far from a novel or recent idea. Consider the following quote from William James: 'The body is the innermost part of the material Self in each of us; and certain parts of the body seem more intimately ours than the rest' (1890, chapter 10). Awareness of the body is for him, too, an essential part of the self. As he continues, when describing our active, conscious awareness of self: 'For this central part of the Self is felt ... just as the body is felt, the feeling of which is also an abstraction, because never is the body felt all alone, but always together with other things' (1890, chapter 10).

What the present discussion adds to these prior discussions is a focus on the mechanisms by which the more primitive forms of bodily awareness might be brought into direct contact and cohesion with the more psychological capacities. These mechanisms are those that realize the heterogeneous class of interactions that I have called 'cooperative integration'. The overall sense of self is not constructed actively by the subject or through rational reflection; nor is it constructed by any dedicated process or system. Instead, it is the result of distributed, flexible rules of cooperative integration. This perspective brings the self into direct alignment with work across a wide range of psychological domains, where such integration problems are widely acknowledged and studied.

The crucial questions to ask here concern the ways in which the bodily self connects with the psychological self (broadly understood). We might expect interaction effects between them, disagreements that need to be rectified, relational involvements with emotions, ideas, narrative, memory, perception, and the like. In other words, the connections between bodily awareness and the overall sense of self will be like everything else: distributed, heterogeneous, occurring at multiple levels and using a variety of principles of integration that are flexible and task sensitive. Bodily self-awareness would not be a completely self-standing component, but one that deeply interacts with, indeed fuses with, the many other self-related systems.

14.5 Conclusion: future directions

Let us summarize the various claims that we have made in this chapter. First, we started with the assumption that we have a unified sense of self, that we are aware of ourselves as unified, coherent agents. This capacity connects many disparate elements: various cognitive capacities, sensory experiences, emotional reactions, desires, and actions. There is an inherent puzzle about the mechanisms by which these heterogeneous, and often conflicting, elements are smoothed out to generate a unified sense of self-awareness. Looking at the example of peripersonal space, I suggested that we already have excellent evidence for the principles by which the brain solves these sorts of integration problems. Rather than a single central mechanism that brings everything together, the brain seems to use flexible, non-centralized rules of integration to solve such problems. The case of peripersonal space is an excellent illustration of how the brain solves these sorts of problems. It is doubly strong because it plays an important role in grounding our sense of bodily self-awareness.

The view, to be clear, is not that anytime you have integration you have a sense of self. It is instead the claim that the sense of self is nothing but the result of integrating, along the same principles used to generate peripersonal space, its constituting elements—its various experiences, beliefs, desires, motivations, emotions, memories, and actions. We focused here on the cohesion and integration of bodily self-awareness, and have only speculated on extensions to these other domains. But we can now, perhaps more clearly, see how these same processes might function to reconcile other potential discontinuities. For instance, we can see the principles of confabulation as operating to reconcile a subject's incompatible beliefs, desires, and experiences. This might function along the lines we've already seen: weights assigned to the various elements, a cooperative exchange ensues, and the most likely, coherent, and adaptive outcome is the one experienced. If this is correct, then a close examination of peripersonal space should have important implications for a wide range of important discussions throughout all levels of our psychology.

Acknowledgements

I am very grateful to Frédérique de Vignemont, Jonathan Cohen, Andrea Serino, and audiences at Kyoto University and UC Santa Barbara for extremely helpful feedback on earlier versions of this chapter.

References

Angelaki, D. E., & Cullen, K. E. (2008). Vestibular system: The many facets of a multimodal sense. *Annual Review of Neuroscience*, 31:125–150.

Baars, B. J. (2005). Global workspace theory of consciousness: Toward a cognitive neuroscience of human experience. *Progress in Brain Research*, 150:45–53.

Bermúdez, J. L. (2018). *The Bodily Self: Selected Essays*. Cambridge, MA: MIT Press.

Campos, J., & Bulthoff, H. (2012). Multimodal integration during self-motion in virtual reality. In *The Neural Bases of Multisensory Processes*, pages 603–628. Boca Raton (FL): CRC Press.

Campos, J. L., Butler, J. S., & Bulthoff, H. H. (2012). Multisensory integration in the estimation of walked distances. *Experimental Brain Research*, 218(4):551–565.

Canzoneri, E., Ubaldi, S., Rastelli, V., Finisguerra, A., Bassolino, M., & Serino, A. (2013). Tool-use reshapes the boundaries of body and peripersonal space representations. *Experimental Brain Research*, 228(1), 25–42.

Carruthers, P. (2011). *The Opacity of Mind: An Integrative Theory of Self-Knowledge*. Oxford: Oxford University Press.

Chandrasekaran, C. (2017). Computational principles and models of multisensory integration. *Current Opinion in Neurobiology*, 43:25–34.

Cole, J. (1995). *Pride and a Daily Marathon*. Cambridge, MA: MIT Press.

Corkin, S. (2013). *Permanent Present Tense: The Unforgettable Life of the Amnesic Patient, H.M.* New York: Basic Books.

Dennett, Daniel (1991). *Consciousness Explained*. New York: Penguin Books.

Dennett, D. C. (1981). *Brainstorms: Philosophical Essays on Mind and Psychology*. Cambridge, MA: MIT Press.

Dennett, D. C., & Kinsbourne, M. (1992). Time and the observer: The where and when of consciousness in the brain. *Behavioral and Brain Sciences*, 15(2):183–201.

de Vignemont, F. D. (2014). A multimodal conception of bodily awareness. *Mind*, 123(492):989–1020.

de Vignemont, F. (2018). *Mind the Body: An Exploration of Bodily Self-Awareness*. Oxford: Oxford University Press.

Derdikman, D., & Moser, E. I. (2010). A manifold of spatial maps in the brain. *Trends in Cognitive Sciences*, 14(12):561–569.

Ernst, M. O., & Banks, M. S. (2002). Humans integrate visual and haptic information in a statistically optimal fashion. *Nature*, 415(6870):429–433.

Farnè, A., & Làdavas, E. (2000). Dynamic size-change of hand peripersonal space following tool use. *NeuroReport*, 11(8), 1645–1649.

Fetsch, C. R., Turner, A. H., DeAngelis, G. C., & Angelaki, D. E. (2009). Dynamic reweighting of visual and vestibular cues during self-motion perception. *Journal of Neuroscience*, 29(49), 15601–15612.

Frissen, I., Campos, J. L., Souman, J. L., & Ernst, M. O. (2011). Integration of vestibular and proprioceptive signals for spatial updating. *Experimental Brain Research*, 212(2):163–176.

Fulkerson, M. (2011). The unity of haptic touch. *Philosophical Psychology*, 24(4):493–516.

Fulkerson, M. (2014). Rethinking the senses and their interactions: The case for sensory pluralism. *Philosophical Psychology*, 5:1426.

Helbig, H. B., & Ernst, M. O. (2007). Optimal integration of shape information from vision and touch. *Experimental Brain Research*, 179(4):595–606.

Holmes, N. P., & Spence, C. (2004). The body schema and multisensory representation(s) of peripersonal space. *Cognitive processing*, 5(2):94–105.

James, W. (1890). *Principles of Psychology, Vol. 1*. New York: Henry Holt and Co. doi:org/10.1037/10538-000

Jones, J. A., & Jarick, M. (2006). Multisensory integration of speech signals: The relationship between space and time. *Experimental Brain Research*, 174(3):588–594.

Kappers, A. M. L. (2007). Haptic space processing–Allocentric and egocentric reference frames. *Canadian Journal of Experimental Psychology/Revue Canadienne de Psychologie Expérimentale*, 61(3):208–218.

Klatzky, R., & Lederman, S. (2010). Multisensory texture perception. In *Multisensory Object Perception in the Primate Brain*, pages 211–230. New York: Springer.

Martin, M. (1992). Sight and touch. In Crane, T. (ed.), *The Contents of Experience*, pages 196–216. Cambridge: Cambridge University Press.

Millar, S. (2008). *Space and Sense. (Essays in cognitive psychology)*. New York: Psychology Press.

Morf, C., & Mischel, W. (2012). The self as a psycho-social dynamic processing system: Toward a converging science of self-hood. In Leary, Mark R., & Tangney, June Price (eds), *Handbook of Self and Identity*, pages 21–49. New York: Guilford Press.

Noel, J.-P., Blanke, O., & Serino, A. (2018). From multisensory integration in peripersonal space to bodily self-consciousness: From statistical regularities to statistical inference. *Annals of the New York Academy of Sciences*, 1426(1), 146–165.

Noel, J.-P., Pfeiffer, C., Blanke, O., & Serino, A. (2015). Peripersonal space as the space of the bodily self. *Cognition*, 144:49–57.

Northoff, G., Qin, P., & Feinberg, T. E. (2011). Brain imaging of the self-conceptual, anatomical and methodological issues. *Consciousness and Cognition*, 20(1):52–63.

O'Callaghan, Casey (2008). Seeing what you hear: Cross-modal illusions and perception. *Philosophical Issues* 18(1):316–338.

O'Callaghan, Casey (2017). Grades of multisensory awareness. *Mind and Language*, 32(2):155–181.

O'Callaghan, C. (2012). Perception and multimodality. In Margolis, E., Samuels, R., & Stich, S. (eds), *Oxford Handbook of Philosophy of Cognitive Science*. Oxford: Oxford University Press.

Rizzolatti, G., Fadiga, L., Fogassi, L., & Gallese, V. (1997). The space around us. *Science*, 277(5323):190–191.

Sacks, O. (2007). The abyss. *The New Yorker*, 24:38–42.

Salomon, R., Noel, J.-P., Łukowska, M., Faivre, N., Metzinger, T., Serino, A., & Blanke, O. (2017). Unconscious integration of multisensory bodily inputs in the peripersonal space shapes bodily self-consciousness. *Cognition*, 166:174–183.

Schechtman, Marya (2011). The narrative self. In Gallagher, Shaun (ed.), *The Oxford Handbook of the Self*. Oxford University Press.

Seilheimer, R. L., Rosenberg, A., & Angelaki, D. E. (2014). Models and processes of multisensory cue combination. *Current Opinion in Neurobiology*, 25:38–46.

Shimojo, S., & Shams, L. (2001). Sensory modalities are not separate modalities: plasticity and interactions. *Current Opinion in Neurobiology*, 11(4):505–509.

Strawson, Galen (2009). *Selves: An Essay in Revisionary Metaphysics*. New York: Oxford University Press.

Wilson, B. A., & Wearing, D. (1995). Prisoner of consciousness: A state of just awakening following herpes simplex encephalitis. In R. Campbell, & M. A. Conway (eds), *Broken Memories: Case Studies in Memory Impairment*, pages 14–30. Blackwell Publishing.

Wu, W. (2011). Confronting many-many problems: Attention and agentive control. *Noûs*, 45(1):50–76.

15

The social dimension of peripersonal space

Yann Coello and Tina Iachini

15.1 Introduction

Interacting efficiently with the environment requires anticipating at every moment what behaviour can be performed depending on the context. Since motor actions are the only way in which organisms can interact with the environment, the cognitive processes flexibly deployed in the interaction with the environment must take into account the state and capabilities of the body whenever it might be relevant for upcoming action in response to external events. Body schema and motor capabilities determine thus whether, in a particular situation, one can consider performing a voluntary action towards a specific object or not. Therefore, before taking any decision, the brain, in a way, must process sensory information from the external world while taking into account information from the body. This implies that the brain retains a functional representation of the environment, which depends not only on the location of objects in space and the body state, but also on the outcome of previous interactions with environmental features (Grüsser, 1983; Hall, 1966; Previc, 1998). Within this functional representation, the peripersonal space (PPS) specifies the limited space encircling the body and dedicated to the interaction with objects located at hand's reachable distance (Bufacchi & Iannetti, 2018; Coello & Iachini, 2016; de Vignemont & Iannetti, 2015; di Pellegrino & Làdavas, 2015; Rizzolatti et al., 1981). As regards the link between PPS and motor actions, objects' processing in PPS involves not only the integration of multisensory information relating to the objects and the body, but also of motor information in relation to the objects' affordances (Graziano, Reiss, & Gross, 1999; Makin, Holmes, & Zohary, 2007; Rizzolatti et al., 1981). This embodied processing of objects is supported by a large subcortical and cortical brain network connecting areas contributing to the processing of sensory and somatosensory information, and also the motor productions (Brozzoli et al., 2012; Cléry, et al., 2015; Holmes & Spence 2004; Serino, Canzoneri, & Avenanti, 2011). Accordingly, PPS must be viewed as a dynamic representation of the space around the body subserving primarily the organization of goal-directed behaviours towards stimuli with the highest reward value. It must also be viewed as a space where potentially harmful stimuli receive specific attention in order to protect the body from the hazards ahead. In the present chapter, we will highlight the anticipatory motor nature of PPS representation and its dynamic properties. We will also show that stimuli in PPS receive particular attention that fosters perceptual and cognitive processes. We finally will propose that PPS serves as a mediation zone between the body and the environment, protecting the body from external threats and, as such, contributing to the organization of the social life.

Yann Coello and Tina Iachini, *The social dimension of peripersonal space* In: *The World at Our Fingertips*. Edited by Frédérique de Vignemont, Andrea Serino, Hong Yu Wong, and Alessandro Farnè, Oxford University Press (2021). © Oxford University Press.
DOI: 10.1093/oso/9780198851738.003.0015.

15.2 The motor nature of peripersonal space

Brain-imaging studies revealed that object processing in PPS activate not only the multisensory (visual, auditory, olfactory ...) brain areas but also the sensorimotor network including the posterior parietal and ventral premotor cortices (Chao & Martin, 2000; Chao, Weisberg, & Martin, 2002; Creem-Regehr & Lee, 2005; Kan et al., 2006; Martin, 2007), as well as the somatosensory cortex (Cardellicchio, Sinigaglia, & Costantini, 2011; Grafton et al., 1997; Matelli, Luppino, & Rizzolatti, 1985). In this regard, Proverbio (2012) reported a modulation of the brain cortical activity when participants observed images of manipulable objects, compared to non-manipulable objects. Time-frequency analysis of EEG signals revealed a desynchronization of the µ rhythm over the centro-parietal region (event-related attenuation of 8–13 Hz cortical oscillation), similar to that observed when executing or observing a voluntary action (Babiloni et al., 1999; Cochin et al.,1999; Llanos et al., 2013; Salenius et al., 1997). More recently, it was found that the µ rhythm desynchronization over the centro-parietal region was present when categorizing manipulable objects according to a perceptual or semantic criterion, but only for objects located in PPS (Kalenine et al., 2016; Wamain, Gabrielli, & Coello, 2016). Moreover, the µ rhythm desynchronization reduced progressively from reachable to unreachable stimuli (Wamain et al., 2016) suggesting a continuity between the representation of the peripersonal space and that of the extrapersonal space (Bufacchi & Iannetti, 2018). Interestingly, it was demonstrated that, in the presence of manipulable objects in PPS, the µ rhythm desynchronization over the centro-parietal region was reduced when the objects evoked not a single one, but multiple motor affordances (Kalenine et al., 2016; Wamain et al., 2018). This is the case when one compares the perception of a bottle (same gesture for manipulation and functional use) to that of a soap-dispenser (different gestures for manipulation and functional use). This finding suggests that the motor encoding of objects in PPS depends on objects' affordances, which includes sorting out the competition between various possible motor activations in the processing of action–object relationship. These observations support the idea that PPS is an action space represented on the basis of motor information similar to that involved in the planning of voluntary actions (Coello & Iachini, 2015). In agreement with this, Coello et al. (2008) found that transcranial magnetic stimulation applied at low frequency to the left motor cortex interferes with perceptual judgements only for stimuli presented in PPS. Likewise, Cardellicchio et al. (2011) stimulated magnetically the left primary motor cortex and recorded motor-evoked potentials (MEPs) while participants were observing manipulable and non-manipulable objects located either in their peripersonal or extrapersonal space. They found higher MEPs during the observation of objects when they were located in PPS.

In order to further explore how motor resources contribute to the processing of objects in PPS, Iachini et al. (2014) used a spatial localization task combined with motor interference. While keeping the dominant hand free or blocked, observers had to judge if manipulable and non manipulable stimuli in peripersonal or extrapersonal space were on the right/left of their own body midline. Consistent with the literature on motor-related effects in object perception, the results showed that manipulable stimuli were more accurately localized with free than blocked arms (Chao & Martin, 2000; Ellis & Tucker, 2000; Iachini, Borghi & Senese, 2008; Tucker & Ellis, 1998, 2001, 2004). However, the crucial finding was a facilitation in peripersonal space when observers located both manipulable and non-manipulable stimuli with their free arms. In contrast, there was no effect of the motor manipulation in extrapersonal space. Therefore, by showing that spatial localization in PPS is facilitated if we can move our hands without hindrance, we can conclude that the nature of this space is intrinsically motor. Any event occurring in PPS, involving either manipulable or non-manipulable stimuli, requires rapid reactions. For example, we naturally withdraw our hand if we hear a strong noise or see a lightening or splinters of glass (e.g. Huang et al., 2012; Symes, Ellis, & Tucker, 2007; for a review, Fischer & Zwaan, 2008). The literature suggests that motor resources may be specified at different degrees of abstraction: from specific grips to more general grasping movements and to generic movements (see Borghi & Binkofski, 2014; Fischer & Zwaan, 2008; Thill, et al., 2013). This leads us to propose that the involvement of motor resources in PPS may also occur at different degrees, with an 'abstract' motor pre-activation as a prerequisite for future actions (e.g. Anderson, Yamagishi, & Karavia, 2002; Bourgeois & Coello, 2012; Phillips & Ward, 2002; Symes, Ellis, & Tucker, 2007), which implyies that limb movements are not constrained (Iachini et al., 2014b). On this basis, a stimulus entering the PPS margin would prime appropriate action programmes, as shown by previous studies (Caggiano et al., 2009; Cardellicchio et al., 2011; Costantini et al., 2010).

The findings described so far are consistent with the idea that the motor nature of PPS may reflect the adaptive need to anticipate what may happen near the body and preparing to react in time. Any stimulus suddenly appearing within the peripersonal boundary would prompt either a quick approach in case of attractive objects or a rapid avoidance in case of threatening stimuli (e.g. Coello & Delevoye-Turrell, 2007; Huang et al., 2012). In any case, the action/safety function of PPS requires rapid reactions and hence the full availability of motor resources (e.g. Graziano, 2017; Graziano & Cooke, 2006; Iachini et al., 2017; Ruggiero et al., 2017). Accordingly, it may be expected that people with a motor deficit also show an altered representation of PPS.

15.3 Alteration of peripersonal representation in populations with sensorimotor deficits

Assuming a causal role of the motor system in the representation of the PPS, we hypothesized that patients with impaired sensorimotor system should be deficient both in the organization of voluntary motor action and the representation of PPS. To address this issue, Bartolo et al. (2014) analysed motor performances and the representation of PPS in neurologic patients with hemiplegia resulting from brain damage localized in either their right or the left hemisphere. They found that right-handed stroke patients with right brain damage associated with left hemiplegia were specifically impaired when performing an actual or imagined sequential motor task with the healthy hand (finger–thumb touch task). Such deficit was not observed in left-brain-damaged patients, even when characterized by the same level of hemiplegia in the contralateral arm. This study also revealed that patients with right hemisphere brain damage had more difficulties in representing their PPS, even when perceptual judgements were provided in reference to the healthy arm, providing further evidence that motor processes associated with voluntary actions contribute to the encoding of PPS (see also Bartolo et al., 2017). Beyond confirming the motor nature of PPS, these results also support the view that the brain hemispheres are specialized in specific components of voluntary motor actions (Schaefer, Haaland, & Sainburg, 2007, 2009). Indeed, according to the hemispheric theory of motor control, the right hemisphere would predominantly contribute to specifying the spatiotemporal features of the motor responses, whereas the left hemisphere would be predominantly involved in the online control of motor actions especially in the context of complex motor skills (Sainburg, 2002, 2005; Sainburg & Kalakanis, 2000; Sainburg et al., 1999).

More recently, Toussaint et al. (2018) highlighted that even transient sensorimotor deficits lasting for several hours can alter the representation of PPS. Using a short-term limb immobilization paradigm consisting in immobilizing the right upper limb for 24 hours, they reported shrinkage of PPS representation after the immobilization period, which was not observed in the non-immobilized control group (see Bassolino et al., 2012, for congruent findings). No effect of arm immobilization was observed in non-spatial perceptual tasks, suggesting herein the absence of specific visual perception or decisional deficits in the limb immobilization group (Toussaint et al., 2018). Restricting upper-limb movements for even a shorter period of time was found to alter cortical excitability of the motor regions dedicated to limb control (Avanzino et al., 2011; Facchini et al., 2002; Huber et al., 2006), resulting in decreased activity in the contralateral motor cortex (Avanzino et al., 2011; Facchini et al., 2002; Huber et al., 2006). Behavioural studies highlighted the negative impact of arm immobilization on the control of motor actions. Specifically, immobilization-induced effects associated with the decrease of input/output signal processing were reflected in a reduction of movement spatial accuracy (Huber et al., 2006) and coordination (Moisello et al., 2008), as well as motor imagery (Meugnot, Almecija, Toussaint, 2014; Meugnot & Toussaint, 2015; Toussaint & Meugnot, 2013). Considered together, these results confirm the crucial role of the motor system in the representation of PPS. They also highlight the plasticity of the motor system resulting in a rapid change of its activity following alteration of the sensorimotor system, with concomitant effect on PPS representation.

15.4 The dynamic properties of peripersonal representation

Overall, the data presented so far strongly suggest that PPS encoding involves the motor system and depends on the ability to anticipate the effects of deployable actions in the environment either to interact with incentive stimuli or to avoid potentially harmful stimuli. Accordingly, PPS should be viewed as a dynamic motor interface between the body and the environment that is continually updated as a function of both the configuration of the body schema and the expected outcome of acting on the available stimuli.

15.4.1 The peripersonal space depends on the body schema

The plastic features of spatial and bodily representations, together with their involvement in motor control, have raised the possibility that the PPS and the body schema are tightly related concepts (Cardinali, Brozzoli, & Farnè, 2009). The idea that the PPS depends on the disposition of the body to interact with the environment is supported by studies on sensorimotor adaptation to tool use in animals (Iriki, Tanaka, & Iwamura, 1996), as well as in humans (Cardinali et al., 2012; Costantini et al.,2011; Farnè, Iriki, & Làdavas, 2005; Witt, Proffitt, & Epstein, 2005). Recently, Bourgeois, Farné, and Coello (2014) demonstrated that the distance at which objects are perceived as reachable with the hand is widened after using a tool artificially extending the length of the arm (a 70-cm long tool), but not after using a short tool providing no functional extension of the arm (a 10-cm long tool). The authors' interpretation was that, through use, the tool was progressively incorporated as a body effector resulting in an elongated representation of arm length in the body schema (Cardinali et al., 2011). As a result of sensorimotor integration, perceptual thresholds change with unreachable objects appearing to be reachable. The selectivity of these findings provided a compelling demonstration that, for the tool to be effective in shaping PPS, a functional benefit to the arm is necessary.

15.4.2 Peripersonal space depends on motor predictive models

Internal models of limb dynamics are used to predict in advance the feasibility and the consequences of performing a voluntary motor action (Wolpert, Ghahramani & Jordan, 1995; Kawato, 1999; Bursztyn et al., 2006; Cullen & Brooks, 2015; Ghez & Sainburg 1995; Tanaka & Sejnowski, 2013). These models operate as predictive systems that anticipate the sensory consequences associated with a particular motor action (Wolpert & Kawato, 1998; Wolpert, Ghahramani & Flanagan, 2001; Pickering & Clark, 2014), and underlie sensorimotor adaptation and learning (Shadmehr, 2017). For instance, whenever the dynamic characteristics of a motor effector change, predictive models are quickly updated so as to maintain a high level of motor performance (Ostry et al., 2010). We may thus expect that changing the predictive models associated with the anticipated outcomes of the actions would modify the representation of PPS. This hypothesis was tested by Bourgeois and Coello (2012), who used an original paradigm consisting in modifying the relation between the predicted and observed outcome of a voluntary motor action. In this study, participants performed a series of manual reaching movements towards a visual target in a situation where the visual target was visible but not the moving hand. Visual feedback was nonetheless given at the end of each manual acquisition of the visual target, to provide participants with estimates of the accuracy of their motor performance in relation to the location of the target, on a trial-by-trial basis. The visual feedback was accurate for one group of participants, but for two other groups it was shifted in depth by 1.5 cm, either further or closer to the actual position of the hand. Throughout movement rehearsals (blocks of 60 trials), participants adapted their responses to the shifted feedback. The same procedure was reiterated until the shift between target location and the feedback about motor performance reached 7.5 cm (requiring thus five successive blocks of adaptation), although not consciously noticed by the participants. The interesting aspect of the study was that the target used in the manual reaching task was also displaced by 1.5 cm in each adaptation block so that the actual amplitude of the hand displacement remained the same at the end of each adaptation period. PPS representation was assessed after each adaptation block throughout a reachability judgement task. The data showed that the motor adaptation was achieved following few trial rehearsals in each block. Furthermore, the representation of PPS followed the motor adaptation in a predictive way. PPS increased when sensorimotor adaptation entailed an increase of target distance for similar movement amplitude, whereas PPS shrank when sensorimotor adaptation entailed a decrease of target distance for similar movement amplitude. Accordingly, modifying the predictive models of limb dynamics with the result of changing the anticipated sensory consequences associated with a particular action affects the representation of PPS. This indicates that not only body schema, but also motor internal models contribute to the representation of PPS (see Leclere et al., 2019, for additional evidence).

15.4.3 Peripersonal space depends on the expected objects' value

It is worth noting that anticipating the possibilities of acting in the environment depends not only on the disposition of the body to objects-oriented actions and motor-related predictive models, but also on the expected benefit from performing these motor actions. Thus, an important factor in the interaction with the environment concerns the physical aspect of objects—that is, their affordances (Gibson, 1979). Another important aspect concerns the benefit expected from acting on those objects (Gawronski, 2005; Lebrecht et al., 2012). In daily life, our attention is oriented towards reward-yielding stimuli and conversely deviated generally from stimuli associated with no reward (Desimone & Duncan, 1995; Akrami et al., 2018). Accordingly, one may expect the neural representation of PPS to be modulated by the value attributed to the stimuli in the environment. In accordance with this, Coello, Bourgeois, & Iachini (2012) showed that the presence of hazardous objects in the proximal space altered the representation of PPS. More precisely, PPS shrank when hazardous objects were located at a reachable distance with their threatening part oriented towards the participants. This finding indicates that PPS representation is influenced by the valence of external stimuli, but according to a specific body–objects functional relationship.

The role of stimuli valence on the representation of PPS was further investigated in a study consisting in modifying the likelihood of getting a positive outcome when performing an object-oriented motor action (Coello et al., 2018). The participants were asked to select on a touch-screen table a series of ten items among a set of random stimuli. Once selected, each stimulus changed its colour by becoming either green (rewarded with one point, 50% of the stimuli) or red (rewarded with no point, 50% of the stimuli). The goal for the participants was to accumulate maximum points throughout the experimental session. Importantly, the touch-screen table was divided into a proximal and distal space of similar size, although this was not visible. In three independent groups, the probability to select a reward-yielding stimulus was either 50%–50%, or 25%–75% or 75%–25% in the proximal and distal spaces of the touch-screen table, respectively. The data revealed that the rate of stimuli selected in the proximal space was determined by the distribution of the stimuli generating a reward, reaching 55% in the 50%–50% condition, 31% in the 25%–75% condition and 73% in the 75%–25% condition. These results indicate that the participants were progressively sensitive, throughout the trials, to the spatial distribution of the reward-yielding stimuli, although not explicitly. When testing PPS representation, the authors observed that although participants on average overestimated what they belived they could reach (8.48% of arm length on average), the extent of PPS was influenced by the spatial distribution of the reward-yielding stimuli: PPS increased after having performed the stimuli-selection task in the 25%–75% condition, while it decreased after having performed the stimuli-selection task in the 75%–25% condition. This study, thus, clearly demonstrated that modifying the probability of getting a positive outcome when acting on objects alters the representation of PPS. This suggests that PPS representation emerges from the combination of information related to the disposition of the body for voluntary actions and the expected benefits from performing those actions. Considered as a whole, these data confirm that PPS operates as an interface between perception and action and, as such, underlies two complementary functions: first, it subserves goal-directed behaviours towards non-threatening stimuli; second, it supports defensive behaviours against threatening and potentially harmful stimuli (Coello et al., 2012; di Pellegrino & Làdavas, 2015; de Vignemont & Iannetti, 2015; Graziano & Cooke, 2006; Hunley & Lorenco, 2018). One must thus expect that the stimuli in PPS receive particular attention that fosters perceptual and cognitive processing.

15.5 The specificity of objects perception in peripersonal space

Taking advantage of the motor nature of PPS, perception (Constantini et al., 2010; Spence et al., 2004) and categorization (Blini et al., 2018) of objects are facilitated when the objects are located in PPS. Constantini et al. (2010), for instance, showed that the motor response triggered by a visual object was 25 ms faster when the visual object had a congruent orientation according to the orientation of the motor response. The perceptual facilitation due to the spatial alignment effect was, however, observed only for stimuli located in PPS. Perceptual facilitation for objects in PPS was also found in tasks where no object-oriented responses were expected. Blini et al. (2018), for instance, showed a persistent shape discrimination advantage (10 ms) for objects in PPS in a task implying the perceptually categorization of 3D visual stimuli as a cube or sphere. The perceptual facilitation was observed even when controlling for retinal size, binocular cues, or upper/lower visual field, firmly indicating that stimuli in PPS benefit from enhanced perceptual processing.

The specific way of processing perceptual information in PPS also emerges when was considered the capacity to predict possible collisions between dynamic stimuli. This capacity is a fundamental survival prerequisite for all moving species and, not surprisingly, people seem to have intuitive preconceptions concerning dynamic events. Studies with healthy adults have shown that collision prediction is based on the processing of various types of perceptual information that physically describe the event, and among them the position over time of the object has a crucial role (Proffitt & Gilden, 1989; Gilden & Proffitt, 1989). Clearly, the capacity to anticipate possible collisions is necessary for acting with moving stimuli, an ability that develops early in infancy (Eilan, Brewer, & McCarthy, 1993; Gray & Thornton, 2001; Zago & Lacquaniti, 2005). This anticipatory capacity should be even more crucial when the dynamic events occur within our PPS.

In a recent study, we directly explored whether the distance between our body and the moving objects may affect the way we process perceptual information in order to predict possible collisions (Iachini et al., 2017). More specifically, a collision judgement task was used to investigate how spatio-temporal information was processed in peripersonal and extrapersonal space in order to predict possible collisions. Collision and non-collision events were devised by manipulating independently the velocity and/or the path of two balls moving one towards the other at two distances from the participants' frontal plane: 30 cm and 120 cm. The results showed that the distance from the observer was crucial: the capacity to predict collisions in PPS was particularly affected by temporal parameters. There was a lower discrimination ability and a more liberal tendency to predict collisions in peripersonal, not extrapersonal space, when the velocity of the balls was different and the path was similar. Moreover, in this specific condition, individuals produced more false alarms—that is, they erroneously predicted a collision even though it was not physically possible. Some studies have shown that it is particularly difficult to keep track of moving stimuli following different velocities (e.g. Fencsik, Klieger & Horowitz, 2007; Pylyshyn, 2004). Therefore, it would be more suitable predicting that a collision event will occur in PPS when it is more difficult to process information. This would provide an adaptive advantage by ensuring more time to prepare for an adequate reaction (Neuhoff, 1998, 2001; Vagnoni et al., 2017). In sum, the overall pattern of results suggested that path and velocity of moving stimuli can influence collision prediction in a different way, depending on whether stimuli are in the peripersonal or extrapersonal space. The specific way of processing spatio-temporal factors may reflect the motor and anticipatory adaptive function of PPS that works like an alert system for preparing timely reactions to events near the body.

15.6 The social dimension of peripersonal space

15.6.1 Social information modulates peripersonal space representation

The defensive role of PPS makes it an important support in the control of social interactions. Indeed, one may assume that the protective role of PPS against harmful stimuli concurrently specifies a private area that could serve as a spatial reference for social interactions. In the recent debate on the nature of PPS, various research lines have investigated whether social information may affect the regulation of PPS. In line with this view, researchers using multisensory integration tasks have shown that the size of PPS is modulated by the presence of confederates. For example, Teneggi et al.(2013) have shown that the size of PPS is reduced when another person is present nearby (see also Cléry et al., 2015; de Vignemont & Iannetti, 2015). However, after a cooperative social task, the representation of the PPS is extended up so as to include the other person. This suggests that the presence of other people around us and the evaluation of their potential benefit for us may play an important role in the representation of the PPS and its role in social interactions.

15.6.2 The relationship between action peripersonal space and social interpersonal space

The observation of how people interact in social contexts suggests a close connection between the motor system, spatial processing, and social interactions. Indeed, in face-to-face social interactions, a too large distance between two confederates would not encourage social communication. On the opposite, a too short distance between two confederates would produce discomfort (Burgoon et al., 2007; Kennedy et al., 2009; Knapp et al., 2013) and trigger defensive behaviour (Evans & Wener, 2007). In the middle of the twentieth century, Hall (1966) and later Hayduk (1978) had the intuition that the space around the body is not only the privileged region of space for reaching and manipulating objects, but also for interacting with conspecifics, which requires accurate control of interpersonal distances. In line with this, Iachini and colleagues (Iachini et al., 2014b, 2015, 2016; Ruggiero et al., 2017) have questioned whether PPS for acting on objects and interpersonal space (IPS) for social interactions refer to a similar or different physical space depending on several social factors. In an immersive virtual reality (IVR) setting, they compared two paradigms typically used in the neurocognitive and proxemics domains to investigate PPS and IPS, respectively: a reachability-distance task (distance at which people perceive a stimulus as reachable) for PPS and a comfort-distance task (distance people prefer when another person is approaching them) for IPS (Aiello, 1987; Dosey & Meisels, 1969; Gessaroli et al., 2013; Hayduk, 1983; Holt et al., 2014). In a first study (Iachini et al., 2014b), participants determined each distance in presence of virtual males and females, a cylinder and a robot under two conditions: while standing still (passive) or walking towards the stimuli (active). The results demonstrated that PPS and IPS had a similar size when individuals could actively walk toward the stimuli: this suggested a common motor nature of these two spaces. Moreover, both spaces were modulated by the social characteristic of stimuli: there was a contraction of distance with humans as compared to objects, and among humans with females as compared to males (see also Hecht et al., 2019). A subsequent study confirmed that both distances were similarly modulated by basic social information such as gender and age (Iachini et al., 2016). Indeed, their size was reduced when interacting with females as compared to males, and enlarged when interacting with adults as compared to children. Remarkably, the effects were similar in an actual laboratory setting and in a IVR setting. Proxemically, participants treated virtual humans as if they were actual humans, thus supporting the ecological validity of the IVR technology as a tool for studying social phenomena (Blascovich, 2001; Bailenson et al., 2003; Slater, 2009).

Further demonstration of the relation between PPS and IPS came from the study by Quesque et al. (2017) who demonstrated that tool use induces not only an enlargement of PPS (Bourgeois et al., 2014), but also an increase of the minimum comfort distance tolerated in dyadic social interactions. They used human-like point-light stimuli in the form of a 1.76 m tall man walking towards the participants with no other information than the body movement, and passing on either the right or left side of the participants with different inter-shoulders distances. The task for the participants was to judge whether the interpersonal distance, at the time the point-light walker reached their level, was comfortable or not. Between two sessions of comfort social distance judgements, the participants manipulated a 70 cm or 10 cm-long tool. The long but not the short tool was assumed to be incorporated into the body schema, resulting in an elongated representation of arm length (Cardinali et al., 2011) and therefore in an extension of PPS (Bourgeois et al., 2014). Following manipulation of the long tool, not the short tool, the interpersonal distance considered as comfortable increased. In other studies, the link between PPS and IPS was found to depend on the experimental method used (Patané et al., 2016; Patané, Farnè, & Frassinetti, 2017). Despite a lack of consistency in the data, the study by Quesque et al. (2017) showed for the first time that the preferred interpersonal distance in social interactions is a function of the representation of PPS. This finding provided thus evidence that interpersonal-comfort space and peripersonal-action space share a common motor nature (Coello & Iachini, 2015).

The above findings also reveal that interpersonal-comfort space and peripersonal-action space are both sensitive to basic social information. However, actual social interactions are complex and involve high-level knowledge about other people. When we encounter unknown persons, we spontaneously form an impression about them. Much research suggests that moral information is fundamental to form a quick impression about others and, consequently, determine the intention to approach or avoid them (Brambilla & Leach, 2014; Goodwin, Piazza, & Rozin, 2014). Has this information a top-down influence on the regulation of the distance between ourselves and others? To answer this question, Iachini, Pagliaro, & Ruggiero (2015) had participants interacting with male and female virtual confederates described by moral, immoral and neutral sentences. The results showed a strong effect of perceived morality on IPS: comfort-distance was particularly large when a person was described as immoral rather than moral or neutral, whereas it was smaller with a moral rather than neutral description. A similar trend characterized PPS, with an increase of distance linked to immoral information. However, only the moral–immoral comparison was significant. These findings show that both spaces are sensitive, at different degrees, to moral evaluation processes. Moreover, they suggest that high-level socio-cognitive processes can have a top-down influence on the way we represent the space near our body.

15.6.3 The influence of emotion on peripersonal space and interpersonal space

Continuing the exploration of social factors on spatial regulation mechanisms, it was necessary to consider the fundamental role of emotional expressions. Studies on emotional expressions have suggested that identifying individuals' intent through their facial expression influences our capacity to approach or avoid others (Ekman, 1999; Keltner et al., 2003). Therefore, we may ask whether emotional expressions have a similar or different effect on peripersonal and interpersonal spaces. To answer this question, Ruggiero et al. (2017) presented participants with virtual characters showing a happy, angry or neutral facial expression. The results showed a general effect of emotional expressions on distances: the latter were larger in the presence of an angry compared to a neutral or happy face, and smaller in the presence of a happy compared to a neutral face. This effect was particularly robust in IPS: comfort-distance was larger with an angry compared to a happy or neutral expression, and smaller with a happy compared to a neutral expression. A similar trend emerged in PPS but only the happy versus angry comparison approached significance. Crucially, however, when participants were standing still and a virtual character looked angry, both IPS and PPS increased. This common increase may represent an automatic avoidance reaction to ensure a larger margin of self-protection and avoid violation of our private area (Dosey & Meisels, 1969; Hayduk, 1983; Siegman & Feldstein, 2014).

Emotional factors mediate visuo-perceptual, psychophysiological and automatic behavioural responses (Vuilleumier & Pourtois, 2007). Further evidence for a PPS-based regulation of inter-individual interactions came from the assessment of the physiological responses associated with PPS invasion (Kennedy et al., 2009), which turned out to be a robust predictor of preferred comfort distance in social contexts. Cartaud et al. (2018) investigated the relation between the emotional responses registered through electrodermal activity (EDA) triggered by human-like Point Light Walkers (PLW) carrying different facial expressions (neutral, angry, happy). These PLWs were presented in either the peri- or extrapersonal space. Following this assessment, the participants performed a comfort-judgement task while the same PLWs were approaching them with different approach angles. The results showed an increase of the phasic EDA when the PLWs with an angry facial expressions were located in the PPS, and a concomitant increase of preferred comfort distance when the same PLWs were approaching the participants. Overall, the findings indicate that comfort social space can be predicted from the level of threat expressed by the facial expressions (Ruggiero et al., 2017), which correlates with the emotional responses triggered by the facial expressions when presented in the PPS (Cartaud et al., 2018). This indicates that peripersonal-action space and interpersonal-social space are coherently sensitive to the emotional expression of conspecifics, which could reflect a common adaptive mechanism shared by these spaces to subtend interactions with both the physical and social environment, along with the protection of the body from potential threats. As a consequence, individuals with enlarged self-representation of PPS reported higher rates of social anxiety (Iachini et al., 2015; Nandrino et al., 2017) and phobia (Lourenco, Longo, & Pathman, 2011). These results are thus consistent with a basic motor function of PPS that underlies its defensive role (Graziano, 2017), the organization of object-directed actions and the regulation of the social life (Coello & Iachini, 2015).

15.7 Conclusion

The overall findings discussed in this chapter suggest that the representation of PPS is based on the action possibilities afforded by our body (depending on its motor status and motor internal models) and by the evaluation of the reward or threat level associated with the stimuli in the environment. Through the integration of these variables, PPS serves the functions of preparing to react to rewarding or potentially harmful stimuli and to socially interact with conspecifics. Accordingly, changing PPS representation modifies preferred distances in social interactions. When our perceptual system signals a potential danger in our physical or social environment, action peripersonal space and social interpersonal space are concurrently updated. Accordingly, the same psychological processes may give rise to different behavioural effects depending on whether there is a need to react to a stimulus located at a predetermined position, or to adjust behaviour to impeding stimuli by regulating distances in a meaningful way. Accordingly, approach/avoidance behaviours are quickly prompted to obtain an optimal regulation of social and action/safety distances (Coello & Iachini, 2016; Coello et al., 2012; Quesque et al., 2017; Iachini et al., 2017; Ruggiero et al., 2017). This organization is in line with the idea that the approach–avoidance regulatory mechanism constitutes a fundamental organizing principle not only at the sensorimotor level, but also at the level of conceptual knowledge acquisition (Solarz, 1960; Phaf et al., 2014; see also Godinho & Garrido, 2016; Meier et al., 2012; Topolinski et al., 2014). In this regard, the need to represent a margin of safety around the body is particularly important when the events occurring in the near space are perceived as potentially harmful (Graziano & Cooke, 2006; Iachini et al., 2015; Kennedy et al., 2009). This suggests that the perception of a potential threat and the need to defend oneself by adjusting the significance of our near space may represent the fundamental function of PPS (Graziano, 2017) and the point of contact between PPS and IPS. This need links together various mechanisms: those devoted to the perceptual analysis of the context, those implied in action planning and spatial regulation, and those evaluating the socio-emotional and psychological meaning of interactions.

References

Aiello, J. R. (1987). Human spatial behavior. In D. Stokols, & I. Altman (eds), *Handbook of Environmental Psychology*. New York: John Wiley and Sons.

Akrami, A., Kopec, C. D., Diamond, M. E., & Brody, C. D. (2018). Posterior parietal cortex represents sensory history and mediates its effects on behaviour. *Nature, 554*, 368–372.

Anderson, S. J., Yamagishi, N., & Karavia, V. (2002). Attentional processes link perception and action. *Proceedings of the Royal Society B, 269*, 1225–1232.

Avanzino, L., Bassolino, M., Pozzo, T., & Bove, M. (2011). Use-dependent hemispheric balance. *Journal of Neuroscience, 31*(9), 3423–3428.

Babiloni, C., Carducci, F., Cincotti, F., Rossini, P. M., Neuper, C., Pfurtscheller, G., et al. (1999). Human movement-related potentials vs. desynchronization of EEG alpha rhythm: a high-resolution EEG study. *Neuroimage, 10*, 658–665.

Bailenson, J. N., Blascovich, J., Beall, A. C., & Loomis, J. M. (2003). Interpersonal distance in immersive virtual environments. *Personality and Social Psychology Bulletin, 29*, 819–833.

Bartolo, A., Carlier, M., Hassaini, S., Martin Y., & Coello, Y. (2014). The perception of peripersonal space in right and left brain damage hemiplegic patients. *Frontiers in Human Neuroscience, 8*(3). doi:10.3389/fnhum.2014.00003.

Bartolo, A., Rossetti, Y., Revol, P., Urquizar, C., Pisella, L., & Coello, Y. (2017). Reachability judgment in optic ataxia: effect of peripheral vision on hand and target perception in depth. *Cortex.* doi:org/10.1016/j.cortex.2017.05.013

Bassolino, M., Bove, M., Jacono, M., Fadiga, L., & Pozzo, T. (2012). Functional effect of short-term immobilization: kinematic changes and recovery on reaching-to-grasp. *Neuroscience, 215,* 127–134.

Blascovich, J. (2001). Social influences within immersive virtual environments. In R. Schroeder (ed.), *The Social Life of Avatars* (pp. 127–145). New York: Springer-Verlag.

Blini, E., Desoche, C., Salemme, R., Kabil, A., Hadj-Bouziane, F., & Farnè, A. (2018). Mind the depth: visual perception of shapes is better in peripersonal space. *Psychological Science, 29*(11), 1868–1877.

Borghi, A. M., & Binkofski, F. (2014). *Words as Social Tools: An Embodied View on Abstract Concepts.* (ed.) Springer New York. doi:10.1007/978-1-4614-9539-0.

Bourgeois, J., & Coello, Y. (2012). Effect of visuomotor calibration and uncertainty on the perception of peripersonal space. *Attention, Perception, & Psychophysics, 74,* 1268–1283.

Bourgeois, J., Farnè, A., & Coello Y. (2014). Costs and benefits of tool-use on the perception of reachable space. *Acta Psychologica, 148,* 91–95.

Brambilla, M., & Leach, C. W. (2014). On the importance of being moral: the distinctive role of morality in social judgment. *Social Cognition, 32,* 397–408.

Brozzoli, C., Makin, T. R., Cardinali, L., Holmes, N. P., & Farnè, A. (2012). Peripersonal space: a multisensory interface for body–object interactions. In M. M. Murray, & M. T. Wallace (eds), *The Neural Bases of Multisensory Processes.* Frontiers in Neuroscience. Boca Raton (FL): CRC Press.

Bufacchi, R. J., & Iannetti, G. D. (2018). An action field theory of peripersonal space. *Trends in Cognitive Sciences, 22,* 1076–1090.

Burgoon, J. K., Stern, L. A., & Dillman, L. (2007). *Interpersonal adaptation: Dyadic interaction patterns.* New York: Cambridge University Press.

Bursztyn, L. L., Ganesh, G., Imamizu, H., Kawato, M., & Flanagan, J. R. (2006). Neural correlates of internal-model loading. *Current Biology, 16*(24), 2440–2445.

Caggiano, V., Fogassi, L., Rizzolatti, G., Thier, P., & Casile, A. (2009). Mirror neurons differentially encode the peripersonal and extrapersonal space of monkeys. *Science, 324,* 403–406.

Cardellicchio, P., Sinigaglia, C., & Costantini, M. (2011). The space of affordances: a TMS study. *Neuropsychologia, 49,* 1369–1372.

Cardinali, L., Brozzoli, C., & Farnè, A. (2009). Peripersonal space and body schema: two labels for the same concept? *Brain Topography, 21,* 252–260.

Cardinali, L., Brozzoli, C., Urquizar, C., Salemme, R., Roy, A.C., & Farnè, A. (2011). When action is not enough: tool-use reveals tactile-dependent access to Body Schema. *Neuropsychologia, 49(13),* 3750–3757.

Cardinali, L., Jacobs, S., Brozzoli, C., Frassinetti, F., Roy, A.C., & Farnè, A. (2012). Grab an object with a tool and change your body: tool-use-dependent changes of body representation for action. *Experimental Brain Research, 218,* 259–271.

Cartaud, A., Ruggiero, G., Ott, L., Iachini, T., & Coello, Y. (2018). Physiological response to facial expressions in peripersonal space determines interpersonal distance in a social interaction context. *Frontiers in Psychology, 9,* 657.

Chao, L. L., & Martin, A. (2000). Representation of manipulable man-made objects in the dorsal stream. *Neuroimage, 12*, 478–484.

Chao, L. L., Weisberg, J., & Martin, A. (2002). Experience-dependent modulation of category-related cortical activity. *Cerebral Cortex, 12*, 545–51.

Cléry, J., Guipponi, O., Wardak, C., & Ben Hamed, S. (2015). Neuronal bases of peripersonal and extrapersonal spaces, their plasticity and their dynamics: Knowns and unknowns. *Neuropsychologia, 70*, 313–326.

Cochin, S., Barthélémy, C., Roux, S., & Martineau, J. (1999). Observation and execution of movement: similarities demonstrated by quantified electroencephalography. *European Journal of Neuroscience, 11*, 1839–1842.

Coello, Y., Bartolo, A., Amiri, B., Devanne, H., Houdayer, E., & Derambure, P. (2008). Perceiving what is reachable depends on motor representations: evidence from a transcranial magnetic stimulation study. *PLoS ONE, 3(8)*, e2862. doi:10.1371/journal.pone.0002862

Coello, Y., Bourgeois, J., & Iachini, T. (2012). Embodied perception of reachable space: how do we manage threatening objects? *Cognitive Processing, 1*, 131–135.

Coello, Y., & Delevoye-Turrell, Y. (2007). Embodiment, space categorisation and action. *Consciousness and Cognition, 16*, 667–683.

Coello, Y., & Iachini, T. (2015). Embodied perception of objects and people in space: towards a unified theoretical framework. In Y. Coello, & M. Fischer (eds), *Foundations of Embodied Cognition* (pp. 198–219). New York: Psychology Press.

Coello, Y., Quesque, F., Gigliotti, M. F., Ott, L., & Bruyelle, J. L. (2018). Idiosyncratic representation of peripersonal space depends on the success of one's own motor actions, but also the successful actions of others! *PloS ONE, 13(5)*, e0196874.

Costantini, M., Ambrosini, E., Sinigaglia, C., & Gallese, V. (2011). Tool-use observation makes far objects ready-to-hand. *Neuropsychologia, 49(9)*, 2658–2663.

Costantini, M., Ambrosini, E., Tieri, G., Sinigaglia, C., & Committeri, G. (2010). Where does an object trigger an action? An investigation about affordances in space. *Experimental Brain Research, 207*, 95–103.

Creem-Regehr, S. H., & Lee, J. N. (2005). Neural representations of graspable objects: are tools special? *Cognitive Brain Research, 22*, 457–469.

Cullen, K. E., & Brooks, J. X. (2015). Neural correlates of sensory prediction errors in monkeys: evidence for internal models of voluntary self-motion in the cerebellum. *The Cerebellum, 14(1)*, 31–34.

de Vignemont, F., & Iannetti, G. D. (2015). How many peripersonal spaces? *Neuropsychologia, 70*, 327–334. doi:org/10.1016/j.neuropsychologia.2014.11.018

Desimone, R., & Duncan, J. (1995). Neural mechanisms of selective visual attention. *Annual Review*, 193–222.

di Pellegrino, G., & Làdavas, E. (2015). Peripersonal space in the brain. *Neuropsychologia, 66*, 126–133. doi:org/10.1016/j.neuropsychologia.2014.11.011

Dosey, M. A., & Meisels, M. (1969). Personal space and self-protection. *Journal of Personality and Social Psychology, 11*, 93–97.

Eilan, N., Brewer, W., & McCarthy, R. (1993). *Spatial representation*. New York: Basil Blackwell.

Ekman, P. (1999). Facial expressions. In Dalgleish, T., & Power, M. J. (eds), *The Handbook of Cognition and Emotion* (pp. 301–320). New York: John Wiley & Sons.

Ellis, R., & Tucker, M. (2000). Micro-affordance: the potentiation of components of action by seen objects. *British Journal of Psychology, 91*, 451–471.

Evans, G. W., & Wener, R. E. (2007). Crowding and personal space invasion on the train: please don't make me sit in the middle. *Journal of Environmental Psychology, 27*(1), 90–94.

Facchini, S., Romani, M., Tinazzi, M., & Aglioti, S. M. (2002). Time-related changes of excitability of the human motor system contingent upon immobilisation of the ring and little fingers. *Clinical Neurophysiology, 113*(3), 367–375.

Farnè, A., Iriki, A., & Làdavas, E. (2005). Shaping multisensory action–space with tools: evidence from patients with cross-modal extinction. *Neuropsychologia, 43*(2), 238–248.

Fencsik, D. E., Klieger, S. B., & Horowitz, T. S. (2007). The role of location and motion information in the tracking and recovery of moving objects. *Perception & Psychophysics, 69*, 567–577. doi:10.3758/BF03193914

Fischer, M. H., & Zwaan, R. A. (2008). Embodied language—a review of the role of the motor system in language comprehension. *Quarterly Journal of Experimental Psychology, 61*, 825–850.

Gawronski, B. (2005). Approach/avoidance-related motor actions and the processing of affective stimuli: incongruency effects in automatic attention allocation. *Social Cognition, 23*(2), 182–203.

Gessaroli, E., Santelli, E., di Pellegrino, G., & Frassinetti, F. (2013). Personal space regulation in childhood autism spectrum disorders. *PLoS ONE, 8*(9): e74959. doi:10.1371/journal.pone.0074959.

Ghez, C., & Sainburg, R. (1995). Proprioceptive control of interjoint coordination. *Canadian Journal of Physiology and Pharmacology, 73*(2), 273–284.

Gibson, J. J. (1979). *The ecological approach to visual perception.* Boston: Houghton Mifflin.

Gilden, D. L., & Proffitt, D. R. (1989). Understanding collision dynamics. *Journal of Experimental Psychology: Human Perception and Performance, 15*, 372–383. doi:10.1037/0096-1523.15.2.372

Godinho, S., & Garrido, M. V. (2016). Oral approach-avoidance: a replication and extension for European–Portuguese phonation. *European Journal of Social Psychology, 46*(2), 260–264.

Goodwin, G. P., Piazza, J., & Rozin, P. (2014). Moral character predominates in person perception and evaluation. *Journal of Personality and Social Psychology, 106*, 148–168.

Grafton, S. T., Arbib, M.A., Fadiga, L., & Rizzolatti, G. (1997). Localization of grasp representations in humans by positron emission tomography: 2. Observation compared with imagination. *Experimental Brain Research, 112*, 103–111.

Graziano, M. (2017). *The Spaces Between Us: A Story of Neuroscience, Evolution, and Human Nature.* Oxford: Oxford University Press.

Graziano, M. S., & Cooke, D. F. (2006). Parieto-frontal interactions, personal space, and defensive behavior. *Neuropsychologia, 44*, 845–859.

Graziano, M. S., Reiss, L. A., & Gross, C. G. (1999). A neuronal representation of the location of nearby sounds. *Nature, 397*(6718), 428.

Gray, R., & Thornton, I. M. (2001). Exploring the link between time to collision and representational momentum. *Perception, 30*, 1007–1022.

Grüsser, O. J. (1983). Multimodal structure of the extrapersonal space. In *Spatially Oriented Behavior* (pp. 327–352). New York: Springer.

Hall, E. T. (1966). *The Hidden Dimension.* New York: Doubleday.

Hayduk, L. A. (1978). Personal space: an evaluative and orienting overview. *Psychological Bulletin, 85*(1), 117–134.

Hayduk, L. A. (1983). Personal space: Where we now stand. *Psychological Bulletin, 94*(2), 293–335.

Hecht, H., Welsch, R., Viehoff, J., & Longo, M. R. (2019). The shape of personal space. *Acta Psychologica, 193*, 113–122.

Holmes, N. P., & Spence, C. (2004). The body schema and the multisensory representation(s) of peripersonal space. *Cognitive Processing*, 5, 94–105.

Holt, D. J., Cassidy, B. S., Yue, X., Rauch, S. L., Boeke, E. A., Nasr, S., . . . Coombs III, G. (2014). Neural correlates of personal space intrusion. *Journal of Neuroscience*, 34, 4123–4134.

Huang, R. S., Chen, C. F., Tran, A. T., Holstein, K. L., & Sereno, M. I. (2012). Mapping multisensory parietal face and body areas in humans. *Proceedings of the National Academy of Sciences*, 109(44), 18114–18119.

Huber, R., Ghilardi, M. F., Massimini, M., Ferrarelli, F., Riedner, B. A., Peterson, M. J., & Tononi, G. (2006). Arm immobilization causes cortical plastic changes and locally decreases sleep slow wave activity. *Nature Neuroscience*, 9(9), 1169.

Hunley, S. B., & Lourenco, S. F. (2018). What is peripersonal space? An examination of unresolved empirical issues and emerging findings. *Wiley Interdisciplinary Reviews: Cognitive Science*, 9(6), e1472.

Iachini, T., Borghi, A. M., & Senese, V. P. (2008). Categorization and sensorimotor interaction with objects. *Brain & Cognition*, 67, 31–43.

Iachini, T., Coello, Y., Frassinetti, F., & Ruggiero, G. (2014). Body space in social interactions: a comparison of reaching and comfort distance in immersive virtual reality. *PLoS ONE*, 9(11), e111511. doi:org/10.1371/journal.pone.0111511

Iachini, T., Coello, Y., Frassinetti, F., Senese, V. P., Galante, F., & Ruggiero, G. (2016). Peripersonal and interpersonal space in virtual and real environments: effects of gender and age. *Journal of Environmental Psychology*, 45, 154–164.

Iachini, T., Pagliaro, S., & Ruggiero, G. (2015). Near or far? It depends on my impression: moral information and spatial behavior in virtual interactions. *Acta Psychologica*, 161, 131–136.

Iachini, T., Ruggiero, G., Ruotolo, F., Schiano di Cola, A., & Senese, V. P. (2015). The influence of anxiety and personality factors on comfort and reachability space: a correlational study. *Cognitive Processing*, 16, 255–258.

Iachini, T., Ruotolo, F., Vinciguerra, M., & Ruggiero, G. (2017). Manipulating time and space: collision prediction in peripersonal and extrapersonal space. *Cognition*, 166, 107–117.

Iriki, A., Tanaka, M., & Iwamura, Y. (1996). Coding of modified body schema during tool use by macaque postcentral neurones. *NeuroReport*, 7(14), 2325–2330.

Kalenine, S., Wamain, Y., Decroix, J., & Coello, Y. (2016). Conflict between object structural and functional affordances in peripersonal space. *Cognition*, 155, 1–7.

Kan, I. P., Kable, J. W., Van Scoyoc, A., Chatterjee, A., & Thompson-Schill, S. L. (2006). Fractionating the left frontal response to tools: dissociable effects of motor experience and lexical competition. *Journal of Cognitive Neuroscience*, 18, 267–277.

Kawato, M. (1999). Internal models for motor control and trajectory planning. *Current Opinion in Neurobiology*, 9(6), 718–727.

Keltner, D., Ekman, P., Gonzaga, G. C., & Beer, J. (2003). Facial expression of emotion. In R. J. Davidson, K. R. Scherer, & H. H. Goldsmith (eds), *Handbook of Affective Sciences* (pp. 415–432). Series in Affective Science. New York: Oxford University Press.

Kennedy, D. P., Gläscher, J., Tyszka, J. M., & Adolphs, R. (2009). Personal space regulation by the human amygdala. *Nature Neuroscience*, 12(10), 1226–1227.

Knapp, M. L., Hall, J. A., & Horgan, T. G. (2013). *Nonverbal communication in human interaction*. Cengage Learning.

Lebrecht, S., Bar, M., Barrett, L. F., & Tarr, M. J. (2012). Micro-valences: perceiving affective valence in everyday objects. *Frontiers in Psychology*, 3, 107.

Leclere, N. X., Sarlegna, F., Coello, Y., & Bourdin, C. (2019). Sensori-motor adaptation to novel limb dynamics influences the representation of peripersonal space. *Neuropsychologia*. doi.org/10.1016/j.neuropsychologia.2019.05.005

Llanos, C., Rodriguez, M., Rodriguez-Sabate, C., Morales, I., & Sabate, M. (2013). Mu-rhythm changes during the planning of motor and motor imagery actions. *Neuropsychologia*, *51*(6), 1019–26

Lourenco, S., Longo, M., & Pathman, T. (2011). Near space and its relation to claustrophobic fear. *Cognition*, *119*(3), 448–453.

Makin, T. R., Holmes, N. P., & Zohary, E. (2007). Is that near my hand? Multisensory representation of peripersonal space in human intraparietal sulcus. *Journal of Neuroscience*, *27*, 731–740.

Martin, A. (2007). The representation of object concepts in the brain. *Annual Review of Psychology*, *58*, 25–45.

Matelli, M., Luppino, G., & Rizzolatti G. (1985). Patterns of cytochrome oxidase activity in the frontal agranular cortex of macaque monkey. *Behavioural Brain Research*, *18*, 125–137.

Meier, B. P., Schnall, S., Schwarz, N., & Bargh, J. A. (2012). Embodiment in social psychology. *Topics in Cognitive Science*, *4*(4), 705–716.

Meugnot, A., Almecija, Y., & Toussaint, L. (2014). The embodied nature of motor imagery processes highlighted by short-term limb immobilization. *Experimental Psychology*, *61*, 180–186. doi:10.1027/1618-3169/a000237

Meugnot, A., & Toussaint, L. (2015). Functional plasticity of sensorimotor representations following short-term immobilization of the dominant versus non-dominant hands. *Acta Psychologica*, *155*, 51–56. doi:10.1016/j.actpsy.2014.11.013

Moisello, C., Bove, M., Huber, R., Abbruzzese, G., Battaglia, F., Tononi, G., & Ghilardi, M. F. (2008). Short-term limb immobilization affects motor performance. *Journal of Motor Behavior*, *40*(2), 165–176.

Nandrino, J. L., Ducro, C., Iachini, T., & Coello, Y. (2017). Perception of peripersonal and interpersonal space in patients with restrictive-type anorexia. *European Eating Disorders Review*, *25*(3), 179–187.

Neuhoff, J. G. (1998). Perceptual bias for rising tones. *Nature*, *395*, 123–124.

Neuhoff, J. G. (2001). An adaptive bias in the perception of looming auditory motion. *Ecological Psychology*, *13*, 87–110.

Ostry, D. J., Darainy, M., Mattar, A. A., Wong, J., & Gribble, P. L. (2010). Somatosensory plasticity and motor learning. *Journal of Neuroscience*, *30*(15), 5384–5393.

Patané, I., Farnè, A., & Frassinetti, F. (2017). Cooperative tool-use reveals peripersonal and interpersonal spaces are dissociable. *Cognition*, *166*, 13–22.

Patané, I., Iachini, T. Farnè, A., & Frassinetti, F. (2016). Disentangling action from social space: tool-use differently shapes the space around us. *PloS ONE*, *11*(5), e0154247.

Phaf, H. R., Mohr, S. E., Rotteveel, M., & Wicherts, J. M. (2014). Approach, avoidance, and affect: a meta-analysis of approach-avoidance tendencies in manual reaction time tasks. *Front Psychol*, *5*: 378. doi:10.3389/fpsyg.2014.00378

Phillips, J. C., & Ward, R. (2002). S-R correspondence effects of irrelevant visual affordance: time course and specificity of response activation. *Visual Cognition*, *9*, 540–558.

Pickering, M. J., & Clark, A. (2014). Getting ahead: forward models and their place in cognitive architecture. *Trends in Cognitive Sciences*, *18*(9), 451–456.

Previc, F. H. (1998). The neuropsychology of 3-D space. *Psychological Bulletin*, *124*, 123–164.

Proffitt, D. R., & Gilden, D. L. (1989). Understanding natural dynamics. *Journal of Experimental Psychology: Human Perception and Performance*, *15*, 384–393. doi:10.1037/0096-1523.15.2.384

Proverbio, A. M. (2012). Tool perception suppresses 10-12 Hz μ rhythm EEG over the somatosensory area. *Biological Psychology*, *91*, 1–7.

Pylyshyn, Z. W. (2004). Some puzzling findings in multiple object tracking: I. Tracking without keeping track of object identities. *Visual Cognition*, *11*, 801–822. doi:10.1080/13506280344000518

Quesque, F., Ruggiero, G., Mouta, S., Santos, J., Iachini, T., & Coello, Y. (2017). Keeping you at arm's length: modifying peripersonal space influences interpersonal distance. *Psychological research*, *81*(4), 709–720.

Rizzolatti, G., Scandolara, C., Matelli, M., & Gentilucci, M. (1981). Afferent properties of periarcuate neurons in macaque monkeys: I. Somatosensory responses. *Behavioural Brain Research*, *2*, 125–146.

Ruggiero, G., Frassinetti, F., Coello, Y., Rapuano, M., Schiano di Cola, A., & Iachini, T. (2017). The effect of facial expressions on peripersonal and interpersonal spaces. *Psychological Research*, *81*(6), 1232–1240.

Sainburg, R. L. (2002). Evidence for a dynamic-dominance hypothesis of handedness. *Experimental Brain Research*, *142*(2), 241–258.

Sainburg, R. L. (2005). Handedness: differential specializations for control of trajectory and position. *Exercise and Sport Sciences Reviews*, *33*(4), 206–213.

Sainburg, R. L., Ghez, C., & Kalakanis, D. (1999). Intersegmental dynamics are controlled by sequential anticipatory, error correction, and postural mechanisms. *Journal of Neurophysiology*, *81*(3), 1045–1056.

Sainburg, R. L., & Kalakanis, D. (2000). Differences in control of limb dynamics during dominant and nondominant arm reaching. *Journal of Neurophysiology*, *83*(5), 2661–2675.

Salenius, S., Schnitzler, A., Salmelin, R., Jousmäki, V., & Hari, R. (1997). Modulation of human cortical rolandic rhythms during natural sensorimotor tasks. *NeuroImage*, *5*(3), 221–8.

Schaefer, S. Y., Haaland, K. Y., & Sainburg, R. L. (2007). Ipsilesional motor deficits following stroke reflect hemispheric specializations for movement control. *Brain*, *130*, 2146–2158.

Schaefer, S. Y., Haaland, K. Y., & Sainburg, R. L. (2009). Hemispheric specialization and functional impact of ipsilesional deficits in movement coordination and accuracy. *Neuropsychologia*, *47*(13), 2953–2966.

Serino, A., Canzoneri, E., & Avenanti, A. (2011). Fronto-parietal areas necessary for a multisensory representation of peripersonal space in humans: an rTMS study. *Journal of Cognitive Neuroscience*, *23*(10), 2956–2967.

Shadmehr, R. (2017). Learning to predict and control the physics of our movements. *Journal of Neuroscience*, *37*, 1663–1671.

Siegman, A. W., & Feldstein, S. (2014). *Nonverbal behavior and communication* (2nd ed.). New York: Psychology Press.

Slater, M. (2009). Place illusion and plausibility can lead to realistic behaviour in immersive virtual environments. *Philosophical Transactions of the Royal Society B*, *364*, 3549–3557.

Solarz, A. K. (1960). Latency of instrumental responses as a function of compatibility with the meaning of eliciting verbal signs. *Journal of Experimental Psychology*, *59*(4), 239.

Spence, C., Pavani, F., Maravita, A., & Holmes, N. (2004). Multisensory contributions to the 3-D representation of visuotactile peripersonal space in humans: evidence from the crossmodal congruency task. *Journal of Physiology-Paris*, *98*(1-3), 171–189.

Symes, E., Ellis, R., & Tucker, M. (2007). Visual object affordances: object orientation. *Acta Psychologica*, *124*, (2) 238–255.

Tanaka, H., & Sejnowski, T.J. (2013). Computing reaching dynamics in motor cortex with Cartesian spatial coordinates. *Journal of Neurophysiology*, *109*(4),1182–1201.

Teneggi, C., Canzoneri, E., di Pellegrino, G., & Serino, A. (2013). Social modulation of peripersonal space boundaries. *Current Biology*, *23*, 406–411.

Thill, S., Caligiore, D., Borghi, A. M., Ziemke, T., & Baldassarre, G. (2013). Theories and computational models of affordance and mirror systems: an integrative review. *Neuroscience and Biobehavioral Review*, *37*, 491–521.

Topolinski, S., Maschmann, I. T., Pecher, D., & Winkielman, P. (2014). Oral approach–avoidance: affective consequences of muscular articulation dynamics. *Journal of Personality and Social Psychology*, *106*(6), 885.

Toussaint, L., & Meugnot, A. (2013). Short-term limb immobilization affects cognitive motor processes. *Journal of Experimental Psychology Learning, Memory & Cognition*, *39*, 623–632. doi:10.1037/a0028942

Toussaint, L., Wamain, Y., Ildei-Bidet, C., & Coello, Y. (2018). Short-term upper-limb immobilization alters peripersonal space representation. *Psychological Research*, doi:org/10.1007/s00426-018-1118-0

Tucker, M., & Ellis, R. (1998). On the relations between seen objects and components of potential actions. *Journal of Experimental Psychology: Human Perception & Performance*, *24*, 830–846.

Tucker, M., & Ellis, R. (2001). The potentiation of grasp types during visual object categorization. *Visual Cognition*, *8*(6), 769–800.

Tucker, M., & Ellis, R. (2004). Action priming by briefly presented objects. *Acta Psychologica*, *116*, 185–203.

Vagnoni, E., Andreanidou, V., Lourenco, S. F., & Longo, M. R. (2017). Action ability modulates time-to-collision judgments. *Experimental Brain Research*, *235*, 2729–2739. doi:10.1007/s00221-017-5008-2

Vuilleumier, P., & Pourtois, G. (2007). Distributed and interactive brain mechanisms during emotions face perception: evidence from functional neuroimaging. *Neuropsychologia*, *45*, 174–194. doi.org/10.1016/j.neuropsychologia.2006.06.003

Wamain, Y., Gabrielli, F., & Coello, Y. (2016). EEG μ rhythm in virtual reality reveals that motor coding of visual objects in peripersonal space is task dependent. *Cortex*, *74*, 20–30.

Wamain, Y., Sahaï, A., Decroix, J., Coello, Y., Kalénine, S. (2018). Conflict between gesture representations extinguishes μ rhythm desynchronization during manipulable object perception: an EEG study. *Biological Psychology*, *132*, 202–211.

Witt, J. K., Proffitt, D. R., & Epstein, W. (2005). Tool use affects perceived distance, but only when you intend to use it. *Journal of experimental psychology: Human Perception & Performance*, *31*(5), 880.

Wolpert, D. M., & Kawato, M. (1998). Multiple paired forward and inverse models for motor control. *Neural Networks*, *11*(7-8), 1317–1329.

Wolpert, D. M., Ghahramani, Z., & Flanagan, J. R. (2001). Perspectives and problems in motor learning. *Trends in Cognitive Sciences*, *5*(11), 487–494.

Wolpert, D. M., Ghahramani, Z., & Jordan, M. I. (1995). An internal model for sensorimotor integration. *Science*, *269* (5232), 1880–1882.

Wolpert, D. M., & Kawato, M. (1998). Multiple paired forward and inverse models for motor control. *Neural Networks*, *11*(7–8), 1317–1329. doi:10.1016/j.neuropsychologia.2004.11.005

Zago, M., & Lacquaniti, F. (2005). Cognitive, perceptual and action-oriented representations of falling objects. *Neuropsychologia*, *43*, 178–188.

16

Action and social spaces in typical development and in autism spectrum disorder

Michela Candini, Giuseppe di Pellegrino, and Francesca Frassinetti

16.1 Introduction

The space close to one's own body represents the region where earliest social interactions, either between infants and individuals, or between infants and objects, occur. The interactions with other individuals enhance the infant's possibilities to experience reciprocal social exchanges, which are fundamental for the subsequent development of communicative and social abilities (Libertus and Violi, 2016). One such behaviour, which is extremely relevant in everyday interactions, is the ability to regulate the physical distance between individuals. Indeed, social distance is critical in determining a successful social interaction and reducing the feelings of discomfort due to interpersonal space violations (Kennedy and Adolphs, 2014).

In this chapter, we primarily focus on the characteristics and plasticity of interpersonal space, reviewing results from different studies on a clinical population characterized by a persistent and profound impairment in social interaction, such as autism spectrum disorders (ASD, DSM-5, APA, 2013; Baron-Cohen et al., 1985; Lord et al., 2000; see, for review, Senju, 2013). Collectively, these studies reveal that the regulation of interpersonal space is impaired, enlarged, or shortened, in individuals with autism. As demonstrated by recent findings (Candini et al., 2019), autism affects interpersonal space regulation, while it seems to leave unaltered peripersonal space regulation, a functionally, at least partially, distinct representation of the space, which is related to the potentiality of acting in the space immediately around the body. Finally, the mechanism and the anatomical correlates of impaired interpersonal space in autism are discussed.

16.2 Deficit in social space regulation in autism

The interpersonal distance that we choose to maintain between ourselves and others during social interactions indicates how close we prefer to stand relative to another individual. This distance has been used by social psychologists to measure interpersonal space, a concept introduced to describe the emotionally tinged area around one's own body, in which another's intrusion may cause a prompt feeling of discomfort or even anxiety. Typically, when these violations occur, individuals may move away to reinstate the margin of safety and to reduce the perceived distress (Sommer, 1959; Hall, 1966; Hayduk, 1978).

In the past years, different studies have shown that the regulation of interpersonal space is altered in people with disorders of social interaction, such as individuals with ASD. Gessaroli

Michela Candini, Giuseppe di Pellegrino, and Francesca Frassinetti, *Action and social spaces in typical development and in autism spectrum disorder* In: *The World at Our Fingertips*. Edited by Frédérique de Vignemont, Andrea Serino, Hong Yu Wong, and Alessandro Farnè, Oxford University Press (2021). © Oxford University Press. DOI: 10.1093/oso/9780198851738.003.0016.

and coworkers (2013b) were the first to provide an ecological measure of interpersonal space in typical and atypical developmental populations. They adopted a comfort-distance task in which the interpersonal space was measured as the distance at which children felt most comfortable when an unfamiliar adult (confederate) approached them, or they approached the confederate (see Figure 16.1). Then, the distance between the confederate's and the participant's body was recorded with a digital laser measurer and considered as a proxy of interpersonal space extent. This procedure allows one to reliably estimate preferred interpersonal distance (permeability) under varied conditions, and represents one of the most frequently used measures of interpersonal space (Hayduk, 1983). Interestingly, the authors found that ASD children chose a larger interpersonal space compared to children with typical development, suggesting that they are less tolerant to the closer proximity of an unfamiliar adult. Furthermore, despite the described differences, an interesting commonality was found across typical development and ASD groups. Indeed, all participants preferred shorter interpersonal distance when they approached the confederate (active approach) compared to when the confederate approached them (passive approach) (Candini et al., 2017). This effect suggests that, since we cannot predict people approaching us, we resolve this unpredictability by maintaining others at a larger distance. This phenomenon might be particularly relevant to assessing potentially threatening social situations, such as people who are too close to us (Lloyd and Morrison, 2008).

Other studies conducted on adult populations reported that ASD individuals are likely to stay closer to others than controls (Pedersen et al., 1989; Pedersen and Schelde, 1997; Parsons et al., 2004; Kennedy and Adolphs, 2014, Asada et al., 2016). For instance, Kennedy and Adolphs (2014) investigated the abnormalities in social distance in individuals with ASD indirectly—that is, by analysing data of a parent-report designed to quantify the severity of autistic symptoms (Social Responsiveness Scale [SRS], Constantino and Gruber, 2005), and directly by measuring the preferred comfort distance from an unfamiliar adult (i.e. a comfort-distance task). Both these measures pointed out that individuals with ASD have the tendency to violate others' interpersonal space (lower scores at SRS compared to

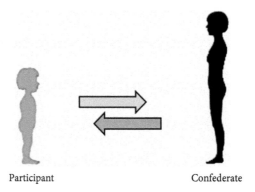

Participant   Confederate

Figure 16.1 Schematic representation of the comfort-distance task. The participant (green) and the confederate (black) are depicted. The arrows indicate the approach directions depending on who performed the movement: the participant (active approach) or the confederate (passive approach).

Adapted from - Candini, M., Giuberti, V., Santelli, E., di Pellegrino, G., and Frassinetti, F. When social and action space diverges: a study in children with typical development and autism. Autism. 2019; 23(7): 1687-1698. https://doi.org/10.1177/1362361318822504.

their unaffected siblings), and to prefer smaller interpersonal space compared to controls, as demonstrated by the comfort-distance task.

Similarly, Asada and coworkers (2016) adopted a similar comfort-distance task to investigate preferred interpersonal space in a Japanese sample of adolescents with and without ASD. In half the trials, the confederate walked toward the participants and, in the other half of the trials, the participants walked toward the confederate. Interestingly, to assess the role of eye contact, in half the trials, the experimenter looked toward the participants' eyes and, in the other half of the trials, the experimenter looked down. Crucially, even if smaller interpersonal space emerged in ASD participants compared to controls, no difference emerged between the two groups when the eye contact condition was considered. Indeed, when the confederate approached the participant, both ASD and controls preferred larger interpersonal distance when there was eye contact, compared to no eye contact condition. This evidence indicates a partially spared ability in ASD individuals to understand social cues, such as other people's gaze, during a social interaction.

By using a virtual reality paradigm, Parsons and colleagues (2004) investigated the use and understanding of *social virtual environments* in ASD adolescents. In this study, participants have to complete a number of tasks in a virtual-café environment that simulates real-life social activities (i.e. pay for food and drink or order a drink from the bar). The authors examined the time spent completing the tasks, the number of errors made, the basic understanding of the virtual environment, and the social appropriateness of performance. The study showed different patterns of results: some ASD participants did not respect the space of virtual characters, walking directly in-between two characters engaged in a conversation, or moving very close to the people at the bar when ordering a drink, whereas other ASD adolescents respected the interpersonal distance rules in the virtual environment. The seemingly controversial results described here might be accounted for by the fact that autism-related social deficits actually fall on a continuum of severity, which may be reflected by different degrees of social space dysfunction in real life. Thus, future research should investigate whether people with ASD, who show deficits in regulating social distance in virtual reality, also have difficulty in understanding the norms and expectations in real-world interactions.

Together the evidence above suggests that autism may have an impact on social distance regulation along two opposite edges: while some research reports that ASD individuals stay too far from other people (Freitag, 1970; Gessaroli et al., 2013b; Candini et al., 2017), other studies conclude that ASD individuals more often violate the space of others (Pedersen et al., 1989; 1997; Parsons et al., 2004; Kennedy and Adolphs, 2014, Asada et al., 2016).

One possible explanation of these opposite results is in terms of the age differences of the samples. Accordingly, an enlargement of interpersonal space was found in children (Gessaroli et al., 2013b; Candini et al., 2017) whereas a reduction of interpersonal space emerged in adolescents and adults (Asada et al., 2016; Parsons et al., 2004; Kennedy and Adolphs, 2014). In accordance with Wing's subclassification scheme (Wing and Attwood, 1987; Wing and Gould, 1979), three subgroups of individuals with autism can be distinguished based on the quality of social interaction: i) the aloof subgroup, characterized by the tendency to reject unsolicited social or physical contact; ii) the passive subgroup, characterized by a lack of spontaneous social approaches to others and by the tendency to engage another person when approached, as long as the other person structures the interaction; and, finally, iii) the active-but-odd subgroup, characterized by a willingness to make social approaches to others. As described by Volkmar and coworkers (1989), aloof children were significantly younger than passive or active-but-odd children. Thus it is possible that, in the mentioned studies, children belonged to the aloof subgroup and consequently they chose a larger interpersonal space from others, whereas adolescents and adults may be character-ized by a passive or an active-but-odd social interaction and, therefore, maintained a shorter interpersonal space. This hypothesis should be investigated in future studies.

Moreover, this different pattern of results can also be ascribed to cultural differences, since some studies were conducted in Europe (Gessaroli et al., 2013b; Candini et al., 2017), while other studies were conducted in the USA (Parsons et al., 2004; Kennedy and Adolphs, 2014), and in Japan (Asada et al., 2016). It is well known that cultural factors may deeply in-fluence social behaviour (Hayduk, 1978), and so it is possible that for some cultures being closer to others is normal, while choosing a larger distance is deemed inappropriate. Further studies will clarify which variables induce an enlargement or a reduction of social distance in autism.

16.3 Anxiety and interpersonal space in autism

Abnormal interpersonal space in ASD children may reflect over-arousal and enhanced fear induced by the presence of other individuals intruding into their social space. Accordingly, previous research has suggested that an excessively functioning amygdala may account for abnormal fears and high level of anxiety in ASD children, thus contributing to impaired social interactions and avoidant behaviours in these patients (Hirstein et al., 2001; Schulkin et al., 2006; Corbett et al., 2006; 2010; Swartz et al., 2011). Supporting this hypothesis, Corbett and coworkers (2010) adopted an ecological paradigm designed to emulate a 'real-life' play-ground in order to determine whether such an environment would be deemed physiolo-gically stressful in a group of children with and without ASD. Results indicated that, during peer-to-peer social interactions, children with autism have significant enhanced levels of cortisol than typical developmental children. Thus, it is reasonable to hypothesize that ASD children, due to increased fear and hyperarousal following other people approaching and the potential violation of interpersonal space, regulate interpersonal space differently, resulting in an inappropriate distance from other individuals.

In this respect, it is important to note that individuals with high level of social anxiety show greater preferred distance from an unfamiliar person compared to those with low social anx-iety traits (Wieser et al., 2010; Rinck et al., 2010). In the same vein, a recent event-related po-tential (ERP) study revealed a positive correlation between levels of social anxiety, measured by using the Liebowitz Social Anxiety Scale (LSAS; Liebowitz, 1987), and preferred inter-personal distance, as measured by using a modified version of the comfortable interpersonal distance paradigm (Duke and Kiebach, 1974; Duke and Nowicki, 1972). In this paradigm, participants are instructed to imagine themselves in the centre of a room visualized on a com-puter screen and to respond to a virtual person approaching them by indicating where they would like the person to stop (Perry et al., 2013). The authors found that individuals with high levels of social anxiety exhibited attenuated early ERP responses (P1 and N1). In con-clusion, the more socially anxious an individual was, the smaller was the early ERP amplitude response, suggesting fewer attentional resources allocated to social stimuli.

A recent systematic review of the field indicates that social anxiety commonly co-occurs with ASD and is associated with poorer social skills, competence, and social motivation (Spain et al., 2018). In line with an avoidance bias, a possible explanation is that individuals with high social anxiety levels during a social engagement may feel earlier distress due to a diminished attention toward social cues, leading them to stand farther away and creating a cycle of less communicative and appropriate social interactions.

This hypothesis has been further verified in a subsequent study with a group of adults with ASD. The authors found that preferred interpersonal space chosen by ASD participants can be explained by differences in social anxiety level: the greater the social anxiety level, the farther away ASD individuals prefer to stand relative to an unfamiliar adult. Interestingly, the preferred interpersonal space can be predicted by the N1 amplitude, an early ERP component related to attention and discrimination processes. These results suggested that complex social behaviours, such as the ability to properly judge the appropriate distance from other individuals, could be influenced by altered early sensory and attentional processes (Perry et al., 2015).

16.4 A reduced interpersonal space plasticity in autism

One relevant characteristic of interpersonal space is that it changes as a function of different social factors, such as age, gender, attachment style, and the degree of intimacy and familiarity between individuals, but also the perceived morality of another person (Dosey and Meisels, 1969; Felipe and Sommer, 1966; Aiello,1987; Remland et al., 1995; Bar-Haim et al., 2002; Iachini et al., 2015; 2016; Pellencin et al., 2018). For instance, Bar-Haim and coworkers (2002) demonstrated how the internal representations of maternal attachment, formed during infancy, is relevant in explaining the variability emerged in interpersonal space regulation in children who experience Israeli kibbutzim. Authors found that children with an ambivalent attachment allowed more intrusion into their interpersonal space, whereas children with insecure-ambivalent attachment exhibited a larger interpersonal space compared with secure attachment children.

It was recently demonstrated that, in children with typical development, it is not necessary that there is a long interaction with another person to modify their interpersonal space, but a brief cooperative interaction with a previous unknown person is sufficient to reduce the distance that we maintain from that person (Gessaroli et al., 2013b; Candini et al., 2017, 2019). The comfort distance was measured in children with typical development and in children with autism before and after a very brief playtime interval during which the child was invited to read an illustrated book together with the confederate. After only ten minutes of a pleasant and cooperative playtime interval with an unfamiliar adult, a rapid reduction of the interpersonal space was observed, indicating that interpersonal space is highly flexible, presumably to facilitate social interactions and communication. Differently, after an uncooperative interaction with the confederate, during which the confederate did not play with the participant and remained silent or occasionally provided criticism, typical development children did not reduce interpersonal space. By contrast, in ASD children, no changes of interpersonal space emerged in response to a brief social cooperative or uncooperative interaction (Gessaroli et al., 2013; Candini et al., 2017). Overall, these results suggest that the plasticity of interpersonal space is altered in ASD children.

Interestingly, a link was provided between interpersonal space plasticity and the severity of social deficit in everyday life in ASD children (Candini et al., 2017). Children's social impairment severity in everyday activities was measured considering the rating on one item from the Wing Subgroups Questionnaire (Castelloe and Dawson, 1993) as an index of appropriate interpersonal space regulation ('When the child is with unfamiliar adults or children, he readily approaches others to interact. His manner of interacting is generally appropriate, not awkward or unusual.'). Considering the frequency of inappropriate social responses on this item, the greater the social impairment severity in social daily interactions, the higher was the impairment observed in interpersonal space plasticity.

16.5 First and third person perspective in interpersonal space regulation

During everyday social interactions, we continuously experience the need to identify other's intentions in order to react with the most appropriate behaviour. This mechanism is present also in animals (McGregor, 1993) and represents a clear advantage during social interactions because it contributes to assessing both the situation and the conspecifics involved, particularly in fearful or threatening context (Olsson et al., 2007).

In this respect, research on social cognition has pointed out that, when ASD individuals observe a social interaction, they exhibit difficulties in adopting another perspective (Frith, 1996). A recent study explored whether ASD individuals are impaired not only when they regulate interpersonal space between themselves and others (first-person perspective) but also when they estimate others' interpersonal space (third-person perspective). The interpersonal space between the participant and an unknown adult was measured in a group of typical developmental children and in two groups of high-functioning ASD children, with a more severe and a less severe social impairment in daily life, respectively (Candini et al., 2017). ASD children preferred larger interpersonal distance compared to typical development children, both when they were asked to adopt a first- or a third-person perspective, suggesting that the adopted perspective did not influence the interpersonal space permeability. Critically, perspective and severity of social impairment influenced the interpersonal space plasticity: following a cooperative social interaction, ASD children with a less severe social impairment reduced the interpersonal space between themselves and the adult (first-person perspective), whereas no such change was found when they were asked to regulate interpersonal distance between others (third-person perspective). Conversely, ASD children with severe social impairment did not show any change of interpersonal space, after compared to before a social interaction, in first- as well as in third-person perspective. Together, these results suggest that the ability to dynamically adapt and appropriately regulate interpersonal space between themselves and others is not sufficient to appropriately estimate interpersonal space between other people.

One interesting question for future studies concerns the relationship between the ability to regulate the distance between the confederate's and the participant's body, and the distinction between self and other. Because there is some evidence to suggest that integration of body-related multisensory signals within the space around the body is a fundamental component of the ability to differentiate self from others, it seems crucial to investigate the link between high-order social cognitive processes and bodily self-representation.

In this respect, Gessaroli and colleagues (2013a) investigated recognition of one's own body in typical development children and in children with ASD. In this study, participants performed a visual matching-to-sample task with pictures depicting self and other people's body parts. Interestingly, both groups showed a facilitation when they visually matched their own, compared to others' body parts. This result is suggestive of a spared bodily self-recognition in children with ASD. A preserved self-other distinction in ASD adults, even if less plastic than healthy controls, was also found in a further study adopting the rubber hand illusion, an experimental manipulation that affects the sense of body ownership and induces the illusion of having a rubber hand (Cascio et al., 2012). The authors reported a marked difficulty in disembodying the bodily self and embodying the bodily other (i.e. the rubber hand) in individuals with ASD. This finding indicates that they are less prone to bodily illusions compared to controls. An intriguing explanation recently proposed by Noel and colleagues suggests the presence of a steeper and less flexible self-other boundary in autism (Noel et al., 2017).

In this perspective, the tendency to maintain a larger interpersonal space between one's own body and others' body in a comfort-distance task can be a manifestation of the sharp boundary and the inflexible distinction between self and other, which affects both the corporeal and social domains, resulting in an impairment of others' emotion recognition, empathy, and theory of mind. Further studies should clarify a possible link between self/other distinction and socio-communicative deficits in ASD individuals.

16.6 Functional dissociation between action and social space

The space close to one's own body is important not only for social interaction but also to reach and grasp objects by using one's own arm. In cognitive neuroscience, a flourishing research literature coined the term 'peripersonal' space by referring to the multisensory interface between the body and the environment, in which stimuli are coded in motor terms for the purpose of voluntary goal-directed actions (Rizzolatti et al., 1981, 1997; Maravita and Iriki, 2004; Makin et al., 2009; Brozzoli et al., 2012; see, for review, Occelli et al., 2011; Makin et al., 2012; di Pellegrino and Làdavas, 2015).

Recently, the functional characteristic of the interpersonal and peripersonal spaces were compared by using the stop-distance paradigm. In this procedure, participants stand facing an unfamiliar person and move toward him/her (Dosey and Meisels, 1969; Hayduk, 1983; Aiello, 1987). In the *reaching-distance task*, participants have to stop when they think they can reach the other person by extending their limb. In the *comfort-distance task*, participants have to stop at the point where they still feel comfortable with the other's proximity. This well-established procedure allows experimenters to ecologically measure the extent of peripersonal and interpersonal space respectively (Iachini et al., 2014; Patanè et al., 2016; 2017; Candini et al., 2017; Pellencin et al., 2018; Candini et al., 2019).

As we have seen before, our interpersonal distance preferences are not immutable, but rather may vary depending on social context. In a similar manner, peripersonal space also has a flexible nature. In fact, human beings can extend their physical action capabilities by using tools (see, for review, Johnson-Frey, 2003, 2004; Maravita and Iriki, 2004; Reynaud et al., 2016), which allows individuals to reach objects located in a far and unreachable space, consequently extending the reaching space's boundary (Berti and Frassinetti, 2000; Longo and Lourenco, 2006; Taffou and Viaud-Delmon, 2014; Hunley et al., 2017; see, for review, Brozzoli et al., 2012; Martel et al., 2016).

Even if the most prominent modulations of peripersonal space described in literature are primarily driven by action-related factors, there is also evidence demonstrating that peripersonal space extent is influenced by high-order social factors. For instance, Ruggiero and colleagues (2016) explored the role of emotional facial expression in defining the peripersonal space in a virtual reality environment. In this study, participants performed the reaching- and the comfort-distance tasks while being approached or walking toward a virtual confederate who exhibited happy, angry, or neutral facial expressions. They reported an enlargement of peripersonal space when participants were approached by an angry confederate. Furthermore, Iachini and colleagues (2015) investigated whether a moral or immoral confederate can influence the reaching and comfort distance chosen by participants. They found that preferred distance chosen by participants actually expanded (shortened) with a confederate described as an immoral (moral) person. This pattern of results was observed both in peripersonal and interpersonal spaces, although it was stronger in the latter one. Overall, this evidence supports the hypothesis that the plasticity of the space around the body allows individuals to dynamically and flexibly react to potentially relevant stimuli, objects, or people, located close the participant's body.

In line with the interpretation proposed by de Vignemont and Iannetti (2015), and further discussed by Bufacchi and Iannetti (2019), the space surrounding the body represents a dynamic interface to safely interact with the physical and social environment. In this respect, it is worth noting that the social modulations of peripersonal space described in literature (Iachini et al., 2015; Ruggiero et al., 2016) emerge particularly when the participant's response has a protective scope, thus emphasizing the defensive nature of the action space.

Several attempts have been made recently to clarify the functional relationship between action and social space (Iachini et al., 2014; 2015; Patanè et al., 2016; 2017; Pellencin et al., 2018; Cartaud et al., 2018). For instance, recent work by Patanè and coworkers (2016) compared the peripersonal and interpersonal space plasticity toward a confederate by measuring the reaching and the comfort distance before and after a cooperative tool-use training. In this training, participant and confederate jointly cooperate to reach objects placed beyond the arm-reaching distance by using a long tool. Interestingly, the training induces opposite effects on peripersonal and interpersonal spaces: peripersonal space extends, whereas interpersonal space reduces as compared to before cooperative tool-use training (Patanè et al., 2017). These findings are in line with the dynamical properties of the space surrounding the body previously described (Berti and Frassinetti, 2000; Longo and Lourenco, 2006; Gessaroli et al., 2013; Candini et al., 2017), but also show that action and social space are subtended by dissociable plastic mechanisms. Indeed, peripersonal space is sensitive to tool use, whereas interpersonal space is sensitive to social cooperation with another person.

In order to verify whether autism affects the plasticity of both interpersonal (Gessaroli et al., 2013b; Candini et al., 2017) and peripersonal space, in a recent study the peripersonal and interpersonal space representations have been explored in a group of children with typical and atypical ASD. Participants performed the reaching- and comfort-distance task in two experimental conditions: facing either an inanimate object or a female unfamiliar adult. Then, to explore the plasticity of these spaces, both tasks were repeated twice: before and after a cooperative tool-use training in which the child used a tool to cooperate with the confederate (see Figure 16.2).

Authors found that both typical development and ASD children preferred larger comfort than reaching distance. This result could be accounted for by the fact that participants were children facing an unfamiliar adult, a situation in which they may have experienced a feeling of discomfort (Dean et al., 1976; Severy et al., 1979). This interpretation is further supported by the fact that the comfort distance was modulated by the type of stimulus approached, with greater interpersonal space found when typical development and ASD children moved toward the person compared to the object. A possible explanation is that the inanimate object was perceived as potentially more safe or predictable compared to an unfamiliar adult, thus allowing children to tolerate a closer distance when facing the object than the person. The intrusion of other people into one's own interpersonal space can increase anxiety, which individuals tend to reduce by keeping other people farther away (Strube and Werner, 1982; Perry et al., 2013; 2015). Conversely, no differences emerged when participants estimated

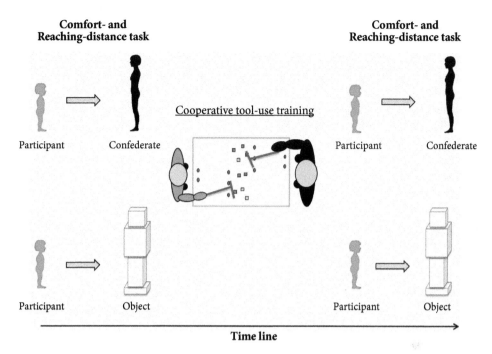

Figure 16.2 Schematic representation of the experimental procedure adopted in Candini et al.'s study (2019). Participants have to stop at the point *where they still feel comfortable with the stimuli's proximity* (comfort-distance task) and *when they think they can reach the stimuli by extending their limb* (reaching-distance task). Both tasks were performed in two experimental conditions: facing either a female unfamiliar adult (confederate) or an inanimate object, before and after a cooperative tool-use training session in which the participant (green) and the confederate (black) jointly cooperate to reach coloured chips placed in the unreachable space by using a rake.

Adapted from - Candini, M., Giuberti, V., Santelli, E., di Pellegrino, G., and Frassinetti, F. When social and action space diverges: a study in children with typical development and autism. Autism. 2019; 23(7): 1687-1698. https://doi.org/10.1177/1362361318822504.

the reaching distance toward the person or the object, thereby suggesting that interpersonal, but not peripersonal, space is selectively shaped by the social significance of stimuli. Regarding the ability to estimate reaching distance, a positive correlation between preferred reaching distance and participants' arm length demonstrates that children from 8 to 13 years old take into consideration their arm length to evaluate stimulus reachability. This finding shows that during developmental age, peripersonal space representation is deeply grounded in the action potentialities of their own body confirming previous developmental studies (Rochat, 1995; Caçola and Gabbard, 2012) (for similar results in adult population, see Witt et al., 2005; Witt and Proffitt, 2008).

Critically, a difference emerged between the plastic properties of peripersonal and interpersonal space in typical development and ASD children. Indeed, in typical development children, following the cooperative tool-use training, peripersonal space extended, both toward the confederate and the object, whereas interpersonal space reduced toward the confederate, but not the object, as compared to before the training. Conversely, in ASD children, both peripersonal and interpersonal spaces extended, regardless of whether they approached the confederate or object. Comparing the effects of cooperative tool-use training in typical development and ASD children, similar results emerged in the reaching- but not in the comfort-distance task: peripersonal space extended in both groups, whereas interpersonal space reduced in typical development but not in ASD children. Thus, in the ASD population the tool use is effective in modulating the peripersonal space, whereas the social interaction, during which the tool is used with the confederate, fails to contract interpersonal space.

Interestingly, this impairment, confined to interpersonal space regulation, correlated with the level of social impairment as measured by the score obtained on the 'Social Interaction' ADOS scale subtest (Autism Diagnostic Observation Schedule, Lord et al., 2000): higher is the severity of social interaction deficit, higher is the deficit of interpersonal space plasticity observed in ASD children. This is in line with previous findings (Candini et al., 2017), and suggests that interpersonal space provides a functional and operational measure of social space. Going further, the demonstration of a preserved peripersonal space regulation in autism is highly relevant because it uncovers an isle of proficiency in reachability estimation, despite the presence of well-documented deficit in planning and executing goal-directed actions in ASD children (Mari et al., 2003; Haswell et al., 2009; Linkenauger et al., 2012; see, for review, Moseley and Pulvermuller, 2018). The intact functional properties of action space found in children with atypical development underlie the specific role of peripersonal space plasticity in order to efficiently adapt our behaviours to a particular context.

16.7 Neural correlates of social space in autism

The neural correlates of social space are the focus of recent neuroimaging studies in healthy subjects and neuropsychological studies in brain-damaged patients.

In a functional magnetic resonance imaging study, Holt and colleagues (2014) found that the activity of a brain network including the dorsal intraparietal sulcus (dIPS) and ventral premotor cortex (vPM) is functionally linked to social behaviours. Indeed, the participants' interpersonal space negatively correlated with the coupling between these two brain regions: the weaker is the dIPS-vPM functional coupling, the larger is the preferred interpersonal space.

Moving to neuropsychological studies, atypical approach-avoidance behaviours are reported in patients affected by amygdala lesion (Harrison et al., 2015; Kennedy et al., 2009). Kennedy and colleagues (2009) described a patient with bilateral amygdala damage, who showed a substantially reduced interpersonal space, both in first- and in third-person perspective, compared to a group of neurologically healthy participants. These findings reveal that bilateral damage to the amygdala results in an abnormally small interpersonal space, thereby suggesting a key role for this subcortical structure in the neural substrate of social space regulation.

A further advance in our understanding of the rules mediating interpersonal space preference has been provided by a recent study by Perry and colleagues (2016). Starting from the idea that interpersonal space strongly relates to social norms and the inhibition of inappropriate social conduct, the authors hypothesized that the orbitofrontal cortex (OFC) may be a critical brain region implicated in social behaviour and interpersonal space. In line with this view, they found that patients with orbitofrontal damage do not keep the expected distance from others, revealing an impaired social distancing. In the comfort-distance task, patients with damage to the OFC preferred very close interpersonal distances regardless of whether they were approaching a stranger or a familiar other, compared to both healthy controls and patients with dorsolateral prefrontal damage.

Linking this anatomical evidence to the reduced tolerance of physical closeness with a stranger and the lack of flexibility of interpersonal space in ASD children, it is reasonable to hypothesize that a cortico-subcortical network, involving amygdala and orbitofrontal cortex is critical in mediating the ability to evaluate the relevance of social stimuli in order to maintain an appropriate interpersonal distance. This hypothesis is supported by recent studies indicating that an excessively functioning amygdala may account for the increased fear and anxiety found in ASD population, thereby leading to withdrawal from social interaction (Kleinhans et al., 2009; Swartz et al., 2013). Additionally, several neuroimaging data of atypical connectivity between frontal and limbic networks in ASD individuals are also reported (Cerliani et al., 2015; Kleinhans et al., 2016; Glerean et al., 2016). Overall, these findings advance our knowledge concerning the neural correlates underlying the space close to the body and provide a foundation for future studies examining the extent to which these representations are flexible both in typical and atypical development.

16.8 Concluding remarks

To sum up, several factors may have a role in influencing the social space in autism: the social anxiety, the nature of social interaction, the severity of social impairment, and the person's perspective. Furthermore, these findings highlight a close link between interpersonal space regulation and social behaviours observed in everyday life. This suggests that the difference in interpersonal space plasticity between typical and atypical development may in part reflect the heterogeneity of cognitive underpinnings characterizing ASD.

References

Aiello, J. R. (1987). Human spatial behavior. In D. Stokols, & I. Altman (eds), *Handbook of Environmental Psychology*. New York: John Wiley and Sons.

APA (American Psychiatric Association) (2013). *Diagnostic and Statistical Manual of Mental Disorders (DSM-5)*. Available at: https://dsm.psychiatryonline.org/doi/book/10.1176/appi.books.9780890425596

Asada, K., Tojo, Y., Osanai, H., Saito, A., Hasegawa, T., & Kumagaya, S. (2016). Reduced personal space in individuals with autism spectrum disorder. *PLoS ONE*, 11(1), e0146306. doi:org/10.1371/journal.pone.0146306

Bar-Haim, Y., Aviezer, O., Berson, Y., & Sagi, A. (2002). Attachment in infancy and personal space regulation in early adolescence, *Attachment & Human Development*, 4(1), 68–83. doi:10.1080/14616730210123111

Baron-Cohen, S., Leslie, A., M., & Frith, U. (1985). Does the autistic child have a 'theory of mind'? *Cognition*, 21, 37–46.

Berti, A., & Frassinetti, F. (2000). When far becomes near: remapping of space by tool use. *Journal of Cognitive Neuroscience*, 12(3), 415–420. doi:org/10.1162/089892900562237

Brozzoli, C. Makin, T. R. Cardinali, L. Holmes, N. P. & Farnè, A. (2012). Peripersonal space: A multisensory interface for body–object interactions. In M. M. Murray, & M. T. Wallace (eds), *The Neural Bases of Multisensory Processes*. Boca Raton, FL: CRC Press.

Bufacchi, R. J., & Iannetti, G. D. (2019). The value of actions, in time and space. *Trends in Cognitive Science*, 23(4), 270–271.

Caçola, P., & Gabbard, C. (2012. Modulating peripersonal and extrapersonal reach space via tool use: a comparison between 6- to 12-year-olds and young adults. *Experimental Brain Research*, 218, 321–330. doi:10.1007/s00221-012-3017-8

Candini, M., Giuberti, V., Manattini, A., Grittani, S., di Pellegrino, G., & Frassinetti, F. (2017). Personal space regulation in childhood autism: effects of social interaction and person's perspective. *Autism Research*, 10(1), 144–154. doi:10.1002/aur.1637

Candini, M., Giuberti, V., Santelli, E., di Pellegrino, G., & Frassinetti, F. (2019). When social and action space diverges: a study in children with typical development and autism. *Autism*, 23(7), 1687–1698. doi:org/10.1177/1362361318822504

Cartaud, A., Ruggiero, G., Ott, L., Iachini, T., & Coello, Y. (2018). Physiological response to facial expressions in peripersonal space determines interpersonal distance in a social interaction context. *Frontiers in Psychology*, 9, 657. doi:10.3389/fpsyg.2018.00657

Cascio, C. J., Foss-Feig, J. H., Burnette, C. P., Heacock, J. L., Cosby, A. A. (2012). The rubber hand illusion in children with autism spectrum disorders: delayed influence of combined tactile and visual input on proprioception. *Autism*, 16(4), 406–419.

Castelloe, P., & Dawson, G. (1993). Subclassification of children with autism and pervasive developmental disorder: a questionnaire based on Wing's subgrouping scheme. *Journal of Autism Developmental Disorders*, 23(2), 229–241.

Cerliani, L., Mennes, M., Thomas, R. M., Di Martino, A., Thioux, M., & Keysers, C. (2015). Increased functional connectivity between subcortical and cortical resting-state networks in autism spectrum disorder. *JAMA Psychiatry*, 72(8), 767–777.

Constantino, J., & Gruber, J. (2005). *Social Responsiveness Scale (SRS) Manual*. Los Angeles: Western Psychological Services.

Corbett, B. A., Mendoza, S., Abdullah, M., Wegelin, J. A., & Levine, S. (2006). Cortisol circadian rhythms and response to stress in children with autism. *Psychoneuroendocrinology*, 31: 59–68. doi:10.1016/j.psyneuen.2005.05.011

Corbett, B. A., Schupp, C. W., Simon, D., & Mendoza, S. (2010). Elevated cortisol during play is associated with age and social engagement in children with autism. *Molecular Autism*, 1(13). doi:org/10.1186/2040-2392-1-13

Dean, L. M., Willis, F. N., & LaRocco, J. M. (1976). Invasion of personal space as a function of age, sex, and race. *Psychological Reports*, 38, 959–965.

de Vignemont, F., & Iannetti G. D. (2015). How many peripersonal spaces? *Neuropsychologia*, 70, 327–334.

di Pellegrino, G., & Làdavas, E. (2015). Peripersonal space in the brain. *Neuropsychologia*, 66, 126–133. doi:org/10.1016/j.neuropsychologia.2014.11.011

Dosey, M. A., & Meisels, M. (1969). Personal space and self-protection. *Journal of Personality and Social Psychology*, 11, 93–97.

Duke, M. P., & Kiebach, C. (1974). A brief note on the validity of the comfortable interpersonal distance scale. *Journal of Social Psychology*, 94(2), 297–298.

Duke, M. P., & Nowicki, S. (1972). A new measure and social-learning model for interpersonal distance. *Journal of Experimental Research in Personality*, 6, 119–132.

Felipe, N. J., & Sommer, R. (1966). Invasions of personal space. *Social Problems*, 14(2), 1. doi:org/10.2307/798618

Freitag, G. (1970). An experimental study of the social responsiveness of children with autistic behaviors. *Journal of Experimental Child Psychology*, 9, 436–453.

Frith, U. (1996). Cognitive explanations of autism. *Acta Paediatrica Supplement*, 85, S416, 63–68.

Gessaroli, E., Andreini, V., Pellegri, E., & Frassinetti, F. (2013a). Self-face and self-body recognition in autism. *Research in Autism Spectrum Disorders*, 7(6), 793–800. doi:org/10.1016/j.rasd.2013.02.014

Gessaroli, E., Santelli, E., di Pellegrino, G., & Frassinetti, F. (2013b). Personal space regulation in childhood autism spectrum disorders. *PLoS ONE*, 8(9): e74959.

Glerean, E., Pan, R. K., Salmi, J., Kujala, R., Lahnakoski, J. M., Roine, U., & Jääskeläinen, I. P. (2016). Reorganization of functionally connected brain subnetworks in high-functioning autism. *Human Brain Mapping*, 37(3), 1066–1079.

Hall, E. (1966). *Distances in Man: The Hidden Dimension*. Garden City, New York: Doubleday.

Harrison, L. A., Hurlemann R., & Adolphs, R. (2015). An enhanced default approach bias following amygdala lesions in humans. *Psychological Science*, 26(10), 1543–55. doi:10.1177/0956797615583804

Haswell, C., Izawa, J., Dowell, L., Mostofsky, S., & Shadmehr, R. (2009). Representation of internal models of action in the autistic brain. *Nature Neuroscience*, 12(8), 970–972. doi:org/10.1038/nn.2356

Hayduk, L. A. (1978). Personal space: an evaluative and orienting overview. *Psychological Bulletin*, 85, 117–134.

Hayduk, L. A. (1983). Personal space: where we now stand. *Psychological Bulletin*, 94, 293–335.

Hirstein, W., Iversen, P., & Ramachandran, V. S. (2001). Autonomic responses of autistic children to people and objects. *Proceedings. Biological Sciences*, 268, 1883–1888. doi:10.1098/rspb.2001.1724

Holt, D. J., Cassidy, B. S., Yue, X., Rauch, S. L., Boeke, E. A., Nasr, S., Tootell, R. B., & Coombs, G. (2014). Neural correlates of personal space intrusion. *Journal of Neuroscience*, 34(12), 4123–4134.

Hunley, S. B., Marker, A. M., & Lourenco, S. F. (2017). Individual differences in the flexibility of peripersonal space. *Experimental Psychology*, 64(1), 49–55. doi:10.1027/1618-3169/a000350

Iachini, T., Coello, Y., Frassinetti, F., & Ruggiero, G. (2014). Body space in social interactions: a comparison of reaching and comfort distance in immersive virtual reality. *PLoS ONE*, 9(11), e111511. doi:10.1371/journal.pone.0111511

Iachini, T., Coello, Y., Frassinetti, F., Senese, V. P., Galante, F., & Ruggiero, G. (2016). Peripersonal and interpersonal space in virtual and real environments: effects of gender and age. *Journal of Environmental Psychology*, 45, 154–164. doi:10.1016/j.jenvp.2016.01.004

Iachini, T., Pagliaro, S., & Ruggiero, G. (2015). Near or far? It depends on my impression: moral information and spatial behavior in virtual interactions. *Acta Psychologica*, 161, 131–136. doi:org/10.1016/j.actpsy.2015.09.003

Johnson-Frey, H. S. (2003). *Cortical Mechanisms of Human Tool Use. Taking Action: Cognitive Neuroscience Perspectives on the Problem Of Intentional Acts*. New York: MIT Press.

Johnson-Frey, H. S. (2004). The neural bases of complex tool use in humans. *Trends in Cognitive Science*, 8(2), 71–78.

Kennedy, D. P., & Adolphs, R. (2014). Violations of personal space by individuals with autism spectrum disorder. *PLoS ONE*, 6(9), 8, e103369.

Kennedy, D. P., Gläscher, J., Tyszka, J. M., & Adolphs, R. (2009). Personal space regulation by the human amygdala. *Nature Neuroscience*, 12, 1226–1227.

Kleinhans, N. M., Johnson, L. C., Richards, T., Mahurin, R., Greenson, J., Dawson, G., & Alward, E. (2009). Reduced neural habituation in the amygdala and social impairments in autism spectrum disorders. *American Journal of Psychiatry*, 166, 467–475. doi:10.1176/appi.ajp.2008.07101681

Kleinhans, N. M., Reiter, M. A., Neuhaus, E., Pauley, G., Martin, N., Dager, S., & Estes, A. (2016). Subregional differences in intrinsic amygdala hyperconnectivity and hypoconnectivity in autism spectrum disorder. *Autism Research*, 9(7), 760–772.

Libertus, K., & Violi, D. A. (2016). Sit to talk: relation between motor skills and language development in infancy. *Frontiers in Psychology*, 7, 475.

Liebowitz, M. R. (1987). Social phobia. *Modern Problems of Pharmacopsychiatry*, 22, 141–173.

Linkenauger, S. A., Lerner, M. D., Ramenzoni, V. C., & Proffitt, D. R. (2012). A perceptual-motor deficit predicts social and communicative impairments in individuals with autism spectrum disorders. *Autism Research*, 5(5), 352–62. doi:10.1002/aur.1248

Lloyd, D. M., & Morrison, C. I. (2008). 'Eavesdropping' on social interactions biases threat perception in visuospatial pathways. *Neuropsychologia*, 46, 95–101.

Longo, M. R., & Lourenco, S. F. (2006). On the nature of near space: effects of tool use and the transition to far space. *Neuropsychologia*, 44(6), 977–981.

Lord, C., Risi, S., Lambrecht, L., Cook, E. H., Jr, Leventhal, B. L., Di Lavore, P. C., et al. (2000). The autism diagnostic observation schedule-generic: a standard measure of social and communication deficits associated with the spectrum of autism. *Journal of Autism and Developmental Disorders*, 30(3), 205–223.

Makin, R., Holmes, N. P., Brozzoli, C., Rossetti, Y., & Farnè, A. (2009). Coding of visual space during motor preparation: approaching objects rapidly modulate corti-cospinal excitability in hand-centered coordinates. *Journal of Neuroscience*, 29, 11841–11851.

Makin, T. R., Holmes, N. P., Brozzoli, C., & Farnè, A. (2012). *Experimental Brain Research*, 219, 421. doi:org/10.1007/s00221-012-3089-5

Maravita, A., & Iriki, A. (2004). Tools for the body (schema). *Trends in Cognitive Science*, 8, 79–86.

Mari, M., Castiello, U., Marks, D., Marraffa, C., & Prior, M. (2003). The reach-to-grasp movement in children with autism spectrum disorder. *Philosophical Transactions of the Royal Society of London. Series B: Biological Sciences*, 358(1430), 393–403.

Martel, M., Cardinali, L., Roy, A. C., & Farnè, A. (2016). Tool use: an open window into body representation and its plasticity. *Cognitive Neuropsychology*, 33(1–2), 82–101. doi:10.1080/02643294.2016.1167678

McGregor, P. K. (1993). Signalling in territorial systems: a context for individual identification, ranging and eavesdropping. *Philosophical Transactions: Biological Sciences*, 340(1292), 237–244.

Moseley, R. L., & Pulvermuller, F. (2018). What can autism teach us about the role of sensorimotor systems in higher cognition? New clues from studies on language, action semantics, and abstract emotional concept processing. *Cortex*, 100, 149–190. doi:10.1016/j.cortex.2017.11.019

Noel, J. P., Cascio, J. C., Wallace, M., & Park, S. (2017). The spatial self in schizophrenia and autism spectrum disorder. *Schizophrenia Research*, 179, 8–12. doi:10.1016/j.schres.2016.09.021

Occelli, V., Spence, C., & Zampini, M. (2011). Audiotactile interactions in front and rear space. *Neuroscience & Biobehavioral Reviews*, 35(3), 589–598. doi:org/10.1016/j.neubiorev.2010.07.004

Olsson, A., Nearing, K. I., & Phelps, E. A. (2007). Learning fears by observing others: the neural systems of social fear transmission. *Social Cognitive and Affective Neuroscience*, 2(1), 3–11. doi:org/10.1093/scan/nsm005

Parsons, S., Mitchell, P., & Leonard, A. (2004). The use and understanding of virtual environments by adolescents with autistic spectrum disorders. *Journal of Autism Developmental Disorders*, 34, 449–66.

Patanè, I., Farnè, A., & Frassinetti, F. (2017). Cooperative tool-use reveals peripersonal and interpersonal spaces are dissociable. *Cognition*, 166, 13–22. doi:org/10.1016/j.cognition.2017.04.013

Patanè, I., Iachini, T., Farnè, A., & Frassinetti, F. 2016. Disentangling action from social space: tool-use differently shapes the space around us. *PLoS ONE*, 11(5), e0154247.

Pedersen, J., & Schelde, J. T. (1997). Behavioral aspects of infantile autism: an ethological description. *European Child & Adolescent Psychiatry*, 6, 2, 96–106.

Pedersen, J., Livoir-Petersen, M. F., & Schelde, J. T. (1989). An ethological approach to autism: an analysis of visual behaviour and interpersonal contact in a child versus adult interaction. *Acta Psychiatrica Scandinavica*, 80, 4, 346–355.

Pellencin, E., Paladino, M. P., Herbelin, B., & Serino, A. (2018). Social perception of others shapes one's own multisensory peripersonal space. *Cortex*, 104, 163–179. doi:10.1016/j.cortex.2017.08.033

Perry, A., Levy-Gigi, E., Richter-Levin, G., & Shamay-Tsoory, S. G. (2015). Interpersonal distance and social anxiety in autistic spectrum disorders: a behavioral and ERP study. *Social Neuroscience*, 10, 354–365.

Perry, A., Rubinstenb, O., Peled, L., & Shamay-Tsoory, S. G. (2013). Don't stand so close to me: a behavioral and ERP study of preferred interpersonal distance. *Neuroimage*, 83, 761–769.

Perry, A., Lwi, S. J., Verstaen, A., Dewar, C., Levenson, R. W., & Knight, R. T. (2016). The role of the orbitofrontal cortex in regulation of interpersonal space: evidence from frontal lesion and frontotemporal dementia patients. *Social Cognitive and Affective Neuroscience*, 1894–1901.

Remland, M. S., Jones, T. S., & Brinkman, H. (1995). Interpersonal distance, body orientation, and touch: effects of culture, gender, and age. *Journal of Social Psychology*, 135, 3, 281–297.

Reynaud, E., Lesourd, M., Navarro, J., & Osiurak, F. (2016). On the neurocognitive origins of human tool use: a critical review of neuroimaging data. *Neuroscience Biobehavioural Review*, 64, 421–437. doi:10.1016/j.neubiorev.2016.03.009

Rinck, M., Rörtgen, T., Lange, W. G., Dotsch, R. Wigboldus, D. H. J., & Becker, E. S. (2010). Social anxiety predicts avoidance behaviour in virtual encounters. *Cognition and Emotion*, 24(7), 1269–1276.

Rizzolatti, G., Fadiga, L., Fogassi, L., & Gallese, V. (1997). The space around us. *Science*, 277, 190–191.

Rizzolatti, G., Scandolara, C., Matelli, M., & Gentilucci, M. (1981). Afferent properties of periarcuate neurons in macaque monkeys. II. Visual responses. *Behavioural Brain Research*, 2(2), 147–163. doi:org/10.1016/0166-4328(81)90053-x

Rochat, P. (1995). Early objectification of the self. In P. Rochat (ed.), *The Self in Infancy*. Advances in psychology book series. Amsterdam: Elsevier.

Ruggiero, G., Frassinetti, F., Coello, Y., Rapuano, M., Di Cola, A. S., & Iachini, T. (2016). The effect of facial expressions on peripersonal and interpersonal spaces. *Psychological Research*. 81, 1232–1240. doi:10.1007/s00426-016-0806-x

Schulkin, J. (2006). Autism and the amygdala: an endocrine hypothesis. *Brain and Cognition*, 65: 87–99.

Senju, A. (2013). Atypical development of spontaneous social cognition in autism spectrum disorders. *Brain Development*, 35(2), 96–101. doi:10.1016/j.braindev.2012.08.002

Severy, L. J., Forsyth, D. R., & Wagner, P. J. (1979). A multimethod assessment of personal space development in female and male, Black and White children. *Journal of Nonverbal Behaviour*, 4(2), 68–86. doi:org/10.1007/BF01006352

Sommer, R. (1959). Studies in personal space. *Sociometry*, 22, 247–260.

Spain, D., Sin, J., Linder, K. B., McMahon, J., & Happé F. (2018). Social anxiety in autism spectrum disorder: a systematic review. *Research in Autism Spectrum Disorders*, 52, 51–68.

Strube, M. J., & Werner, C. (1982). Interpersonal distance and personal space: a conceptual and methodological note. *Journal of Nonverbal Behaviour*, 6, 163–170.

Swartz, J. R., Wiggins, J. L., Carrasco, M., Lord, C., & Monk, C. S. (2011). Amygdala habituation and prefrontal functional connectivity in youth with autism spectrum disorders. *Journal of American Academic Child Adolescence Psychiatry*, 52, 84–93.

Swartz, J. R., Wiggins, J. L., Carrasco, M., Lord, C., & Monk, C. S. (2013). Amygdala habituation and prefrontal functional connectivity in youth with autism spectrum disorders. *Journal of American Academic Child Adolescence Psychiatry*, 52, 84–93.

Taffou, M., & Viaud-Delmon, I. (2014). Cynophobic fear adaptively extends peri-personal space. *Frontiers in Psychiatry*, 5, 122. doi:10.3389/fpsyt.2014.00122

Volkmar, F. R., Cohen, D. J., Bregman, J. D., Hooks, M. Y., & Stevenson, J. M. (1989). An examination of social subtypologies in autism. *Journal of the American Academy of Child and Adolescent Psychology*, 28, 82–86.

Wieser, M. J., Pauli, P., Grosseibl, M., Molzow, I., & Mühlberger, A. (2010). Virtual social interactions in social anxiety—the impact of sex, gaze, and interpersonal distance. *Cyberpsychology, Behaviour, and Social Networking*, 13(5), 547–554.

Wing, L., & Attwood, A. (1987). Syndromes of autism and atypical development. In D. J. Cohen, & A. Donnelan (eds), *Handbook of Autism*. New York: Wiley.

Wing, L., & Gould, J. (1979). Severe impairments of social interaction and associated abnormalities in children: epidemiology and classification. *Journal of Autism and Developmental Disorders*, 9, 11–29.

Witt, J. K., & Proffitt, D. R. (2008). Action-specific influences on distance perception: a role for motor simulation. *Journal of Experimental Psychology: Human Perception and Performance*, 34, 1479–1492.

Witt, J. K., Proffitt, D. R., & Epstein, W. (2005). Tool use affects perceived distance, but only when you intend to use it. *Journal of Experimental Psychology: Human Perception and Performance*, 31, 880–888.

17

Risk-taking behaviour as a central concept in evolutionary biology

Anders Pape Møller

17.1 Introduction

Peri-personal space (PPS) is defined as the space immediately surrounding the body (this volume). This space is determined by the actions that an individual is taking to optimize the distance from humans or other organisms that may pose an immediate threat. This space may appear threatening because any actions affecting it may result in an increased probability of injury or even death. Individuals belonging to different species vary in their PPS. This may have consequences for individuals with different kinds of flight responses to the presence of predators with such behaviour having different ontogenies, proximate mechanisms, functions, and phylogenetic histories (Tinbergen, 1963). If we accept this classical definition, we have to take into account that these four levels of understanding apply to all these disparate aspects of behaviour. Thus, different kinds of behaviour may affect size and shape of PPS.

Flight initiation distance (FID), PPS and IPS are different concepts with different evolutionary histories and even different levels of understanding. While FID deals with both proximate and ultimate aspects of this phenomenon, current studies of PPS and IPS are almost entirely proximate in approach. This implies that they largely ignore the roles of ontogeny, function, and phylogeny as the basis for fully understanding these concepts (Tinbergen, 1963). Experimental neuroscience has shown that the sense of one's own body influences the visual perception of the world. For example, Van der Hoort et al. (2012) showed that manipulation of the size of perceived body affected the perception of objects as being larger and farther away, while perception of a small body resulted in objects being perceived as being smaller and nearer. These effects resembled studies of FID showing large effects of body size on anti-predator response (Samia et al., 2015). Similarly, a perceptual illusion of having an invisible body reduced the anxiety response like that when standing 'unprotected' in front of an audience (Guterstam et al., 2015). If an individual experienced invisibility, this has been shown to result in the contraction of IPS without affecting the perceived space surrounding the body (D'Angelo et al., 2017). In contrast, modifying the boundaries of PPS left IPS unaffected, supporting a close relationship between IPS and the representation of the body. These observations could suggest that there are two different representations of space surrounding the human body. Although these neuroscience experiments on humans may be linked to FID and other types of escape behaviour, further and more specific experiments dealing with the function of such differences in behaviour are required.

There has been an explosion in the number of citations of papers dealing with both FID and PPS (Figure 17.1). While papers dealing with these concepts only had a handful of

Anders Pape Møller, *Risk-taking behaviour as a central concept in evolutionary biology* In: *The World at Our Fingertips*.
Edited by Frédérique de Vignemont, Andrea Serino, Hong Yu Wong, and Alessandro Farnè, Oxford University Press (2021).
© Oxford University Press. DOI: 10.1093/oso/9780198851738.003.0017.

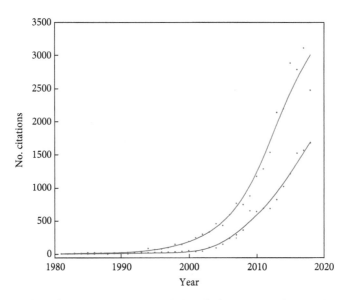

Figure 17.1 Number of citations on papers including flight initiation distance (FID; blue lines and dots) and peripersonal space (PPS; red lines and dots).

citations in the early 1980s, both fields of research recently reached more than 1,000 citations per year. Hence, both fields could potentially gain considerably from integration into a single area of research.

17.2 The history of biological studies of flight response

Almost unsurprisingly, Darwin (1839) listed a number of ideas and reflections on flight distance, mainly in domestic animals, but also in wild animals that he encountered during his journeys. He observed that animals on uninhabited islands often had lost their fear of humans compared to the level of fear in sites inhabited by humans.

Subsequently, Hediger (1934) distinguished between flight distance (run boundary), critical distance (attack boundary), personal distance (distance separating members of non-contact species, such as a pair of swans or swallows sitting on a wire), and social distance (defined as intraspecific communication distance). These different components of flight behaviour have subsequently been reduced to starting distance (the distance at which a human starts the approach of an animal), alert distance (the distance at which an animal is alerted by an approaching human), flight initiation distance (the distance at which an animal takes flight when approached by a human) and distance actually fled (the difference in distance between the position where an animal takes off and FID). Hediger (1934) made the first generalizations about flight distance and other flight responses, and the selective agents that could have shaped the evolution of such behaviour. Numerous species have become 'tame' or developed short flight distances in their natural environment, while the appearance of humans may change that situation during just a few generations (Darwin, 1839, 1868; Hediger, 1934; Cooper et al., 2014). For example, crabs on Cook Island have changed from being diurnal to completely nocturnal in just a few generations after the appearance of humans. Hediger (1934) showed that flight response is species-specific albeit with a significant component of variation among individuals. He also noticed that flight response may evolve as a compromise between different selective forces. For example, Adélie penguins *Pygoscelis adeliae* show a balance between the direction of flight and aversion of humans and avoidance of predatory mammals such as leopard seals *Hydrurga leptonyx* in the nearby ocean, resulting in optimization of flight distance (Gain, 1913). Several studies have suggested that flight reactions depend on the ability of locomotion and hence on the size of locomotor organs. Hediger (1934) showed, based on behavioural observations, that tame mammals and birds appear in the absence of humans, implying that flight distance can change rapidly. This suggests that flight behaviour is optimized as a response to different components of threat.

In another seminal paper, Hemmingsen (1951) noted that a number of factors, such as body size, the size of the locomotor apparatus, and reproductive rate, could all account for differences in FID among species. Likewise, the mode of approach to a wild animal, such as speed of walking or running, direction of approach, and size of the target prey, but also the larger muscles required to escape from a predator by large animals compared to small-sized ones, may also affect FID. These hypotheses and predictions have all subsequently been shown to apply.

FID may not only depend on the risk that the approach of a potential predator such as a human implies, but also on the colouration of an individual and hence its degree of camouflage against the background. Hemmingsen (1951, p. 76) noted that large birds 'do not possess the possibilities of developing skulking habits like grasshopper warblers ([…] especially *Locustella certhiola* [Pall.] and *Locustella lanceolata* [Temm.]), which like the quails or quail-like birds ([…] *Turnix tanki blanfordii* Blyth and *Coturnix c. japonica* [Temm. & Schleg.]) are sometimes nearly trodden upon even in open country with low grass […] before being detected.' These observations imply that not only flight responses, but also the link between behaviour and crypsis against the background, independently and in combination contribute to affect flight responses. These early findings by Hemmingsen have interesting links to studies of perceived body size in humans (van der Hoort et al., 2012), showing that manipulation of the size of perceived objects affects the perception of objects as being larger and farther away while perception of a small body results in objects appearing to be smaller and nearer than they actually are.

While Darwin (1839) wondered about flight distance in environments with the presence of humans and other predators, Hediger (1934) and Hemmingsen (1951) for the first time identified a large number of factors accounting for intra- and interspecific variation in flight distance, and they correctly identified the presence of a hereditary component, but also phenotypic plasticity that would allow rapid change in behaviour.

Domestication is the process by which organisms adapt to the proximity of humans (Clutton-Brock, 1987). The components of domestication include ability to live in the proximity of humans and their domesticated animals such as dogs and cats, increased sociality and large flock size, loss of aggression and loss of fear responses, reduced corticosterone response, and increased rate of reproduction at the cost of reduced survival prospects (Darwin, 1839, 1868; Clutton-Brock, 1987). Interestingly, domesticated animals can rapidly become 'wild' again following release or escape from humans (Krieg, 1925).

17.3 Peri-personal and inter-personal space in birds and other animals

Numerous organisms maintain a quasi-fixed distance from other individuals of the same species and to individuals of other species. Surprisingly, there are hardly any empirical studies of PPS and IPS in organisms other than humans. Such individual distances are reported as being constant in the literature although there are clear differences among individuals. Extensive studies of birds during adverse weather in winter or during migration result in the complete disappearance of individual distances. For example, when more than one million barn swallows *Hirundo rustica* and house martins *Delichon urbica* were stranded in very early snow in the Alps during the normal migration season in late October 1974, huge numbers of birds formed dense aggregations of individuals until weather improved and temperatures increased again (Grubb, 1973; Meservey and Kraus, 1976; Gilbert et al., 2010; Møller, 2011). However, many individuals did not survive. Such huddling results in the complete disappearance of PPS and IPS (Gilbert et al., 2010). Many birds reduce PPS and IPS during the breeding season. For example, zebra finches *Taeniopygia guttata* intermittently may change their PPS and IPS during breeding, although this situation disappears again outside the breeding season when 'fixed' PPS and IPS re-emerge. The basis for these changes remains poorly understood.

PPS and IPS may also change in response to disease. When individuals have fever, or if fever is induced by injection with an antigen such as lipo-poly-saccharid (LPS; an extract from *Escherichia coli*), PPS and IPS are reduced, or may disappear completely. The almost constant individual distance in birds implies that this distance is under strong selection with intermediate distances being favoured. Surprisingly, there is only anecdotal evidence for this phenomenon.

Another example of the loss of peri-personal and inter-individual distance occurs when migrating birds are 'lost' during adverse weather. When birds land on a boat during migration, sometimes hundreds or even thousands of individuals settle resulting in PPS and IPS decreasing or even disappearing. T. Duch (pers. comm.) reported an observation onboard a ship during thick fog in the Baltic Sea in October 2015, when hundreds of birds alighted. While most of these birds were small passerines, there were also several buzzards *Buteo buteo*, sparrowhawks *Accipiter nisus*, and long-eared owls *Asio otus*. Numerous thrushes were sitting next to each other and physically touching each other. For example, several kinglets *Regulus regulus*, robins *Erithacus rubecula*, and other small birds were sitting on the shoulders or even on the heads of the raptors, and some of these raptors killed and even ate a prey, although many just remained perched as if this was not a dangerous situation. No alarm calls were emitted despite all small birds habitually producing such alarms when encountering a predator. Neither T. Duch nor I have ever seen one single similar situation during the past 50 years despite having spent thousands of days in the field. This highly abnormal situation appears as if normalcy breaks down under abnormal weather conditions, and neither predator nor prey behave as in normal situations when both prey and predator react as if there is no danger present. The close proximity, the predators involved, and the reduction in distance between individuals makes this situation very similar in several respects to that when peri-personal space (PPS) and inter-personal space (IPS) are changing.

A slightly similar situation emerged when humans for the first time appeared on islands where there were no predators. This situation allowed such island species to approach their prey with no flight initiation distances, inter-personal distances, or peri-personal distances. When the first individuals showed fear reactions to humans, it was already too late. Very soon afterwards, the solitaire *Pezophaps solitaria*, the dodo *Raphus cucullatus*, and the moas disappeared, and numerous 'tame' insular bird species followed in their footsteps.

17.4 Fear, flight initiation distance, and peri-personal space

Response to proximity of individuals being either predators, parasites, or conspecifics may all be costly in terms of fitness loss, and be described as peri-personal behaviour. As an example, let us take the proximity of a potential predator such as a lion *Panthera leo* to humans. Humans have a long history of being attacked by lions for millennia with the number of fatal cases during recent decades still exceeding 120 attacks per year (Ikanda, 2009). In the presence of lions, humans run away reflecting increased FID, or seek shelter in trees or inside buildings, reducing the risk of predation (Packer et al., 2005). PPS and FID should be larger in the presence than in the absence of lions, and in areas with recent cases of death caused by lions. There is no requirement for predator-induced mortality because the mere presence of a predator suffices to suppress fecundity and reduce survival rate (Abrams, 1984; Lima, 1998; Lima and Dill, 1990). Hence the presence of lions that mainly hunt bush pigs *Potamochoerus larvatus* and antelopes should have a negative impact on humans even in the absence of any direct interactions between humans and lions. Here we use the term 'flight distance' as originally proposed by Hediger (1934) despite Blumstein (2006) and later Cooper and Blumstein (2015) defining FID as the distance at which animals take flight when approached directly by a human being.

This example of lions and humans is no different from examples on anti-predator behaviour in common free-living organisms such as the European blackbird *Turdus merula*. Because there are many fewer predators in urban than in rural habitats, blackbirds run smaller risks of predation when living in cities, but not in the original forest habitat, and we predicted that they should thus have shorter FID in urban than in rural habitats, as observed. Because juveniles experience a much larger risk of mortality than adult blackbirds, juveniles should take greater risks when searching for food than should adult blackbirds. Again, this difference in mean FID between prey and non-prey is as expected, with adults being subject to directional selection for longer FID compared to non-prey. Similar data exist for many other species of common small birds. A number of studies have shown phenotypic plasticity in flight distances, either because of differences in underlying quantitative genetics, or because environmental conditions are variable. For example, animals should take greater risks when environmental conditions deteriorate. Likewise, Rand (1964) showed that FID was inversely related to body temperature in a lizard, implying that FID varied with risk of predation, which is the highest at low temperatures.

Animals and humans alike make decisions about when to flee (Samia et al., 2015). Prey make decisions on when to flee with immediate flight minimizing attentional costs involved in monitoring the exact location of a dangerous predator and its immediate behaviour. Relative brain size reflects the ability to monitor the whereabouts of potential predators, while habitat complexity, but also social complexity as reflected by a large number of individuals, may contribute to the difficulty of this task. Predators monitor the whereabouts of prey through the use of eyes, and by processing such visual information by the brain (Møller and Erritzøe, 2015). Therefore, prey species with relative large brains for their body size should be better able to monitor the intentions of predators, they should delay flight for longer, and hence postpone or entirely avoid energetically costly displacement and also be able to feed for a longer time. Thus, such individuals with relatively larger brains should have shorter FID than individuals with smaller brains. In fact, FID increased with relative eye size and decreased with relative brain size among species of birds (Møller and Erritzøe, 2015). In addition, FID increased independently with size of the cerebellum, which plays a key role in motor control in predator–prey interactions. These findings are consistent with cognitive monitoring as an antipredator behaviour that delays and hence results in the energetically least expensive escape flight. Therefore, antipredator behaviour may have co-evolved with the size of sense organs (as, in this case, eyes), brains, and compartments of the brain involved in responses to risk of predation.

Next I will briefly describe the underlying theory of studies of FID and its component parts. First, I will develop life history theory behind FID including optimal FID. Second, I will describe pace-of-life syndromes (POLs) and FID. Third, I provide a brief review of empirical studies of FID in relation to different aspects of human disturbance. Finally, I list a number of perspectives that may facilitate merging the different fields of PPS, IPS, and FID and their importance for the study of animal and human behaviour.

17.5 When humans rapidly cause flight initiation distance to change

The world is rapidly changed by humans through the effects of urbanization, but also through ecotourism that brings humans to even the most remote spots on earth (Blumstein et al., 2017). Free-living organisms respond to the proximity and the approach of potential predators by showing variable distances of FID. This has resulted in human impact on wild animals having been the focus of many studies.

There is an extensive literature on FID and hunting showing that hunted birds take flight earlier when approached by humans (Madsen, 1995, 1998a, 1998b; Madsen and Fox, 1995; Weston et al., 2012). For example, Laursen et al. (2005) showed a large increase in FID linked to hunting across 19 species of waterbirds in Denmark. However, this literature on the effects of hunting on FID is heterogeneous. In a comparative study, Møller (2008a) found no significant difference in FID between hunted and non-hunted species of birds in an analysis of 55 species during the hunting season. Likewise, there was little difference in FID between species of birds in inhabited and uninhabited areas on the tropical island of Hainan in China in an area with relatively little hunting (Møller and Liang, 2013).

Humans are a source of disturbance, but wild animals have also adapted to such an increase in disturbance regime through tolerance of humans (Frid and Dill, 2002; Samia et al., 2015). A recent study showed that disturbed populations of lizards, birds, and mammals were more tolerant of humans than less disturbed populations (Samia et al., 2015). The best predictors of such tolerance of disturbance were the type of disturbed habitat (urban vs rural habitat), and body size with large-sized species being more tolerant of disturbance than small species. Therefore, animals living in areas with urban habitats and large-sized species should on average be more readily managed than species living in areas with rural habitats. Vehicles of various kinds differ in their effects on flight responses of birds with such effects being highly consistent (Hemmingsen, 1951; McLeod et al., 2013).

17.6 Theory: optimal flight initiation distance

There is a clear theoretical basis for FID and therefore potentially also for PPS and IPS. Here, I first develop optimal flight initiation distance theory. Second, I describe the link between FID and life history theory. Third, I explain how FID may relation to POL theory.

How can we account for differences in FID among individuals? Optimal flight initiation distance theory was initially developed by Ydenberg and Dill (1986), who showed that the optimal FID occurs at the distance when the cost of not fleeing and the cost of fleeing are equal. In contrast, Cooper and Frederick's (2007) optimality model of escape behaviour predicts that prey select the FID that maximizes expected lifetime fitness, when the encounter between a predator and a prey comes to an end. Furthermore, if we first determine fitness of an individual, this has to be increased by any fitness gains acquired during the approach by not fleeing. Finally, this fitness estimate has to be adjusted for the probability of survival if the prey individual survives at a given distance (Cooper and Frederick, 2007).

Fitness as just described above does not only depend on FID, but also on the use of refuges. The use of refuges (a 'shelter' that protects against predators) and the time spent inside them should be tuned to actual predation risk, in order to avoid wasting resources at the expense of other activities. Because prey have shorter FID than predators, and because birds in urban habitats have shorter FID than birds in rural habitats, we consider that prey in predator refuges in urban habitats have gained a 10-fold increase in refuges in urban compared with rural habitats (Møller, 2012). Theoretical models in which prey make explicitly optimal escape decisions have added to our understanding of escape and provided quantitative predictions that can be tested experimentally (Cooper and Frederick, 2007). These optimality models predict how close prey will permit predators to approach before starting to flee. In addition, FID is determined by a trade-off between predation risk, costs of fleeing (i.e. the benefits lost by fleeing), and the prey's fitness while monitoring the predator's approach. In the end, optimality models of escape behaviour predict that prey selects the FID that maximizes expected lifetime fitness at the end of the encounter between predator and prey.

17.7 Theory: FID and why we do not live forever

Life history theory explains the evolution of suites of characters such as when and how often to reproduce, the number and size of offspring, and the association between rate of reproduction and survival. Therefore, life history theory plays a crucial role for understanding the evolution of flight responses of animals. Studies have shown that FID correlates with life history traits as expected from the fact that animals must trade the fitness benefits from one life history component against the benefits of another. Thus short FID are associated with early start of reproduction, high fecundity, and low survival prospects, while long FID are associated with start of reproduction at old age, low fecundity, and high survival prospects.

Møller and Garamszegi (2012) showed for different species of birds that relative age at first reproduction was positively related to FID, annual fecundity was negatively correlated with FID, and adult survival rate was positively correlated with FID. These findings are as expected from life history theory when these life history components are related to FID.

If we investigate the relationship between FID in different species of birds and risk of predation by a common predator such as a sparrowhawk, we find a strong negative relationship. This implies that species with relatively longer FID for their body size run a smaller risk of getting killed than species with a relatively short FID (Møller et al., 2008). This finding may sound surprising. Why do prey individuals belonging to species with short FID not simply fly up at longer distances and thereby avoid getting killed? However, the observed pattern is exactly what we would expect because not all components of fitness can be maximized simultaneously. There is no Malthusian demon that survives indefinitely, produces large families continuously, and still shows no signs of being worn out and suffering from senescence

(Cole, 1954). The cost of reproduction reflected in the tear and wear that all parents know so well is a function of investment in reproduction. Such high reproductive rates have among other consequences a shortening of telomeres at the ends of chromosomes, a deficiency of DNA repair, and a deficit of essential antioxidants that help prevent premature damage to the machinery of life. A reproducing individual experiencing such costs will have fewer resources available for the next reproductive event, and, therefore, less incentives for further reproduction. While some individuals may survive better than others, this always occurs at the cost of reduced reproduction.

Birds show signs of having developed some of the characteristics of a Malthusian demon by having high survival rates, combined with high levels of metabolism, high blood glucose levels, and high body temperatures that render such individuals prime candidates for significant aging (Holmes et al., 2001). Surprisingly, such signs of senescence are generally absent from birds, or they occur much later in life of birds than, for example, in mammals like ourselves. Flying birds can escape from predators in two dimensions while most mammals are only able to escape in two dimensions, bats being a notable exception (Pomeroy, 1990; Møller, 2010). Blumstein et al. (2007) showed that horizontal and vertical components of escape differed among species of birds. However, vertical escape is energetically more costly than horizontal escape (Møller, 2010). The ability to escape in three dimensions as in birds should thus result in greater survival and slower rates of senescence than in most mammals that only escape in two dimensions. The slope of the relationship between horizontal distance and FID increases with body mass across species, whereas the slope of the relationship between vertical distance and FID decreases with body mass. Across bird species, there is a negative relationship between the extent of horizontal and vertical escape. The horizontal slope decreases with increasing density of vegetation from grass over shrub to trees, while that is not the case for the slope of vertical escape. Adult survival rate increases while rate of senescence decreases with increasing vertical, but not with horizontal, slope. These findings are consistent with the prediction that vertical escape indeed reduces the rate of senescence, while providing a means for reducing the impact of predation because of reduced survival prospects.

A central premise of this theory is the existence of a fast–slow continuum of life history variation, which reflects the impossibility to simultaneously maximize survival and fecundity (Cole, 1954; Stearns, 1983). Thus, individuals may maximize their survival prospects by only laying a single egg every second year as in the wandering albatross *Diomedea exulans*, or a blue tit *Cyanistes caeruleus* may lay a clutch of 15 eggs. If adult albatrosses are forced to invest more by having their clutch or brood size experimentally increased, this results in a dramatic reduction in survivorship. Hence, current interest in extending longevity prospects in humans through artificially improved survivorship is theoretically impossible because an improvement in one component of life history will automatically result in deterioration of others (Cole, 1954).

17.8 Theory: risk taking and pace-of-life syndromes

Interestingly, a recent study showed that FID increased along the fast–slow continuum for rural, but not for urban, birds of the same species (Sol et al., 2018). The discovery of POLs associated with risk-taking behaviour across species of birds (Sol et al., 2018) provides a suggestion for how behaviour and life history interact to influence the response of animals to changes in their altered urban environment. While it is unclear whether changes in POLs reflect plastic and/or evolutionary adjustments, the findings reported by Sol et al. (2018) highlight the need to integrate behaviour into life history theory for understanding how animals tolerate human-induced environmental changes.

POLs are characterized by suites of life history traits. For example, a central premise of this theory is the fast–slow continuum of life history variation, which reflects the impossibility to simultaneously maximize survival and fecundity (Cole, 1954). The fast–slow continuum aligns organisms along a POL axis from a highly reproductive (fast-lived) strategy at one end to a survival (slow-lived) strategy at the other end. As slow-lived animals prioritize future over current reproduction (Cole, 1954), they should generally be more risk-averse compared to those at the fast extreme (Martin et al., 2000). In contrast, fast-lived animals should prioritize behaviours that enhance current reproductive effort, even when doing so involves taking some risks. Therefore, POL theory explains why selection should favour behaviours ensuring higher adult survival in slow-lived animals and behaviours that enhance reproductive effort in fast-lived animals.

17.9 Future prospects

Genetic and epigenetic effects may both contribute to adaptation to novel environments impacted by humans, and may thereby affect both IPS and PPS. First, personality, defined as the individual consistency in behaviour across settings or environments (Budaev, 1997; Gosling, 2001; Sih et al., 2004; Wolf et al., 2007), is widespread in the animal kingdom. Such consistent differences in behaviour are sometimes linked to habitat as reflected by urbanization. Consistent types of behaviour in birds appear to have underlying genetic polymorphisms that may be decoupled in urban, but not in rural, habitats (Riyahi et al., 2017). Indeed, novelty-seeking or bold behaviour seems to have a genetic basis, but this behaviour can also be modified by epigenetic mechanisms that affect the expression of genes (Riyahi et al., 2015). Similarly, individual black swans *Cygnus atratus* with more common DRD4 genotypes coding for wariness were differentially associated with non-urban habitats (Guay et al., 2015). However, there are so far no studies of the genomics of PPS, IPS, or associated phenomena.

Second, there may be ways in which disease in humans, but also in other organisms, can be linked to FID, IPS, and PPS. Here IPS is defined as the safety zone that humans, but also all other organisms, maintain around their body during social interactions (Hediger, 1934). This distance is called 'individual distance' in the animal behaviour literature (Manning, 1998). IPS may vary with nature and nurture. Extreme proximity, which occurs in humans but also in other organisms, is perceived as a medical condition (Hall, 1966; Hayduk, 1983). PPS is defined as that surrounding the body when encoded in a body-centered frame.

Third, while there are numerous studies of IPS and PPS, why are there no similar studies across populations of humans? If there are fitness costs of such behaviour, there should be evidence of IPS and PPS being linked to specific environments, behaviour, or life history.

Fourth, there is now a multitude of automatic devices that can be used to monitor behaviour that constitute components of IPS and PPS. These devices can also be used for monitoring locomotion, heartbeat, body temperature, and a multitude of other aspects of health.

Fifth, there is a deficit of records of predation on humans as caused by lions, hippopotamuses *Hippopotamus amphibius*, tigers *Panthera tigris*, and great white sharks *Carcharodon carcharias*. Recording such events and the circumstances such as FID, studies of behaviour linked to IPS and PPS could help reduce the number of fatalities. Recording near-death experiences in humans and how they are related to fear responses and flight behaviour in animals may also be useful.

17.10 Conclusions

Flight and other components of escape reactions to the proximity of predators have evolved in response to evolutionary interactions between prey and predators. Such behaviour may be linked to IPS and PPS, and hence be under selection of a similar strength and direction, as does affecting FID. Future studies of humans and animals alike should investigate the correlations between such disparate components of behaviour, but also investigate the extent to which they are coevolving and their consequences for all living beings.

Acknowledgements

F. de Vignemont kindly invited me to contribute to this interesting volume and provided a number of suggestions.

References

Abrams, P. A. (1984). Foraging time optimization and interactions in food webs. *American Naturalist*, 124, 80–96.

Blumstein, D. T. (2006). Developing an evolutionary ecology of fear: How life history and natural history traits affect disturbance tolerance in birds. *Animal Behaviour*, 71, 389–399.

Blumstein, D. T., Geffroy, B., Samia, D. S. M., and Bessa, E. (eds). (2017). *Ecotourism's Promise and Peril: A Biological Evaluation*. New York: Springer.

Budaev, S. V. (1997). 'Personality' in the guppy (*Poecilia reticulata*): A correlational study of exploratory behavior and social tendency. *Journal of Comparative Psychology*, 111, 399–411.

Clutton-Brock, J. (1987). *A Natural History of Domesticated Mammals*. Austin, TX: University of Texas Press.

Cole, L. E. (1954). The population consequences of life history phenomena. *Quarterly Review of Biology*, 29, 103–137.

Cooper Jr, W. E., and Frederick W. G. (2007). Optimal flight initiation distance. *Journal of Theoretical Biology*, 244, 59–67.

Cooper Jr., W. E., Pyron, R. A., and Garland Jr, T. (2014). Island tameness: Living on islands reduces flight initiation distance. *Proceedings of the Royal Society of London B Biological Sciences*, 281, 20133019.

Cooper, W. E., and Blumstein, Jr, D. T. (eds). *Escaping From Predators: An Integrative View of Escape Decisions*. Cambridge, UK: Cambridge University Press.

D'Angelo, M., di Pellegrino, G., and Frassinetti, F. (2017). Invisible body illusion modulates interpersonal space. *Scientific Reports*, 7, 1302.

Darwin, C. (1839). *Journal of Researches into the Geology and Natural History of the Various Countries visited by H. M. S. Beagle, under the Command of Captain Fitzroy, R. N. from 1832–1836*. Henry Colburn.

Darwin, C. (1868). *The Variation of Animals and Plants under Domestication*. London: John Murray.

Frid, A., and Dill, L. (2002). Human-caused disturbance stimuli as a form of predation risk. *Conservation Ecology*, 6 (1), 11.

Gain, L. (1913). La vie et les moeurs du pingouin Adélie. *Congres Internationale de Zoologie de Monaco, Section* 3, 9, 501–521.

Gilbert, C., McCafferty, D., Le Maho, Y., Martrette, J.-M., Giroud, S., Blanc, S., and Ancel, A. (2010). One for all and all for one: The energetic benefits of huddling in endotherms. *Biological Reviews*, 85, 545–569.

Gosling, S. D. (2001). From mice to men: What can we learn about personality from animal research? *Psychological Bulletin*, 127, 45–86.

Grubb, T. C. Jr. (1973). Absence of 'individual distance' in the Tree Swallow during adverse weather. *Auk*, 90, 432–433.

Guay, P.-J., Robinson, R. W., Weston, M. A., Mulder, R. A., and van Dongen, W. F. D. (2015). Variation at the DRD4 locus is associated with wariness and local site selection in urban black swans. *BMC Evolutionary Biology*, 15, 253.

Guterstam, A., Abdulkarim, Z., and Ehrsson, H. H. (2015). Illusory ownership of an invisible body reduces autonomic and subjective social anxiety response. *Scientific Reports*, 5, 9831.

Hall, E. T. (1966), *The Hidden Dimension*. London: Macmillan.

Hayduk, L. A. (1983). Personal space: Where we now stand. *Psychological Bulletin*, 94(2), 293.

Hediger, H. (1934). Zur Biologie und Psychologie der Flucht bei Tieren. *Biologische Zentralblatt*, 54, 21–40.

Hemmingsen, A. M. (1951). The relationship of shyness (flushing distance) to body size. *Spolia Zoologica Musei Hauniensis*, 11, 74–76.

Holmes, D. J., Flückiger, R., and Austad, S. N. (2001). Comparative biology of aging in birds: An update. *Experimental Gerontology*, 36, 869–883.

Ikanda, D. K. (2009). *Dimensions of a Human–Lion Conflict: The Ecology of Human Predation and Persecution of African Lions* Panthera leo *in Tanzania*. PhD thesis. Trondheim, Norway: Norwegian University of Science and Technology, Faculty of Natural Sciences and Technology, Department of Biology.

Krieg, H. (1925). Studien über Verwilderung bei Tieren und Menschen in Südamerika. *Archiev für Rassen- und Gesellschaft-Biologie*, 16, 241–267.

Laursen, K., Kahlert, J., and Frikke, J. (2005). Factors affecting escape distances of staging waterbirds. *Wildlife Biology*, 11, 13–19.

Lima, S. L. (1998). Nonlethal effects in the ecology of predator–prey interactions: What are the ecological effects of antipredator decision- making? *BioScience*, 48, 25–34.

Lima, S. L., and Dill, L. M. (1990). Behavioural decisions made under the risk of predation: a review and prospectus. *Canadian Journal of Zoology*, 68, 619–640.

Madsen, J. (1995). Impacts of disturbance on migratory waterfowl. *Ibis*, 137, S67–S74.

Madsen, J. (1998a). Experimental refuges for migratory waterfowl in Danish wetlands. I. Baseline assessment of disturbance effects of recreational activities. *Journal of Applied Ecology*, 35, 386–397.

Madsen, J. (1998b). Experimental refuges for migratory waterfowl in Danish wetlands. II. Tests of hunting disturbance effects. *Journal of Applied Ecology*, 35, 398–417.

Madsen, J., and Fox, A. D. (1995). Impacts of hunting disturbance on waterbirds: A review. *Wildlife Biology*, 1, 193–207.

Manning, A. (1998). *An Introduction to Animal Behavior*. Cambridge, UK: Cambridge University Press.

Martin, T. E., Martin, P. R., Olson, C. R., Heidinger, B. J., and Fontaine, J. J. (2000). Parental care and clutch sizes in north and south American birds. *Science*, 287, 1482–1485.

McLeod, E. M., Guay, P.-J., Taysom, A. J., Robinson, R. W., and Weston, M. A. (2013). Buses, cars, bicycles and walkers: The influence of the type of human transport on the flight responses of waterbirds. *Public Library of Science One*, 8, e82008.

Meservey, W. R., and Kraus, G. F. (1976). Absence of 'individual' distance in three swallow species. *Auk*, 93, 177–178.

Møller, A. P. (2008). Flight distance of urban birds, predation and selection for urban life. *Behavioral Ecology and Sociobiology*, 63, 63–75.

Møller, A. P. (2010). Up, up, and away: Relative importance of horizontal and vertical escape from predators for survival and senescence. *Journal of Evolutionary Biology*, 23, 1689–1698.

Møller, A. P. (2011). Behavioral and life history responses to extreme climatic conditions: Studies on a migratory songbird. *Current Zoology*, 57, 351–362.

Møller, A. P. (2012). Urban areas as refuges from predators and flight distance of prey. *Behavioral Ecology*, 23, 1030–1035.

Møller, A. P. (2015). Birds. In: W. E. Cooper, and D. T. Blumstein (eds). *Escaping from Predators: An Integrative View of Escape Decisions*, pp. 88–112. Cambridge, UK: Cambridge University Press.

Møller, A. P., and Erritzøe, J. (2015). Predator-prey interactions, flight initiation distance and brain size. *Journal of Evolutionary Biology*, 27, 34–43.

Møller, A. P., and Garamszegi, L. Z. (2012). Between individual variation in risk-taking behavior and its life history consequences. *Behavioral Ecology*, 23, 843–853.

Møller, A. P., and Liang, W. (2013). Tropical birds take small risks. *Behavioral Ecology*, 24, 267–272.

Møller, A. P., Nielsen, J. T., and Garamszegi, L. Z. (2008). Risk taking by singing males. *Behavioral Ecology*, 19, 41–53.

Packer, C., Ikanda, D., Kissui, B., and Kushnir, H. (2005). Lion attacks on humans in Tanzania. *Nature*, 436, 927–928.

Pomeroy, D. (1990). Why fly? The possible benefits of lower mortality. *Biological Journal of the Linnean Society*, 40, 53–65.

Rand, A. S. (1964). Inverse relationship between temperature and shyness in the lizard *Anolis lineatopus*. *Ecology*, 45, 863–864.

Riyahi, S., Sánchez-Delgado, M., Calafell, F., Monk, D., and Senar, J. C. (2015). Combined epigenetic and intraspecific variation of the DRD4 and SERT genes influence novelty seeking behavior in great tit *Parus major*. *Epigenetics*, 10, 516–525.

Riyahi, S., Björklund, M., Mateos-Gonzalez, F., and Senar, J. C. (2017). Personality and urbanization: Behavioural traits and *DRD4* SNP830 polymorphisms in great tits in Barcelona city. *Journal of Ethology*, 35, 101–108.

Samia, D. S. M., Nakagawa, S., Nomura, F., Rangel, T. F., and Blumstein, D. T. (2015). Increased tolerance to humans among disturbed wildlife. *Nature Communications*, 6, 8877.

Sih, A., Bell, A. M., Johnson, J. C., and Ziemba, R. E. (2004). Behavioral syndromes: An integrative overview. *Quarterly Review of Biology*, 79, 241–277.

Sol, D., Maspons, J., Gonzalez-Voyer, A., Morales-Castilla, I., Garamszegi, L. Z., and Møller, A. P. (2018). Risk-taking behavior, urbanization and the pace of life in birds. *Behavioral Ecology and Sociobiology*, 72, 59.

Stearns, S. C. (1983). The influence of size and phylogeny on patterns of covariation among life-history traits in the mammals. *Oikos*, 41, 173–187.

Tinbergen, N. (1963). On aims and methods of ethology. *Zeitschrift für Tierpsychologie*, 20, 410–433.

Van der Hoort, B., Guterstam, A., and Ehrsson, H. H. (2012). Being Barbie: The size of one's own body determines the perceived size of the world. *Public Library of Science*, 6(5), e20195.

Weston, M. A., McLeod, E. M., Blumstein, D. T., and Guay, P.-J. (2012). A review of flight-initiation distances and their application to managing disturbance to Australian birds. *Emu*, 112, 269–286.

Wolf, M., van Doorn, G. S., Leimar, O., and Weissing, F. J. (2007). Life-history trade-offs favour the evolution of animal personalities. *Nature*, 447, 581–584.

Ydenberg, R. C., and Dill, L. M. (1986). The economics of fleeing from predators. *Advances in the Study of Behavior*, 16, 229–249.

18

Human emotional expression and the peripersonal margin of safety

Michael S.A. Graziano

18.1 Introduction

This chapter describes how smiling, laughing, and crying may have evolved. It does not try to explain the origin of the underlying emotional states such as happiness, humour, or sadness. Instead, the hypothesis focuses on the possible evolutionary origin of the overt, motoric components, the body and face movements that are used by humans during the expression of these emotions. Why do we make such bizarre, quirky gestures to convey specific internal states? Why should leaking lubricant from the eyes become a means of soliciting psychological support? Even supposing we could understand what humour is and how it emerged in humans, why should it be expressed by baring the teeth and crinkling the skin around the eyes?

The hypothesis arose from previous work on a network of neurons in the primate brain that processes the space around the body, the so-called 'peripersonal space', and that may be related to defending a margin of safety (Cooke and Graziano, 2004; de Vignemont and Iannetti, 2015; di Pellegrino and Làdavas, 2015; Duhamel at al., 1998; Graziano, 2018; Graziano et al., 1994; Rizzolatti et al., 1981). Each peripersonal neuron responds to tactile stimuli within a specific, receptive field on the body, and to visual stimuli in the space near that tactile receptive field. Some neurons also respond to auditory stimuli, and are sensitive to sound sources in the space near the body. Given these properties, each neuron appears to monitor a multisensory region of space anchored to the body surface. In aggregate, the population of neurons could act almost like an air-traffic radar system, monitoring the location and movement of nearby objects relative to different parts of the body. In studying these neurons in two specific regions of the primate cerebral cortex, the ventral intraparietal area in the parietal lobe and a polysensory zone in the frontal lobe, my colleagues and I discovered that artificial electrical stimulation of the neurons typically evoked a consistent suite of movements (Cooke and Graziano, 2004; Cooke et al., 2003; Graziano et al., 2002). The stimulation-evoked movements closely mimicked defensive or protective actions.

For example, if a site in cortex—a cluster of neurons about half a millimeter in diameter—responded to a touch on the right hand and to the sight of objects looming toward the hand, then electrical stimulation of that site, artificially activating the local neurons with a train of pulses for half a second, would evoke a characteristic, fast withdrawal of the arm behind the back into a guarding posture. If a site in cortex responded to a touch on the right cheek and to the sight of objects in the space near the right cheek, then stimulation would evoke an even richer, characteristic set of movements. Both eyes would squint, with stronger muscular contraction on the right. The facial skin would fold in a manner that appeared to protect the

Michael S.A. Graziano, *Human emotional expression and the peripersonal margin of safety* In: *The World at Our Fingertips*. Edited by Frédérique de Vignemont, Andrea Serino, Hong Yu Wong, and Alessandro Farnè, Oxford University Press (2021). © Oxford University Press. DOI: 10.1093/oso/9780198851738.003.0018.

eyes. The upper lip would lift, caused by a contraction of muscles in the face that mobilized the skin on the cheek upward toward the eye. The ears would fold back against the head. The head would duck down and turn toward the left. The shoulders would pull up. The torso would turn and the right arm would lift, the hand moving rapidly into the space beside the head, as if blocking an impending impact toward the right side of the face. Stimulating other sites in cortex evoked many other defensive movement sets, each one specific to the region of the body monitored by the multisensory neurons that had been stimulated.

My colleagues and I conducted a series of studies on these multisensory neurons, not just using electrical stimulation, but also chemical stimulation and single neuron recording, to examine their relationship to defensive actions (Cooke and Graziano, 2004; Cooke et al., 2003; Graziano et al., 2002). We also studied protective movements evoked by natural stimuli such as an airpuff or a ping pong ball fired from an air gun (Cooke and Graziano, 2003). These studies suggested that the peripersonal neurons were part of a sophisticated input-output system that could transform sensory information about the space near the body into coordinated protective actions.

In the course of these studies, I spent years watching protective actions unfold in real time and in slow motion on recorded media, dissecting them into types and components. It was during these many observations that I began to notice the similarity between defensive actions and emotional expressions. The similarity suggested a possible evolutionary path, in which some human emotional expressions were mimics of a set of behaviours originally evolved for the protection of the body surface.

Evolutionary hypotheses can be highly speculative, and I acknowledge that the present proposal is no exception. It is extremely difficult to design a decisive experiment to test how a specific trait evolved. Nonetheless, over the past decade, I have written about the hypothesis, explaining the rationale behind it and how it fits with relevant data (Graziano, 2008, 2015, 2018). In this chapter, I summarize the case for the hypothesis. I argue that the similarity between defensive actions and emotional expressions is not a superficial one. It is not simply that human emotional expressions involve contraction of the facial muscles, and defensive actions do too. Instead, the relationship may be more specific. I describe how protective actions may have been co-opted by evolution and turned into social displays, and then further shaped and modified into human smiles, laughter, and crying. In this hypothesis, the brain mechanisms of peripersonal space are the origin of at least some of our most characteristic human emotional expressions.

18.2 Defensive actions

The pioneering studies of Strauss (1929) first showed that an unexpected loud sound (a gunshot behind the head, in Strauss's experiments) causes an extremely fast, consistent reflex in people. The reflex has been studied extensively since then (Davis, 1984; Koch, 1999; Landis and Hunt, 1939). Figure 18.1 illustrates some of the components in humans from classic studies by Landis and Hunt (1939). The startle response includes the following components:

> The torso curves forward and the knees and hips bend, reducing the person's height.
>   The arms are drawn forward and pulled close around the stomach or chest.
>   The shoulders are lifted and the head pulled down and forward, in effect blocking external access to the neck.
>   The eyes blink, and the musculature around the eyes contracts to cause a squint.

(a)   (b)

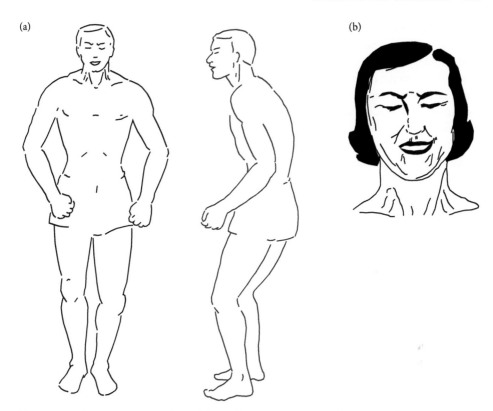

Figure 18.1 Startle response, adapted from classic illustrations in Landis and Hunt (1939). a. Body components. b. Facial components. In the startle response to a loud sound, the torso curves forward and the knees and hips bend, reducing the person's height. The arms are drawn forward and pulled close around the stomach or chest. The shoulders are lifted and the head pulled down and forward, reducing external access to the neck. The eyes blink, and the musculature around the eyes contracts to cause a squint. The muscles in the cheeks contract, pulling the facial skin up toward the eyes and increasing the folds of skin that pucker around and potentially protect the eyes. A consequence of this upward mobilization of facial skin is that the teeth are exposed—especially the upper teeth. The strongest, most consistent part of the action is the facial component. The action resembles the human smile.

The muscles in the cheeks contract, pulling the facial skin up toward the eyes and increasing the folds of skin that pucker around and potentially protect the eyes. A consequence of this upward mobilization of facial skin is that the teeth are exposed — especially the upper teeth, and especially the "eye teeth" or the cuspids located laterally in the mouth.

Although Strauss never observed it, another characteristic component of the startle reaction is a centering of the eyes. In studies from my own lab (Cooke and Graziano, 2003), we found that these eye movements are not the normal saccades a person might make to look at objects. They are slower and have a distinctive curved trajectory. They are probably caused by the co-contraction of all six extra-ocular muscles, a reflex that pulls the eyeball back into the head by a millimeter or two.

The acoustic startle reflex could be described as a generalized defensive stance. It is not tuned to the specifics of the stimulus. Whether the stimulus comes from the left or right, or has a specific timbre or meaning, the initial startle reaction is essentially the same. Crouching down reduces the exposure of the body to predators. Pulling the arms over the torso protects both the soft abdomen and the hands. Raising the shoulders and ducking the head protects the neck, one of the most vulnerable body parts to predation. The facial muscle movements conspire to protect the eyes.

The protection of the eyes is the strongest part of the reaction. As the startle response becomes weaker, perhaps in reaction to lower amplitude sounds, or perhaps due to habituation during repeated stimulus presentation, the components from the neck down begin to drop out, while the facial components are the last to remain. The final reaction to drop out involves only a blink and some tension in the cheek muscles. Moreover, the contraction of the orbicularis muscle around the eyes has the fastest reaction time of any startle component, as fast as 12 milliseconds from the onset of the sound to the initial rise in muscle activity (Cooke and Graziano, 2003). When a video of a startle reaction is slowed and shown frame by frame, the most obvious components, the ones that jump out visually, are the closure of the eyes and the flashing of teeth as the upper lip pulls up.

This flashing of the teeth during a generalized defensive stance can easily be misinterpreted. It is easy to imagine the action as readying the teeth to bite, or perhaps to warn off an attacker. But that interpretation is incorrect. The muscles involved are different from the biting, attacking, or snarling muscles that ring the mouth, and consequently the shape of the mouth is different. Instead, the muscle contraction is mainly in the cheeks, bunching the flesh of the cheeks upward in a manner that helps protect the eyes. Imagine walking from a dark indoor space into an ultra-bright, sun-saturated summer day. Your whole face contracts into a kind of sun smile, or maybe a sun grimace, exposing your upper teeth, bunching your cheeks upward, wrinkling the skin around your eyes and protecting them from the excess light. You are not preparing to bite anything. That pseudo-smile is a byproduct of protecting the eyes.

A defensive reaction to a loud sound or to a looming object can be broken down into two phases. The first is the generalized protective stance, which I have just described. It is fast, preliminary, and takes into account essentially none of the specifics of the threatening stimulus. The second is a more stimulus-specific response. In our own studies of defensive reactions to airpuff on the cheek (Cooke and Graziano, 2003), we saw an initial phase that began in the facial muscle activity at about 10 to 20 milliseconds. The face and body shaped into a generalized protective stance. Then, by about 50 milliseconds, the stance began to evolve into a spatially directed one, stronger on the side of the airpuff, the head turning away, the arm and hand on that side rising up in a blocking movement. Figure 18.2 shows some of the facial components of a standard defensive movement in a species of monkey, *Macaca fascicularis*, illustrating especially the muscle tension around the eyes and the lifting of the upper lip, in this case on the side of the face relevant to the threat.

This second, spatially specific phase can be extremely complicated, as we saw in our studies. A looming object or airpuff threatening the top of the head causes the head to pull down and both hands to rise. A threat to the side of the face causes the head to turn aside and the hand and arm to shoot out laterally, as if thrusting away or blocking the potential threat. A threat to the side of the torso causes the elbow to move rapidly to a blocking position near the waist, and the body to shift away. A threat to the forearm causes the arm to pull rapidly in toward the abdomen and the upper body to hunch protectively over the arm. A threat to the hand causes the arm to whip behind the back. All these reactions are slower than the initial startle phase, and yet so fast that they preclude any cognitive component. Within 50

(a)   (b)

Figure 18.2 Defensive response in a monkey (*Macaca fascicularis*) to an activation of peripersonal mechanisms monitoring the left side of the face (from Cooke and Graziano, 2004). a. Resting state. b. State approximately 100 ms into the defensive reaction. The musculature around the eye and in the upper face contract, drawing skin protectively toward the eye. A consequence of this upward mobilization of facial skin is that the upper teeth are exposed. Again, the defensive reaction resembles a normal affiliative gesture (such as a human smile), at least on one side of the face where the protective action occurs. It differs in that, in an affiliative gesture, although the muscles around the eyes contract, the eyes usually do not close entirely.

milliseconds, the spatial computations are evidently in progress and the body can react in a directed manner.

In our studies of cortical mechanisms, we found evidence that this second phase, the spatially specific phase, is controlled by and depends on the peripersonal neurons that we studied in cortex (Cooke and Graziano, 2004). The initial phase, the generalized protective stance, may involve other brain mechanisms. Work on the acoustic startle reflex suggests that it is coordinated at least partly by the pontine reticular formation (Davis, 1984; Koch, 1999). Other aspects of a defensive reaction, such as computing which objects should be recognized as safe and which are intrinsically suspicious or threatening, may involve many structures and networks, notably the amygdala and other systems involved in emotional valence (e.g. LeDoux, 2007).

18.3 How defensive movements might evolve into social signals

A defensive reaction, as fast as it may be, is not a simple, unvarying reflex. It changes depending on context, and those changes can reveal a great deal about the inner state of a person. For example, if a person is put on edge through a series of weak electrical shocks, and then hears an unexpected, loud sound, that person's defensive startle reaction will be greatly exaggerated (Grillon et al., 1991). Irritants like unpleasant pictures or odours can cause the same exaggeration of the startle reaction (Ehrlichman et al., 1995; Lang et al., 1990; Patrick et al., 1996). People who suffer from anxiety disorders have measurably enhanced defensive reactions (Grillon, 2008; Grillon et al., 1996; McTeague and Lang, 2012). If your child suddenly lunges at you, you'll react one way—maybe putting out your hands to catch him.

If a large dog that you have only just met suddenly lunges, you are already nervous and suspicious, probably already partly in a protective stance with respect to the animal, and your defensive reaction to the lunge will be exaggerated. Mood, thought, attention, emotion, and expectation sift through cortical and subcortical circuitry, and modulate the mechanisms that govern the defensive reaction. The defensive reaction is, in turn, visible to anyone else watching.

If an antagonist is watching you, and you wish to avoid broadcasting clues about your internal state, you can suppress or delay a great deal of behaviour, but a defensive response is urgent. It cannot safely be suppressed. You have no good option except to make the movement, protect yourself from the threat of the moment, and very possibly reveal something of your internal state to your watchful antagonist. This obligatory throughput from internal state to visible display is what makes defensive movements a good starting point for evolution to shape social signals. Defensive movements are, in effect, a data breach. They are a conduit through which information about your internal states, especially your emotional vulnerabilities, leak out to anyone watching. The information can be used to predict your behaviour in the near future. Evolution can go to work on this situation, shaping the brains of animals to automatically perceive and take advantage of the streams of information leaking out of other nearby animals.

For example, we all intuitively recognize the body language of confidence. A person stands tall, his back straight. His shoulders are down and his head is up. His arms are at his sides, or even spread out expansively. He is showing a kind of negative image, an exaggerated opposite to the defensive stance shown in Figure 18.1. We also intuitively recognize the body language of timidity. The person has a slight hunch, the head is ducked down, the shoulders slightly raised, and the arms tend to be pulled in across the front, perhaps clasped together over the chest or stomach. We can read something of a person's internal state from the extent to which a generalized defensive stance is active or absent in that person.

My contention here is not that peripersonal space and defensive movements shaped the evolution of emotion. Much has been written about the evolution of emotions and the commonalities in the emotional mechanisms of human and many non-human animals (e.g. de Waal, 2011; Panksepp, 2007). The evolution of emotion, however, is not at issue here. My contention is that the specific, quirky, physical actions by which we communicate internal emotional states have been profoundly influenced by peripersonal space and defensive movements.

18.4 The origin of smiling

The evolution of a social signal is easily misunderstood. The reason is that it is easy to pay too much scientific attention to the sending of the signal. One is tempted to pose the evolutionary question: how did the sender evolve to use that specific signal as a means of expressing itself? For example, how did people evolve to use a smile to express happiness? However, it is now widely accepted that, with respect to social signals, evolution shapes the receiver first, then the sender (Dawkins and Krebs, 1978; Fridlund, 1994; Godfray and Johnstone, 2000; Grafen and Johnstone, 1993; Schmidt and Cohn, 2001). The receiver evolves to react in a specific way when it observes a specific stimulus. As a result of that first evolutionary step, the sender has been given a lever by which to manipulate the behaviour of the receiver. The sender then evolves to control or exaggerate that triggering stimulus in a strategic way. To help illustrate that hypothesized process, in the following paragraphs I will tell a story, a hypothetical step-by-step account of how a defensive movement might turn into a smile, starting with the originating stimulus, then progressing to the evolution of the receiver, and finally moving to the evolution of the sender. This account is a fiction meant to clarify the concepts. In reality, the components probably co-evolved in a highly interactive manner.

Although my account of the origin of smiling emphasizes protective movements more than some other accounts, it is nonetheless close to the current, widely accepted explanation of the evolutionary origin of the smile, or the affiliative gesture called the 'silent bared teeth display' (Beisner and McCowan, 2014; De Marco and Visalberghi, 2007; Preuschoft, 1992; Thierry et al., 1989; Von Hooff, 1962). In subsequent sections, I will extend the argument and propose an analogous account for laughing and of crying.

18.5 The original stimulus

To explain how a smile might have evolved from a defensive gesture, imagine a scenario in which you and I are primates, perhaps as many as 50 million years ago, before the evolution of the smile but after the evolution of the standard defensive stance against a looming threat. Suppose I am a large, aggressive monkey and you are smaller. I stride past you.

Since I am a looming object with high negative valence, your peripersonal neurons respond to me, monitoring my trajectory, coordinating signals that adjust your posture. You lean away from me. Your torso hunches. Your arms pull in to protect your hands and your abdomen. Your head lowers and your shoulders lift to protect your neck, more so on the side that faces me. The muscles around your eyes contract. It is useful to keep your eyes open and your face turned partly toward me, so that you can maintain a close watch. But, even though your eyes are open, the surrounding muscles in the brow and on the side of the snout contract to form a protective puckering of the skin toward the eyes. As a consequence of this facial contraction, your upper lip is pulled up, exposing your upper teeth. The baring of teeth is not a prelude to biting. The shape of the mouth is associated with the mobilization of the nasolabialis muscle, wrinkling the skin on the side of the face upward toward the eyes.

Your reaction inadvertently broadcasts information about you—information about how you perceive me. I could, in principle, infer that you are non-aggressive toward me and a hierarchical underling. I could guide my own behaviour toward you partly on the basis of that information. At this early moment in primate evolution, however, I lack the specialized neural pathways to process that information and use it to my advantage. Your defensive stance is simply a matter of pragmatic self-protection, and not yet acting as a social signal.

18.6 The receiver

Suppose, again, that you and I are monkeys, but evolution has further shaped the brain, giving me the tools to take advantage of the available visual cues. I now have a set of reactions wired into me. The reactions do not stem from explicit cognition. I cannot look at you and logically deduce the relevant information. I have something more like a cortical reflex or an instinct, like a rabbit reacting instinctively to a shadow passing overhead. When I see your stance, your hunched posture, raised upper lip, and squinted eyes, that stimulus acts as an automatic trigger. It makes me treat you as a non-threatening conspecific, and therefore makes me less aggressive toward you. The brain has evolved to take advantage of useful, available information. The receiver has evolved.

18.7 The sender

Now that we have a stimulus (the defensive reaction) and a receiver whose behaviour is affected by that stimulus, evolution can go to work shaping the sender to manipulate that signal. Suppose a brain system evolves that can *mimic* the defensive stance. Even when I am not directly looming into your personal space, even when your peripersonal neurons are not triggered and your defensive mechanisms are not recruited, you now have a capacity to flash a mimic defensive stance in my direction, thereby altering my behaviour. By tapping into pre-existing wiring in me, the stimulus makes me less likely to attack you. Your behaviour is, again, not the result of explicit cognition. You do not cleverly reason out that, if you generate a pretend cringe, it will make me think you are non-threatening, thereby altering my behaviour toward you. The interaction between us lies deep beneath the level of explicit cognition. Evolution has given us these behaviours and reactions because they confer a survival advantage. In a similar way, a stick insect does not know that it is mimicking a stick or that the mimicry has the useful effect of camouflaging it from predators.

Just as the stick insect is a mimic, your socially generated defensive stance is a mimic. It is not a real defensive reaction. I do not know what specific mechanisms in the brain might generate this social signal, although there has been some speculation. There is no reason to think that the peripersonal networks are responsible. The two behaviours are quite different. A true defensive stance protects you from a potentially dangerous object looming into personal space. It is fast and exquisitely tuned to protect the most vulnerable parts of your body, especially the eyes. In contrast, the mimic defensive stance is a way to manipulate another monkey even at a distance. It is much slower and tuned to be easily visible. It probably involves turning your face directly toward the other monkey and exaggerating the facial components of the action, while at the same time keeping your eyes open. It is a separate, modified behaviour that evolved on the back of an older, defensive behaviour.

Here we finally have a true social signal. You produce it and I receive it. The signal is known as the 'silent bared teeth display' and has been documented in many species of primate (Beisner and McCowan, 2014; De Marco and Visalberghi, 2007; Preuschoft, 1992; Thierry et al., 1989; Von Hooff, 1962). They cringe down, duck the head, and raise the shoulders. The skin puckers and crinkles around the eyes but the eyes remain open. The upper lip pulls up and the upper teeth show. The display is a signal of non-aggression and is believed to be the origin of the human smile.

The term 'silent bared teeth display' is misleading because it refers only to the teeth and in that way misses the connection to a standard defensive stance. The epicentre of the action is the face, and in a passing, quick example of the display, one sees the upper lip pull up, flashing the upper teeth. But a full gesture of non-aggression can recruit a larger set of muscles. In humans, the epicentre of the smile is not the teeth but the contraction of musculature around the eyes, as the nineteenth-century neurologist Duchenne pointed out (Duchenne, 1990). Although a human smile is usually limited to the face, other components around the body can also appear. Think of the new intern, with low status in the company, smiling at a vice-president far up in the hierarchy. He grins, teeth on display, face crinkled painfully around the eyes, body slightly hunched, knees slightly bent, shoulders slightly raised, hands pulled inward and curled over the abdomen or chest. At least in its more extreme manifestations, a smile retains the echo of the defensive cringe.

I am not suggesting that the human smile *is* a defensive cringe. A smile does not protect the eyes or express fear. I am suggesting that the evolutionary *precursor* of a smile is a defensive cringe that protects the eyes in folds of skin. In this proposal, a smile is an evolutionary mimic that has lifted free of its original context.

The evolutionary origin of smiling is different from its psychological origin. Most people assume that we smile because we feel happy. Many scientists have studied the origin of emotional states and the commonality of emotional mechanisms in humans and non-human animals (e.g. de Waal, 2011; Panksepp, 2007). Here, however, the question is much more specific: not why do we have certain emotional states, but why do we produce a specific motoric action? Why, when making an affiliative gesture, do we bare the upper teeth and crinkle the eyes? Ultimately, the survival advantage of a smile is the same as the advantage of any social signal: it manipulates the behaviour of the receiver.

18.8 The origin of laughter

Ethologists have described a gesture called the 'open-mouth play face' (Cordoni et al., 2016; Darwin, 1872; Henry and Herrero, 1974; Jolly, 1966; Palagi, 2008, 2009; Preuschoft, 1992; Ross et al., 2010; von Hooff, 1962). It is common among many mammals. Anyone with a pet dog knows it well. When playing, the dog opens its mouth slightly in a characteristic way. When mammals play, they gently bite, and that mouth action may have evolved into a communicative gesture to help regulate the play. The great apes, like most primates, have an open-mouth play face (Darwin, 1872; Ross et al., 2010; von Hooff, 1962). In addition to the visual display, the great apes add a sound. For example, when a chimpanzee is tickled, it opens its mouth and makes a series of huffing sounds. Bonobos, gorillas, and orangutans do the same. Darwin (1872) discussed this remarkable ape huffing sound, and it has been studied systematically more recently by Ross et al. (2010). They find similarities in the sound spectrum between the huffing in apes and human laughter. The more genetically related a species of ape is to humans, the more similar the sound spectrum. By implication, at least part of human laughter may have evolved first in the common ancestor to apes and humans. Other scientists have argued that an analogue of play laughter can even be found in rats, who emit high-frequency sounds as part of their social interactions (Panksepp, 2007).

In my view, the human version of play huffing goes beyond the open mouth and the 'ha ha' sound. Consider what extreme laughter looks like. The skin wrinkles and puckers around the eyes. The muscles in the cheeks mobilize the skin upward, further protecting the eyes in puckered folds. As the cheeks bunch upward, the upper lip is pulled up, exposing the teeth. Tears are secreted. The shoulders lift and pull forward, the torso curls forward, and the arms pull in, curling around the abdomen. In humans, laughter includes what appears to be a mimic defensive reaction. In other great apes, during the play display, the components of a defensive stance are not obviously displayed. The eyes are not closed or puckered. Something may have happened in evolution, after our separation from the great apes, that shaped our uniquely human style of laughter.

In 2008, I suggested a possible explanation (Graziano, 2008). Just like the commonly accepted evolutionary explanation for the smile, my suggested explanation for laughter depends on the reflexive actions that protect us when peripersonal space is invaded. The speculation begins with tickle-evoked laughter, a reaction to an intrusion into personal space.

Consider a time five or six million years ago, after the human split with chimpanzees. Our ancestors have already evolved the open-mouthed play face and the huffing sound seen in all great apes, but have not yet evolved our specific human variety that includes a defensive stance. Suppose you and I are Australopithecines and we are play fighting. Many animals play fight with their mouths, trying to land gentle bites. As primates, we also play fight with our hands. Suppose you are an adult and I am a child. Your goal in the play fight is to penetrate my defences and make contact with a vulnerable body part. My goal is to block you and protect myself.

When your hands intrude into the defended buffer of space around me, my peripersonal neurons become active and trigger a defensive reaction. My body curls, my arms move into blocking postures, my shoulders lift to protect my neck, my facial muscles contract to protect my eyes. As your attacking hand looms farther into my peripersonal space, my defensive reaction becomes stronger. If you make contact with my skin, my peripersonal neurons fire at peak activity and my defensive reactions become frantic. If you land a blow or a scratch near my eyes, even a gentle one, my tear ducts leak lubricant to protect my eyes.

In the context of the play fight, this reflexive, defensive behaviour inadvertently broadcasts information about me. It demonstrates that you have won that moment in the fight. You have scored a point, so to speak. The defensive set broadcasts information content that could be roughly translated as 'touché.' At this point, however, we do not yet have a social signal. We have only a normal defensive reaction. There is, so far, no reason for you to interpret my behaviour in any specific way, or for me to deploy that behaviour strategically as a communicative signal.

Imagine we fast-forward, perhaps a million years. Evolution has had time to further shape systems in the brain to take better advantage of the available information. Suppose again we are human ancestors engaged in a play fight. I produce a defensive set as your hand penetrates my protected spaces. You now have pathways wired into your brain to react to that defensive set. Your reaction is not the result of any explicit cognition. An instinctive behaviour has evolved, and you react automatically. My defensive set demonstrates that your hand action has just been successful, and therefore it reinforces your behaviour, shaping your ability to win the play fight. My defensive set becomes a specific reward to you. But my defensive squirm also has the effect of causing you to pause. In a play fight, it is counterproductive to push too far or scratch too deep. In these ways, my defensive set provides useful information to regulate your behaviour in the play fight. At this point, we *still* do not have a true social signal. I am merely defending my body, and you are adjusting your behaviour based on your observations of my defensive actions.

Fast-forward once again, another million or so years. The brain has evolved, not only to receive the signal, but to send it strategically. By deploying a mimic defensive set, I can manipulate your behaviour. Although I can still generate a normal defensive set when your hand looms in, I can now also generate a mimic behaviour. The mimic behaviour is not a real defensive reaction because I can produce it even when the defence is not urgent. Even if you lightly touch my skin, the touch can trigger an exaggerated reaction from me. I'm sending out a touché signal. I'm saying, 'You got me! I'm dispensing your reward! Now give it a moment's rest and don't go too far!' The behaviour I emit has lifted free of the original behaviour and turned into a true social signal.

And so we have tickle-evoked laughter, a social signal that evolved to regulate a particular kind of human interaction. It acts as a social reward and as a mediating signal.

The explanation I have offered here is quite narrow. It does not explain where the huffing sound comes from, or the open-mouth play face (see section 18.8). It does not explain how tickle-evoked laughter might have branched into the many kinds of human laughter used in a vast range of social contexts. It narrowly focuses on how some of the motoric components of human laughter mimic a natural defensive set typical of an intrusion into personal space. Laughter may have evolved originally from play fighting in which one player attacks the defended spaces of the other player. It is now a social reward that one person can give to another person as part of playful interaction, and it retains some physical characteristics of a defensive set.

18.9 The origin of crying

Crying is a difficult behaviour to study from an evolutionary perspective because only humans do it. Other animals make distress cries. We may call it 'crying' when a puppy whimpers, but generating a distress sound differs from human crying. Most attempts to explain the origin of human crying focus on the tears, but the tears are only one out of a large set of components. Human crying in its most intense form includes a secretion of tears, a pursing of skin around the eyes, a bunching upward of the cheeks, a lifting of the upper lip, a lowering of the head, a raising of the shoulders, a hunching of the torso, a pulling of the arms into a blocking posture around the abdomen or chest or face, and a repeated aspiration that is sometimes voiced. Many of these components match a normal defensive set.

Other animals solicit comfort by making noise. No other animal, as far as I know, solicits comfort by partially mimicking the actions that normally protect the face from a collision. Why do humans cry like this?

Darwin's explanation (Darwin, 1872) begins with babies screaming in order to express negative emotion. In his speculation, the extreme forcing of air through the windpipe excites blood flow to the face. That extra blood flow risks rupturing the blood vessels in the eyes. To protect the eyes, the facial muscles contract, packing the eyeballs in a tight, protective cushion. The squeezing of muscles around the eyes, along with the air pressure from the screaming, forces fluid out of the tear ducts.

Another influential account of crying was proposed nearly a hundred years later by Andrew (1963). He argued that crying mimics a case of contaminants in the eyes. It evolved as a way to express distress.

These previous accounts focus on the sender of the signal—on why the physical act of crying would be used to express sadness or distress. As I noted earlier, it is now generally accepted that social signals evolve because of the impact they have on the receiver (Dawkins and Krebs, 1978; Fridlund, 1994; Godfray and Johnstone, 2000; Grafen and Johnstone, 1993; Schmidt and Cohn, 2001). There may be many reasons why we evolved the emotional state of sadness, but the specific external signal is a different matter. Crying, like any other social signal, is likely to be a display for others. It is a means of manipulating a receiver. Crying solicits comfort from others. To understand the evolutionary origin of crying, we must start with its specific impact on the receiver.

Although other animals do not cry in the human sense, they do provide comfort to each other. Most commonly across animal species, adults comfort infants. Infants therefore have a range of distress calls that can solicit help from their parents. But my own proposed account begins with circumstances in which adults comfort adults. Suppose you and I are both chimpanzees and we belong to the same family group. One day you badly beat me in a dispute over food. After the fight, you comfort me. Other chimps from the same group might also comfort me by grooming or touching me. In bonobos, the comforting sometimes takes the form of make-up sex (Clay and de Waal, 2013; Furuichi, 2011). Underlying these instances of adult-on-adult comforting is an initial burst of aggression that threatens social amity. The social amity is crucial in a highly cooperative species. Because fights are inevitable, it is adaptive to have a mechanism for comforting the victim afterward.

Given this social dynamic, here is my proposed evolutionary account of the quirky motoric components of crying.

Sometime after hominins split from the chimpanzee lineage, when our ancestors were Australopithecines three or four million years ago, we lived in cooperative social groups. But we were prone to fighting. An analysis of the bone structure of Australopithecines (Morgan and Carrier, 2013; Carrier and Morgan, 2014) suggests how our ancestors might have fought. In one interpretation, the facial bones of hominins, from Australopithecus to modern humans, are buttressed to withstand the stress of a blow, much like the facial bones of a bighorn sheep are buttressed to withstand the stress of a head collision. Moreover, according to the same authors, the bones of the Australopithecus hand are shaped to optimize curling the fingers into a fist and delivering a forceful punch. The implication is that Australopithecines engaged in ritual fighting by making fists and punching each other in the face. Many species have unique methods of fighting. Deer lock antlers. Giraffes swing and bang their necks together. Hippopotamus fight with wide-open mouths. Many species of monkey bite and also scratch each other with their fingernails. Humans, evidently, ball their hands into a boney club and hit each other in the face, and may have been doing so for millions of years, pre-dating our modern species. It is a species-typical behaviour. I know of no other species that engages in that specific mode of fighting. Perhaps it is one of the many evolutionary factors in flattening our snouts into vertical faces.

Suppose you and I are Australopithecines in a fight and I win. I punch you hard on the nose. I've penetrated your peripersonal space and made violent contact with your face. All your usual defensive reflexes are deployed. Your eyes water in a rapid autonomic response, protecting your eyeballs from potential scratch or contaminants. (If you have ever accidentally struck yourself on the nose, note how much your eyes water.) The skin purses around your eyes and your upper lip pulls up hard, further wrinkling the skin protectively around your eyes. Your head ducks down, your shoulders rise, your arms pull across your torso or into a blocking posture across your face.

As the aggressor, I need a mechanism for recognizing when I have won the fight, and especially when I have gone too far and hurt you. That mechanism should automatically trigger me to reduce my aggression and offer comfort. In that way, I can repair the social amity after the fight. Others in the family group also need a mechanism for recognizing when to offer comfort to you, if you are in distress after the fight.

Your extreme defensive reaction offers the most obvious signal. In this hypothesis, the brain evolved to receive that particular signal. When I see you enact an extreme defensive set, the kind normally triggered by a violent punch to the nose, it triggers an instinctive reaction in me. I reduce aggression and give comfort. The adaptation is a simple, effective way to help preserve social amity after a fight.

Now we reach the Machiavellian part of the story. Given that I have evolved that reaction, you can take advantage of my wiring. If you mimic that particular type of defensive reaction, especially if you exaggerate it and extend it in time over seconds or minutes, you should be able to extract comfort from me. Maybe I never fought you, and have shown no aggression toward you. Maybe nobody has hurt you. Your peripersonal neurons are not involved and you are not making an actual defensive movement. Nonetheless, if you approach me and display that particular kind of behaviour, it will press my built-in buttons and extract comfort from me. I'm wired to dispense comfort, or at least to cease any aggression toward you, when I see you produce that signal.

Once again, the process is not an explicit, cognitive one. You have no need to figure out the causes and effects intellectually, in order to deploy the behaviour. Evolution has built the behaviour into the brain. When you need comfort, that behaviour is triggered in you instinctively.

In this hypothesis, crying is not a facial protective action. It is a mimic. The mimic roughly resembles, but is not exactly the same, as the original. In my lab, I have watched hours of video of people and other primates hit in the face with ping pong balls and airpuffs and the reaction is brief, efficient, and not nearly as dramatic as crying. In contrast, the mimic behaviour is exaggerated, extended in time, and noisy. Perhaps the noise helps attract attention. The mimic behaviour is tuned not to protect the body, but to evoke a reaction in the receiver. Crying, in this proposal, is a distortion and exaggeration of a defensive set, deployed strategically to elicit a comfort reaction, or at least a rapid de-escalation of aggression, in others.

18.10 Summary

I acknowledge that the hypotheses proposed in this chapter are speculative. My argument is that a standard defensive stance, typically triggered by intrusions into peripersonal space, was co-opted by evolution and modified to become a set of situation-specific social signals. The argument rests on the point-by-point similarity among smiling, laughing, and crying. All three expressions resemble a standard defensive behavioural set. They share the contraction of the obicularis muscle around the eyes; the bunching of the cheek muscles upward toward the eyes, causing the upper lip to pull up, exposing the upper teeth; and, in extreme forms, the ducking of the head, lifting of the shoulders, hunching of the torso, and pulling of the arms over the front of the torso. This behavioural set is different from other emotional expressions (Ekman and Friesen, 1972). For example, in human anger, the musculature around the eyes is not contracted, but instead the eyes are opened wide; the cheeks are not mobilized upward; the teeth are exposed by a retraction of local muscles in the lips rather than in the cheeks; and the shoulders do not lift protectively around the neck. Only some human expressions mimic the defensive stance.

Protective peripersonal space is easily overlooked. It invisibly surrounds the body and the mechanism for it lies mainly beneath the surface of consciousness. Yet it coordinates some of the most important interactions between self and world. It clears a margin of safety and protects the body. As research into peripersonal space expands, one emerging lesson is that the mechanism has an outsized impact on almost all aspects of life, well beyond the narrow scope of space within a metre or so of the skin. Peripersonal space has an unexpected relevance to tool use, social spacing, and perhaps even self-awareness (e.g. Graziano, 2018; Làdavas and Serino, 2008; Pellencin et al., 2018; Salomon et al., 2017). Here I suggest that peripersonal space, through evolutionary time, has even shaped our most characteristic human facial expressions—a thought that makes this peripersonal-space researcher smile.

References

Andrew, R. J. (1963). The origin and evolution of the calls and facial expressions of the primates. *Behaviour* 20, 1–107.

Beisner, B. A., and McCowan, B. (2014). Signaling context modulates social function of silent bared-teeth displays in rhesus macaques (Macaca mulatta). *Am. J. Primatol.* 76, 111–121.

Carrier, D., and Morgan, M. (2014). Protective buttressing of the hominin face. *Biol. Rev. Camb. Philos. Soc.* 90, 330–346.

Clay, Z., and de Waal, F. B. M. (2013). Bonobos respond to distress in others: Consolation across the age spectrum. *PLoS ONE* 8, e55206.

Cooke, D. F., and Graziano, M. S. A. (2003). Defensive movements evoked by air puff in monkeys. *J. Neurophysiol* 90, 3317–3329.

Cooke, D. F., and Graziano, M. S. A. (2004). Super-flinchers and nerves of steel: Defensive movements altered by chemical manipulation of a cortical motor area. *Neuron* 43, 585–593.

Cooke, D. F., Taylor, C. S. R., Moore, T., and Graziano, M. S. A. (2003). Complex movements evoked by microstimulation of the ventral intraparietal area. *Proc. Natl. Acad. Sci. U.S.A.* 100, 6163–6168.

Cordoni, G., Nicotra, V., and Palagi, E. (2016). Unveiling the 'secret' of play in dogs (*Canis lupus familiaris*): Asymmetry and signals. *J. Comp. Psychol.* 130, 278–287.

Darwin, C. (1872). *The Expression of the Emotions in Man and Animals*. London: John Murray.

Davis, M. (1984). The mammalian startle response. In: *Neural Mechanisms of Startle*. Edited by R. C. Eaton. New York: Plenum Press.

Dawkins, R., and Krebs, J. R. (1978). Animal signals: Information or manipulation? In: *Behavioral Ecology: An Evolutionary Approach*, pp. 282–309. Edited by R. Krebs, and N. B. Davies. Oxford: Blackwell.

De Marco, A., and Visalberghi, E. (2007). Facial displays in young tufted Capuchin monkeys (*Cebus apella*): Appearance, meaning, context and target. *Folia Primatol. (Basel)* 78, 118–137.

de Vignemont, F., and Iannetti, G. D. (2015). How many peripersonal spaces? *Neuropsychologia* 70, 327–334.

de Waal, Frans, B. M. (2011). What is an animal emotion? *Ann N. Y. Acad. Sci.* 1224, 191–206.

di Pellegrino, G., and Làdavas, E. (2015). Peripersonal space in the brain. *Neuropsychologia* 66, 126–133.

Duchenne G.-B. (1990). *The Mechanism of Human Facial Expression*. New York: Cambridge University Press. Edited and translated by R. Andrew Cuthbertson. Original published in 1862.

Duhamel, J. R., Colby, C. L., and Goldberg, M. E. (1998). Ventral intraparietal area of the macaque: Congruent visual and somatic response properties. *J. Neurophysiol.* 79, 126–136.

Ehrlichman, H., Brown, S., Zhu, J., and Warrenburg, S. (1995). Startle reflex modulation during exposure to pleasant and unpleasant odors. *Psychophysiol.* 32, 150–154.

Ekman, P., and Friesen, W. V. (1972). *Emotion in the Human Face: Guidelines for Research and an Integration of Findings*. Oxford: Pergamon Press.

Fridlund, A. (1994). *Human Facial Expression: An Evolutionary View*. New York: Academic Press.

Furuichi T. (2011). Female contributions to the peaceful nature of bonobo society. *Evol Anthropol.* 20, 131–142.

Godfray, H. C. J., and Johnstone, R. A. (2000). Begging and bleating: The evolution of parent-offspring signaling. *Phil. Trans. R. Soc. Lond. (Biol.)* 355, 1581–1591.

Grafen, A., and Johnstone, R. A. (1993). Why we need ESS signalling theory. *Phil. Trans. R. Soc. Lond. (Biol.)* 340, 245–250.

Graziano, M. S. A. (2008). *The Intelligent Movement Machine*. New York: Oxford University Press.

Graziano, M. S. A. (2015). A new view of the motor cortex and its relation to social behavior. In: *Shared Representations: Sensorimotor Foundations of Social Life*. Edited by S. S. Obhi, and E. S. Cross. Cambridge, UK: Cambridge University Press.

Graziano, M. S. A. (2018). *The Spaces Between Us: A Story of Neuroscience, Evolution, and Human Nature*. New York: Oxford University Press.

Graziano, M. S. A., Taylor, C. S. R., and Moore, T. (2002). Complex movements evoked by microstimulation of precentral cortex. *Neuron* 34, 841–851.

Graziano, M. S. A., Yap, G. S., and Gross, C. G. (1994). Coding of visual space by pre-motor neurons. *Science* 266, 1054–1057.

Grillon, C. (2008). Models and mechanisms of anxiety: Evidence from startle studies. *Psychopharmacol.* 199, 421–437.

Grillon, C., Ameli, R., Woods, S. W., Merikangas, K., and Davis, M. (1991). Fear-potentiated startle in humans: Effects of anticipatory anxiety on the acoustic blink reflex. *Psychophysiol.* 28, 588–595.

Grillon, C., Morgan, C. A., Southwick, S. M., Davis, M., and Charney, D. S. (1996). Baseline startle amplitude and prepulse inhibition in Vietnam veterans with posttraumatic stress disorder. *Psychiat. Res.* 64, 169–178.

Henry, J. D., and Herrero, S. M. (1974). Social play in the American black bear: Its similarity to canid social play and an examination of its identifying characteristics. *Am. Zool.* 14, 371–389.

Jolly, A. (1966). *Lemur Behaviour: A Madagascar Field Study.* Chicago: University of Chicago Press.

Koch, M. (1999). The neurobiology of startle. *Prog. Neurobiol.* 59, 107–128.

Làdavas, E., and Serino, A. (2008). Action-dependent plasticity in peripersonal space representations. *Cogn Neuropsychol.* 25, 1099–1113.

Landis, C., and Hunt, W. A. (1939). *The Startle Pattern.* New York: Farrar and Rinehart Inc.

Lang, P. J., Bradley, M. M., and Cuthbert, B. N. (1990). Emotion, attention, and the startle reflex. *Psychol. Rev.* 97, 377–395.

LeDoux, J. (2007). The amygdala. *Curr Biol* 17, R868–R874.

McTeague, L. M., and Lang, P. J. (2012). The anxiety spectrum and the reflex physiology of defense: From circumscribed fear to broad distress. *Depress Anxiety* 29, 264–281.

Morgan, M. H., and Carrier, D. R. (2013). Protective buttressing of the human fist and the evolution of hominin hands. *J. Exp. Biol.* 216, 236–244.

Palagi, E. (2008). Sharing the motivation to play: The use of signals in adult bonobos. *Animal Beh.* 75, 887–896.

Palagi, E. (2009). Adult play fighting and potential role of tail signals in ring-tailed lemurs (*Lemur catta*). *J. Comp. Psychol.* 123, 1–9.

Panksepp, J. (2007). Neuroevolutionary sources of laughter and social joy: Modeling primal human laughter in laboratory rats. *Behav. Brain Res* 182, 231–244.

Patrick, C. J., Berthot, B. D., and Moore, J. D. (1996). Diazepam blocks fear-potentiated startle in humans. *J. Abnorm. Psychol.* 105, 89–96.

Pellencin, E, Paladino, MP, Herbelin, B, Serino, A. (2018). Social perception of others shapes one's own multisensory peripersonal space. *Cortex* 104, 163–179.

Preuschoft, S. (1992). 'Laughter' and 'smile' in Barbary Macaques (*Macaca sylvanus*). *Ethology* 91, 220–236.

Rizzolatti, G., Scandolara, C., Matelli, M., and Gentilucci, M. (1981). Afferent properties of periarcuate neurons in macaque monkeys. II. Visual responses. *Behav. Brain Res* 2, 147–163.

Ross, M. D., Owren, M. J., and Zimmermann, E. (2010). The evolution of laughter in great apes and humans. *Commun. Integr. Biol.* 3, 191–194.

Salomon, R., Noel, J.-P., Łukowska, M., Faivre, N., Metzinger, T., Serino, A., and Blanke, O. (2017). Unconscious integration of multisensory bodily inputs in the peripersonal space shapes bodily self-consciousness. *Cognition* 166, 174–183.

Schmidt, K. L., and Cohn, J. F. (2001). Human facial expressions as adaptations: Evolutionary questions in facial expression research. *Am. J. Phys. Anthropol.* Suppl. 33, 3–24.

Strauss, H. (1929). Das Zusammenschrecken. *Journal fur Psychologie und Neurologie* 39, 111–231.

Thierry, B., Demaria, C., Preuschoft, S., and Desportes, C. (1989). Structural convergence between silent bared-teeth display and relaxed open-mouth display in the Tonkean macaque (*Macaca tonkeana*). *Folia Primatol. (Basel)* 52, 178–184.

Von Hooff, J. A. R. A. M. (1962). Facial expression in higher primates. *Symp. Zool. Soc. Lond.* 8, 97–125.

Index

For the benefit of digital users, indexed terms that span two pages (e.g., 52–53) may, on occasion, appear on only one of those pages.

The editors would like to express their special thanks to Selina Guter, Malte Hendrickx, and Ruben Schwaben for their work on the index.

Tables and figures are indicated by *t* and *f* following the page number.